AF779303

Tobias Alexander Jopp

Insurance, Fund Size, and Concentration

JAHRBUCH FÜR WIRTSCHAFTSGESCHICHTE
BEIHEFT 16

Im Auftrag der Herausgeber des Jahrbuchs für Wirtschaftsgeschichte
herausgegeben von Reinhard Spree

Tobias Alexander Jopp

Insurance, Fund Size, and Concentration

Prussian Miners' Knappschaften in the Nineteenth- and Early Twentieth-Centuries and Their Quest for Optimal Scale

Akademie Verlag

Bibliografische Information der Deutschen Nationalbibliothek
Die Deutsche Nationalbibliothek verzeichnet diese Publikation
in der Deutschen Nationalbibliografie;
detaillierte bibliografische Daten sind im Internet über
http://dnb.d-nb.de abrufbar.

© Akademie Verlag GmbH, Berlin 2013
Ein Wissenschaftsverlag der Oldenbourg Gruppe

www.akademie-verlag.de

Das Werk einschließlich aller Abbildungen ist urheberrechtlich geschützt. Jede Verwertung außerhalb der Grenzen des Urheberrechtsgesetzes ist ohne Zustimmung des Verlages unzulässig und strafbar. Das gilt insbesondere für Vervielfältigungen, Übersetzungen, Mikroverfilmungen und die Einspeicherung und Bearbeitung in elektronischen Systemen.

Einbandgestaltung: Entwurf hauser lacour
Druck und Bindung: Beltz Bad Langensalza GmbH

Dieses Papier ist alterungsbeständig nach DIN/ISO 9706

ISBN 978-3-05-006370-6
eISBN 978-3-05-006371-3

Table of Contents

List of Tables

List of Figures

List of Abbreviations

AKP	Allgemeine *Knappschafts*-Pensionskasse
ASPD	Average sick pay per day
ATP	Average total pension
AW	Average annual wage
CR	Contribution rate
DP	Probability of dying
EAP	Economically active population
GDP	Gross domestic product
GLM	Generalized linear model
GNP	Gross national product
HHI	Hirschman-Herfindahl index
ICR	Invalids-to-contributors ratio
IP	Probability of becoming invalid
KV	*Knappschaft* fund (according to the German term *Knappschaftsverein*)
MC	Miscellaneous costs
MES	Minimum efficient size
MR	Miscellaneous revenues
N	Number of contributors
NNP	Net national product
OCR	Orphans-to-contributors ratio
OMP	Probability of out-migrating
OP	Operating costs
OSD	Overall sick days
P	Pensioners
PAYG	Pay-as-you-go
PCR	Pensioners-to-contributors ratio
RR	Replacement rate
RVO	Reich insurance regulation (according to the German term *Reichsversicherungsordnung*)

SCR	Survivors-to-contributors ratio
SD	Standard deviation
SDCR	Sick days-to-contributors ratio
VAR	Variance
WCR	Widows-to-contributors ratio
YOR	Young contributors-to-old contributors ratio

Preface

This study is a revised and shortened version of my doctoral thesis, accepted by the Department of Economics and Social Sciences at *Universität Hohenheim* in March 2012. Written while I was working at the Chair of Economic and Social History, my study has to be considered as a part of the interdisciplinary research project entitled "Vergangenheit und Zukunft sozialer Sicherungssysteme am Beispiel der Bundesknappschaft und ihrer Nachfolger" funded by the Leibniz Association. I worked as a project assistant on the economic sub-project which was led by the *Rheinisch-Westfälisches Institut für Wirtschaftsforschung* (RWI) located in Essen in cooperation with *Universität Hohenheim* and Yale University. To my great delight, this study has been awarded the *Friedrich-Lütge-Preis* 2013 for outstanding dissertations in the fields of economic history by the *Gesellschaft für Sozial- und Wirtschaftsgeschichte* (GSWG). I would like to thank the GSWG very much for acknowledging my study in this way.

In fact, I am not the only one who contributed to the study – several persons did whose valuable professional and emotional support I highly appreciate, of course. To begin with, I would particularly like to acknowledge and express my gratitude for my doctoral supervisor, Prof. Dr. Jochen Streb, on the one hand, for enabling me to conduct research on the questions addressed according to my particular ideas and, on the other hand, for his dedicated mentoring at any time during the origination process. I am truly indebted to him as my academic teacher. Moreover I am very grateful to Prof. Timothy Guinnane, PhD, for discussing with me the subject matter on various occasions and giving me a lot of suggestions, which likewise were invaluable for preparing this study. Additionally, I owe special thanks to Prof. Dr. Christoph M. Schmidt and Prof. Dr. Manuel Frondel for their valuable support – funding inclusive – directly coming from the RWI. Besides, I would like to thank Prof. Dr. Jörg Schiller very much for acting as co-supervisor providing critical advice and important suggestions for revisions, too, and Prof. Dr. Gert Kollmer-von Oheimb-Loup for acting as chairman of the examining board.

In particular, I would also like to highlight the contribution that the amicable atmosphere among my appreciated colleagues provided for preparing my study. For human warmth and the willingness to discuss the subject matter at any time, I offer appreciation and many thanks to Nicole Waidlein, Stefanie Werner, and Harald Degner. Besides, a handful of student assistants generously supported me by collecting and preparing data, namely Jana Schrödlen, Meike Thumm, Patrick Baldes, Reiner Frommer, and Niklas Roming.

I am also deeply indebted to Prof. Dr. Carsten Burhop for setting the course to proceed with the project "doctorate" after I finished my studies by mentoring me as an external doctoral candidate. Moreover, I am very grateful to Prof. Dr. Mark Spoerer for having given me a considerable amount of freedom in fine-tuning my study and completing my doctorate after moving to *Universität Regensburg*. Finally, I would like to thank Prof. Dr. Reinhard Spree, who – as editor of the series *Beihefte des Jahrbuchs für Wirtschaftsgeschichte* – recommended incorporating my study, as well as the *Akademie Verlag* for publishing it.

Above all, I have to give my greatest thanks to my mother for her constant backing, encouragement and support for my project "doctorate" in all respects. To her and my grandparents Klara and Willibald I dedicate my study.

Tobias A. Jopp
Regensburg, April 2013

1. Introduction

Anyone who is to some extent familiar with the German welfare state probably knows about the *Bundesknappschaft*, established in 1969, and its successor, the *Deutsche Rentenversicherung Knappschaft-Bahn-See*, established in 2005. While the former provided a package of social insurance benefits exclusively to miners – and, thus, was one of the few strictly sector-specific carriers of statutory health and pension insurance – the latter is now free to contract with everybody, not only with miners, railroaders, and shippers as its name implies.[1] Presumably less known are both of these social insurance institutions' direct organizational forerunners, to which they can be traced for their very origins in mediaeval times. This study examines one of these forerunners: the various *Knappschaftsvereine* (*Knappschaft* funds, or simply KVs[2] in the following) of the second half of the nineteenth and the early twentieth centuries.[3] These mining-specific insurance funds, which had emerged from purely charitable associations of religious origin, existed throughout the various German mining areas. These funds insured their

---

1 Actually, there is no natural translation of the terms *Bundesknappschaft* and *Deutsche Rentenversicherung-Knappschaft-Bahn-See*. As for non-German readers, however, we might call them "Federal *Knappschaft*" (for the term *Knappschaft* itself see the following footnote) and "German Mining-Railway-Sea Pension Fund." Precisely, the *Deutsche Rentenversicherung Knappschaft-Bahn-See* is now the second pillar of the German statutory old-age insurance, besides the *Deutsche Rentenversicherung Bund* ("German Federal Pension Fund"). It was recently, in 2005, that the *Bundesknappschaft* merged with the *Bahnversicherungsanstalt* (social insurance for the railroad sector) and the *Seekasse* (health insurance for the shipping sector) into the new provider; see Bartels et al. (2009), pp. 195–208, and Klenk (2008), pp. 125–126.

2 To clarify: *Knappschaften* is plural, and *Knappschaft* is singular. I use the abbreviation "KV" (singular) or "KVs" (plural) according to the German term *Knappschaftsverein(e)*. It is not possible – at least not to the best of my knowledge – to translate the term into an accurate English expression; "miners' fund" would probably do it. The term itself relates to the mediaeval German term for miner, the *Knappe*. When I speak of "the *Knappschaft*," I always refer to the *Knappschaft* institution – the basic idea *behind* the fund(s) and not a particular fund taken as an organization.

3 The timely gap between the KVs and the *Bundesknappschaft* is closed, on the one hand, by the *Reichsknappschaft*, which was established in 1923 and continued to operate until 1945, and, on the other hand, by the few self-standing *Bezirksknappschaften* (district *Knappschaft* funds), into which the *Reichsknappschaft* was split during the immediate post-war years; see Lauf (1994), pp. 201–204.

members against the major risks of life, including invalidity, sickness, survivorship, and old-age (i.e., longevity). Related insurance benefits – lifetime pensions and short-term payments such as daily sick pay, among others – were combined in a single package.[4]

The origin of the *Knappschaft* institution lies in the Middle Ages, when miners began to form mutual associations to collectively deal with contingencies like temporary or permanent incapacity to work and, thus, to earn one's living and to provide for one's household. As nineteenth- and early twentieth-century contemporaries of the KVs believed, the emerging collective action directed towards risk provision was due, essentially, to the unique character of the profession itself and to the highly developed sense of solidarity among miners, isolating mining from other economic activities.[5] The German miners' funds, increasingly spreading over areas with resource deposits and characterized by the principle of charity, had then been under the patronage of the sovereign's administration during the absolutist-mercantilist era. Finally, KVs were converted from charitable or, respectively, state-run organizations into insurance funds thanks to the Prussian mining reform of 1851–1865, which significantly contributed to establishing a new liberal economic order regarding all kinds of mining activities.[6] The seminal New *Knappschaft* Law of 1854 and, later, the Prussian General Mining Law of 1865 essentially shaped the miners' job-related social insurance system and re-established the KVs as publicly-licensed, but self-administered insurance providers. Membership in a fund would provide a miner with a conditional promise that he would receive a certain insurance benefit of pre-defined amount in the event of certified sickness or invalidity. Ensuring that this exchange took place was the KV's – or, more accurately, the KV's management's – responsibility. In return, a miner paid regular social insurance contributions to complete the transfer of insured risks to the KV. The KVs were still kind of mutual insurance organizations in that the various local memberships, together with mine owners, managed the funds, though within the limits of the states' regulations and without owning property rights in the funds.[7]

---

4 I prefer the term "invalidity" over the term "disability." The terms may, however, be used interchangeably.

5 See, for example, von der Heyden-Rynsch (1881), p. 259, and Köhne (1915), p. 5.

6 See Fischer (1961a).

7 As emphasized, the underlying definition of what "insurance" might be is chosen to be fairly specific here. In a broader economic sense, the essence of "insurance" certainly is that it enables a risk-averse individual to exchange an uncertain and potentially high loss in the future for a small certain loss in the present; see, for example, Arrow (1996) on this. Whaples and Buffum (1991), for example, provide an overview of the different ways of provision against risks. Moreover, the coexistence of several different forms of provision against the risks of sickness or invalidity – through the state or through governmental agencies, individual care, collective mutual action, and the market – has been termed in the literature "the mixed economy of welfare." Surely, such a mixed economy of welfare existed in Germany at the time, and the KVs were part of it; see, for example, Harris and Bridgen (2007), p. 1, and Pons Pons and Vilar Rodriguez (2011), p. 72.

Given their purpose and character, KVs might remind the reader of friendly societies and other related mutual funds, more intensively discussed in the economic-historical literature. KVs, however, were different in some important aspects. First, miners were obligated to join the local KV that was in charge of the area in which their workplace was situated – that is, membership was mandatory; this, in particular, meant that adverse selection as a basic insurance problem *did not* play a role for KVs unless, of course, it was a matter of choice of profession.[8] Second, the benefit package that the KVs were obligated by law to provide was, at least, qualitatively more extensive in the sense that it combined daily sick pay, medical treatment and funeral benefits with lifetime invalidity and survivorship pensions. Third – and this fact provides the basis for the "connecting tissue" of this study – contemporary observers of the KVs seem to have focused on whether financial stability could be improved by exploiting economies of scale – all else being equal –, rather than, as contemporary observers of friendly societies in Britain and North America suggested, by improving the actual pricing techniques used at the time. The central research question of this study – which will be outlined below – is essentially motivated by the fact that there was this intense discussion among observers of the KVs – and that the importance of scale was at its centre.

Several scholars have, so far, emphasized the KVs' importance as an early – presumably the earliest – *social insurance* scheme. The KVs had already been in operation for more than two decades when Bismarckian worker insurance came into existence in 1883 as their "counterpart" on the national level. According to Varnas (1947), Köhler and Zacher (1982), Tampke (1982) and Geyer (1985, 1992), for example, the great achievement of Bismarckian insurance was to make the advantages of insurance available to a larger circle of workers, selectively directed at first towards those in industry. It was, however, not the *Gesetz betreffend die Krankenversicherung der Arbeiter* (Health Insurance Law) of June 15, 1883 that formally invented and established social insurance – or the insurance business in general, for that matter.[9] Instead, "insurance," in the narrow sense of the word, already existed as an economic innovation and a technology to satisfy people's and businesses' needs for future financial security in the face of omnipresent uncertainty. As such, it emerged, at the latest, around the middle of the nineteenth century, in the private commercial (life insurance) sphere, as well as in the sphere of sector-specific social insurance (mining, railroad, and shipping).[10] Initially the focus was on protection against sickness and incapacity to work, including the conse-

8 For the idea behind adverse selection, see Akerlof's seminal 1970 paper on the "market for lemons." Applications to economic history include, for example, Murray (2005) and Gottlieb (2007).

9 See Varnas (1947), p. 59, Köhler and Zacher (1982), p. 22, Tampke (1982), pp. 71–74, Geyer (1985), p. 96, and Geyer (1992), pp. 1 046–1 047.

10 See Borscheid and Drees (1988), pp. 3–49, for a short history of the insurance business in Germany, including social insurance. Furthermore, see Montz (2010) and Sulzer (2010) for the most recent studies about the occupational systems for the railroad and shipping sectors. Pearson (1997) provides an international perspective.

quences of work-related accidents; protection against the consequences of unexpected longevity, of survivorship and of unemployment appeared on the national agenda only in the early twentieth century.[11] Moreover, besides those scholars mentioned, Wagner-Braun (2002) brings particular attention to the KVs' historical and economic importance as the definitive pioneers of *compulsory* membership.[12] Finally, according to Hennock (2006) who wrote, among other things, on the emerging German and British welfare states in comparative perspective, entrepreneur Carl Ferdinand Stumm recommended several times, in the late 1870s, that politicians should seriously consider adopting the miners' insurance model for all employees in the industrial sector.[13] Obviously, all countries that, to one extent or another, adopted the tradition of the Bismarckian welfare state have been influenced somewhat indirectly by the German miners' insurance model.

Against the historical background of the KVs' significance in the emergence of the German welfare state, researchers have made only limited efforts to understand this welfare institution in more detail and, thus, have neglected a crucial piece of German insurance history. This impression manifests all the more when we consider, in particular, approaches that would apply economics' methodology to filter out important insights into the KVs' historical development and its broader implications. This study helps fill that research gap by focusing on a particular issue that to explore demands some baseline (insurance) economics: the observable process of internal and, especially, external concentration among the KVs that began in the earlier 1870s and that culminated in the foundation of the *Reichsknappschaft* in 1923, which replaced the locally-organized and state-regulated funds with one national-level institution.

Before proceeding with the central research question that picks up this point, I like to address questions that may immediately have come to the reader's mind after these first introductory words: What precisely was the KVs' nature? Were those funds really carriers of a social insurance system – thus, may be appropriately called "social insurance funds" – or were they not, rather, kind of mutual insurance organization (*Gegenseitigkeitsversicherer*) in a private market? These questions touch on an important, yet – at this very early stage – not satisfactorily assessable, issue. Clearly, the nature of the KVs' insurance determines the theoretical and empirical concepts that do or do not matter in this particular context. Thus, one needs to know more about the KVs' nature before the way the analysis can be or should not, under all circumstances, be conducted can be properly specified. Unfortunately, I can answer these questions properly only

---

11 This holds for Germany – or the German *Kaiserreich* – as well as for a set of states that had begun to implement social policies for the masses, shortly after Germany made the initial step in 1883. Lindert (1994) and Cutler and Johnson (2004) explain the emergence of the late nineteenth- and early twentieth-century welfare state(s) with reference to a set of theories and quantitative methods.

12 See Wagner-Braun (2002), pp. 32–33.

13 See Hennock (2007), pp. 87–88.

after having introduced the institutional design KVs were embedded in. What I can do right now, however, is to stress my opinion about the KVs' nature: In my view, they can indeed be called the carriers of an early social insurance system, which distinguishes them in some way from what "mutual insurance organization" commonly means, at least in my perception. The remaining part of this introductory chapter might evoke a picture of KVs that is in contradiction to what is commonly seen in connotation with "social insurance," So, for example, I will provide some anecdotal evidence on the fact that KVs seem to have specified risk loadings as part of the contribution payment. This instrument of risk policy may rather be associated with commercial insurance, but not social insurance. Just one remark: KVs seem, indeed, to show up as hybrids. Yet, a basic characteristic that makes the mutual insurance organization "mutual" – policy-holders own the organization such that they are at the same time the supplier of their insurance coverage – was definitely not fulfilled in the KVs' case.

1.1. Central research question and scope

Concentrating on the *Prussian* miners' KVs' experience over the formative period of the German welfare state, 1854 to 1923, the central research question is straightforward: Was there a minimum efficient or even optimal scale of operation, and, depending on the nature of economies of scale, was the proclaimed "cleaning of the market" by absorptions of KVs efficient in retrospect? The broader aim of this research is to evaluate the merger and liquidation wave among the Prussian *Knappschaften* beyond what historical research has hitherto done. Once we have examined what information on scale (dis)advantages is hidden in the data, we can relate this information to three important questions in this context: First, what does the evidence tell us about the efficiency of the insurance arrangement carried by KVs? Second, how does the evidence relate to the important discussion in the contemporary literature on the costs and benefits of *Knappschaft* insurance – a discussion centering on the claim that, ultimately, efficiency and survival of the miners' insurance was a function of KV size? Third, how does the evidence relate to the concentration process among KVs as it factually took place? Put differently, is there a picture emerging that is consistent with the traditional reading of the KVs' history, or does this fresh view on the KVs' struggle for optimality suggest a different reading? As to provide the reader with a prospect, in my opinion, evidence will suggest the latter.

This study's observation period covers the years during which the landscape of *Knappschaft* insurance was characterized, on the one hand, by the switch to insurance in the more narrow sense of the concept (see above) and, on the other hand, by many operating funds, or fragmentation. It has to be pointed out, though, that the "statistical observation period" – i.e., the period for which data are available – is 1861–1920. The reason why to focus on the Prussian KVs primarily is fourfold: First, it was the Prussian state that initiated the influential reform of the mining law that subsequently spread across almost every German state; second, the Prussian KVs covered roughly 90 percent of German miners at the time; third, they were, statistically and with respect to contemporary secondary sources, the relatively best-documented funds; and, fourth, these KVs showed a clear consolidation process until 1923 that, nonetheless, leaves enough variation – basically in fund size – to perform serious, quantitative, mass data-based tests. Although Bavarian and Saxon *Knappschaft* insurance saw notable consolidation as well, variation with respect to size was smaller.

It seems helpful here to illustrate the stylized historical facts that this study takes as its starting point for research. Therefore, Figure 1.1 depicts the (statistical) beginning of KVs in 1861. Reported are the number of KVs, the number of working members ("contributors"), and the number of pensioners as consisting of invalids and survivors. All quantities are depicted by classes of KV size.[14]

Figure 1.1: The Prussian KVs' starting position in 1861: Number of funds, of contributors, and of pensioners by size class

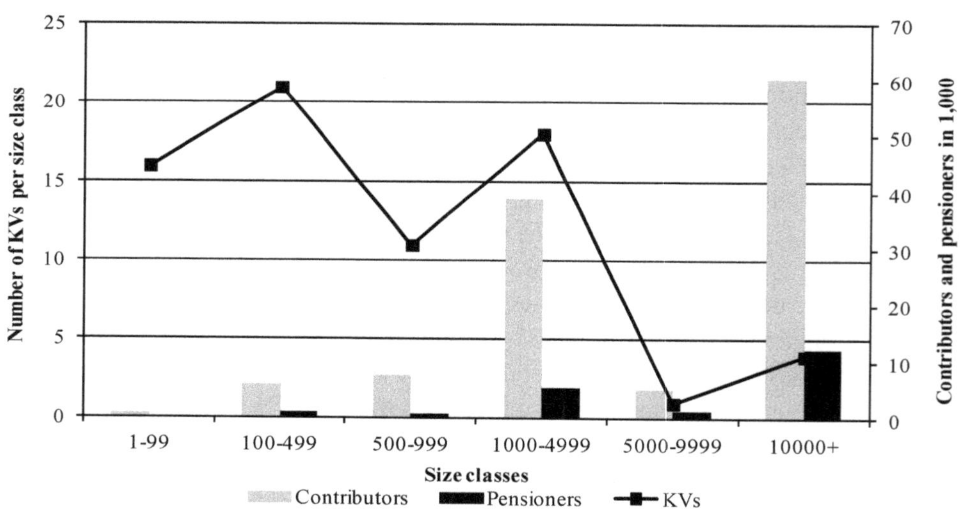

Sources: My own calculations based on Ministerium für Handel, Gewerbe und öffentliche Arbeiten (1861–1878).

As of 1861, two stylized facts stand out: First, Prussian KVs initially covered around 119 000 working members and 21 000 pensioners. Second, KVs were distributed very unequally over size classes; most KVs were rather small, and the overwhelming number – namely 48 of 71 funds – operated below a size of 1 000 working miners. However,

[14] KV size is measured in terms of contributors; see Chapter 3.

the largest five KVs alone accounted for about 55 percent of contributors (or 56 percent of overall members including inactive pensioners).

In addition, Figure 1.2 illustrates the consolidation process among Prussian, as well as Bavarian and Saxon, KVs. The net reduction in the number of operating Prussian funds, which set in at the beginning of the 1870s and lasted until the formation of the *Reichsknappschaft*, resulted from a combination of mergers (absorptions may be a more accurate term) and liquidations, thus apparently a kind of selection process.[15] At the same time, however, the number of working miners insured in Prussian, Bavarian and Saxon KVs climbed from about 119 000 (from 1861 onwards), 5 000 (from 1871 onwards) and 9 700 (from 1868 onwards), respectively, to more than one million in Prussian, 20 600 in Bavarian and 51 400 in Saxon KVs; counting insured pensioners, too, would, of course, push covered membership further up (around 1 340 000 or, respectively, 25 000 and 69 500 in 1920). Hence, against the background of a continuously growing aggregate membership, directly reflecting the mining sector's growth in terms of employment, more and more insurants were obviously distributed among fewer insurance carriers, and the average size among KVs climbed up.

Figure 1.2: The development of the number of Prussian, Bavarian and Saxon KVs

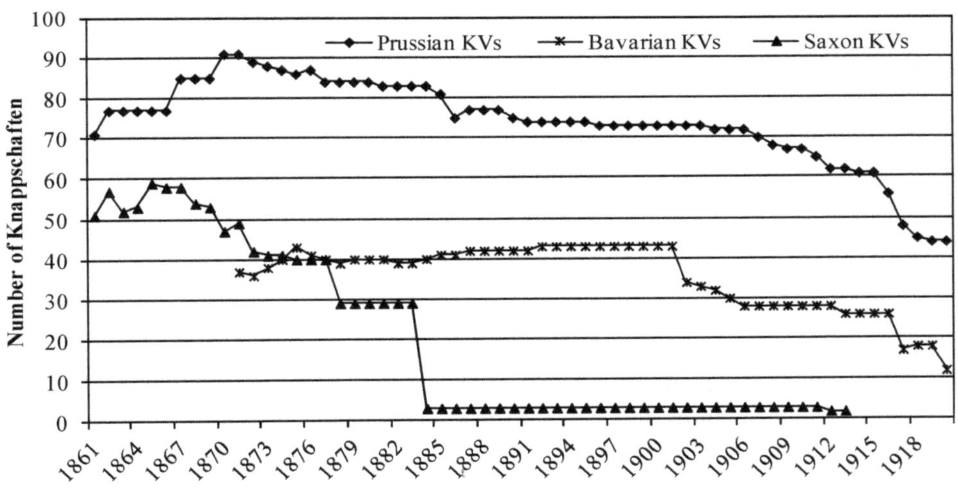

Sources: My own calculations based on Ministerium für Handel, Gewerbe und öffentliche Arbeiten (1861–1878), Ministerium für öffentliche Arbeiten (1879–1889), Ministerium für Handel und Gewerbe (1890–1922), and Oberbergamt München (1873–1921), and Königliches Finanzministerium Sachsen (1870–1872) and (1873–1921).

15 According to Kling (2006), who addresses the first merger wave among German firms before the First World War, we might speak in this context of the first social insurance fund merger wave in German history.

The one issue that motivates this study, besides the concentration phenomenon that we do not yet fully understand, is the historical debate in the literature among contemporary observers of the KVs, who discussed the funds' operations primarily in the context of some perceived major "design flaws." In this respect, the contemporary literature mirrors the steady attempt to not only identify those flaws, but to also seek solutions or improvements to the implemented institutional order – above all, external concentration. To make a long story short: Too many KVs were perceived as too small to provide sustainable insurance. In addition, complicating the whole story, the bundling of pension insurance and sickness insurance in one fund per location was perceived as inferior to separate provision in two self-standing funds. This, argument goes, prevented KVs from establishing the optimal-sized fund for each class of insurance – sizes that were proclaimed to be different for sickness and pension insurance. Contemporary critics *unanimously* supported the observable concentration process, which supposedly sorted out financially or actuarially unviable funds and, perhaps, led to the rescue of the funds that deserved to be rescued. Throughout the observation period, these critics even went beyond defending the observable process and demanded a still higher degree of concentration beyond what had been actually realized.

Although the KVs' quantitative account clearly reveals that the KVs' history at the time was, essentially, a history of concentration, the compiled statistical frameworks on them have – to my astonishment – not been used to date as an academic source to trace and evaluate the KVs' and, what is more, their observers' "quest for optimal scale."

There was actually one circumstance, also obvious from the statistics, that forced KVs into action and observers into thinking, and that, thus, initialized and motivated the search for the optimal institutional design of the entire insurance system: Economic dependency continually increased; put differently, the KVs' memberships aged. To illustrate this stylized fact, Figure 1.3 portrays the development of the number of economically dependent pensioners per 100 contributors, derived as annual cross-KV averages.

The average number of invalids (survivors) per 100 contributors per KV increased from about four (14) dependent pension recipients to 14 (more than 25) in 1913; this equals an overall growth by 357 percent regarding invalids and around 181 percent, at the least, regarding survivors. There is reason to conclude that KVs were under massive pressure from a continuously rising proportion of pensioners, precisely because they applied what is commonly known as pay-as-you-go (PAYG) financing (in contrast to funding; *Umlage* and *Kapitaldeckung*). This underlying circumstance was all the more relevant because KVs used to pay out pensions until death in the event of both invalidity of a miner and survivorship of a miner's wife. Thus, KVs effectively covered the risk of longevity and, therefore, implicitly provided old-age coverage.[16]

[16] The KVs' collective solidarity – within their respective memberships, but also across memberships – "was going to be tested", so to say; see Pearson (2002) and Pearson (2003), p. 51. I refer here to the title of Pearson's paper on English mutual fire insurance.

Figure 1.3: Average number of invalids and survivors per 100 contributors (Prussia)

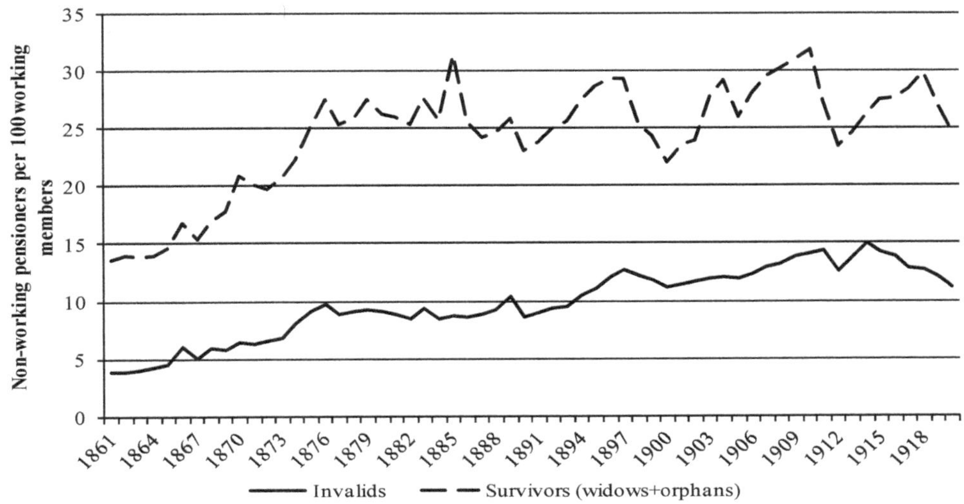

Sources: My own calculations based on Ministerium für Handel, Gewerbe und öffentliche Arbeiten (1861–1878), Ministerium für öffentliche Arbeiten (1879–1889), and Ministerium für Handel und Gewerbe (1890–1922).

This study should not be understood as arguing that the economic, social, and political conditions in the nineteenth- and early twentieth-century, in which the KVs were embedded, were the same as in today's industrialized economies with their advanced – or should we, at least, say matured – social security systems. Since the KVs were job-related, mining-specific factors probably were an important source of financial challenges (e.g., financial shocks due to massive accidents; exhaustion of resource deposits and, thus, structural decline of a mining area; and worsening geological conditions because of an increasing average depth). Nonetheless, by answering the central research question as it stands, this empirical study identifies parallels to modern economies in terms of the basic fundamental economic problem of creating a viable social insurance system that would provide efficient and lasting insurance coverage.[17] The question of what might be the optimal scale at which to target insurance operations – whether or not that optimal scale could be realized or whether it existed at all – worried KVs and con-

[17] That KVs still are a quite neglected piece of German insurance history becomes exemplarily clear when taking a look at Arps (1965), whose book *Auf sicheren Pfeilern – Deutsche Versicherungswirtschaft vor 1914* still seems to be the most comprehensive and, thus, standard historical account on the subject matter. He mentions the *Knappschaft* only four times – first and foremost in the context of Stumm's proposition of factory worker insurance along the lines of the KVs. This fact would not be that astonishing if he did not address Bismarckian social insurance either. But, indeed, there are separate chapters on *Reich* sickness, accident, and invalidity insurance.

temporary observers, including the regulator. Today, that same question tends to worry researchers, politicians, regulators, and lobbyists for various, perhaps conflicting, reasons.

1.2. The historical discussion about the appropriate fund size

A closer look at what I call the "historical discussion of the appropriate fund size," a look that goes beyond the notes in the previous subchapter, is inevitable for the subject matter.[18] The following concise overview discusses some anecdotal evidence; the overview is inevitably selective. However, I believe that the quotes presented are representative enough to provide an understanding of the discussion's bottom line. What might strike the reader after few lines as problematic is the fact that the historical discussion lacks counter-arguments. In fact, it was quite one-sided. Contemporary observers did not conduct a true discussion in the sense of weighing advantages and disadvantages. Rather, they – whether researchers, *Knappschaft* officials, or some kind of members' advocates – agreed with one other that their basic arguments about "design flaws" were correct. As a result, the contemporary literature does contain significant repetition and redundancy, not to mention a set of unproven assertions.

This overview begins with the "essence" of the contemporary observers' complaints, namely the important design flaws or structural deficiencies identified by them. Note that, for the most part, design flaws were identified with reference to Prussian KVs' experience, although they were of practical concern to the German *Knappschaft* institution as a whole. The essential flaws identified by observers were:

(1) KVs were too many in number and reflected a highly unequal size distribution ("fragmentation" design flaw);

(2) KVs were used to provide pension and sickness insurance "out of one hand," which prevented them from creating efficient KV sizes in each insurance section ("different efficient size" or "compound insurance" design flaw);

(3) If KVs engaged in mergers to partly overcome flaw (1), either voluntarily or under pressure from the industry regulator, their engagement was not sufficient to contribute to the stability of the whole insurance system ("insufficient merger activity" design flaw);

---

[18] This subchapter contains sections of the articles "Old times, better times? German miners' Knappschaften, Pay-as-you-go Pensions, and Implicit Rates of Return, 1854-1913" and "Insurance, size, and exposure to actuarial risk: Empirical evidence on nineteenth- and twentieth-century German Knappschaften;" see Jopp (2011b) and (2012a). Moreover, this section, as well as other parts of the study, includes a number of direct quotations. All translations from German are my own; see this study's long version (Jopp 2012b) for the German originals.

(4) Due to their choice of PAYG financing – or, alternatively, the absence of a strong regulator who would have prescribed full pre-funding of present and future liabilities – KVs were seen as incapable of keeping benefit promises to the ready-to-retire membership once made to them during their working years ("securing of benefits" or "PAYG" design flaw);

(5) The different treatment of members with more rights and those with fewer rights that many KVs perpetuated was judged unjust and misleading ("unestablished miners" design flaw);

(6) Since the bottom line of KV operations was that any working member who was willing to relocate to another KV's area lost his accumulated entitlements, free spatial movement of labor was *de facto* restricted ("portability of entitlements" design flaw); and

(7) The design of the KVs often offered members incentives to feign illness and to claim sickness benefits in excess of what was actuarially appropriate ("simulation" design flaw).

All of these complaints – which had already been voiced at the beginning of the KVs' new insurance era after the Prussian General Mining Law of 1865 and were steadily revived thereafter – were finally legally addressed by the important 1906 reform of the basic regulations, but with a delay of four decades (see Chapter 2 for details). In the following, I concentrate on design flaws (1) to (3), but address also flaw (7).[19]

As one of the earliest observers, Julius Hiltrop argued in 1869 that many KVs were, on the one hand, too small to ensure actuarial stability of their pension insurance scheme and, on the other hand, too large to successfully control for moral hazard in their health insurance scheme:

> "Of greatest importance for a KV's usefulness and efficiency, however, is its size. The more members a KV has, [...], the more solid will it become in view of granting benefits and overcoming challenges. [...] The basic evil, rather, is the preposterous fusion of health and pension insurance; a KV's size may be too large for it as provider of health insurance and too small for it as provider of pension insurance."[20]

---

19 Regarding point (4), which has, in the opinion of contemporaries, something to do with KV size as well, I refer the reader to the long version of this study, where I consider that point in more detail; see Jopp (2012b). Furthermore, due to the fact that many of the selected quotes contain implications on more than just one issue, it seems not appropriate to me to introduce separate subsections (indicated explicitly by four subheadings) and process the anecdotal evidence strictly separately. Besides, I am aware that the reader may wish at one point or another to have more institutional knowledge at hand to better classify the evidence. However, since I faced a "problem of sequence" in order to avoid too much repetitions, I decided to provide here only a minimum of institutional information. Since the only intention of this section is to convey an impression of what contemporaries thought were the bottom lines regarding the KVs' problems, this lack of additional information may be acceptable. Hirsch (1875), for example, offers an early contemporary listing of the KVs' deficiencies.

20 Hiltrop (1869), pp. 223–225.

In his opinion, this was so because of the assessment that:

> "Sickness and pensions funds are so much different regarding their underlying basic principles, such that a fusion of both [in one scheme] is inappropriate and would not serve the purpose."[21]

Which factors or requirements that establish such differences had he in mind? Actually, this is difficult to say. What we can say, in my opinion, is that a major and long-standing claim made early in dealing with the KVs' operations had been that small funds were actuarially unviable regarding pension finance because of their inappropriate size, and that the funds' performance would considerably improve alongside a growing collective of insurants. The favored formula was simple: The larger the KV, the better – if pension provision was addressed; however, if sickness insurance was addressed, then the formula would have to be re-stated as: The smaller the KV, the better. This interpretation is in line with Hiltrop's views, expressed a few years after the enactment of the basic regulations.

But Hiltrop was not alone in this view. Others, such as Albert Caron, Harry Karwehl, Heinrich Imbusch, and Ferdinand Bertrams, expressed similar thoughts.[22] Caron (1882), for example, combined the implications of the "law of large numbers" – that the relative frequency of an event *x* would steadily approach its true, but hitherto unknown probability of occurrence the more observations (i.e., in this case: insurants) are available – with pricing considerations. He did so by considering the implicit degree of risk loading as an important component of a premium, in addition to the expected value of costs and the administrative overhead loading. He actually established a relationship between the degree of risk loading and a KV's size, such that a smaller KV would require a higher risk loading:

> "The first fundamental condition for a restructuring of Knappschaft insurance would be to form large Knappschaft areas to bring into effect the law of large numbers. We would best achieve this by merging all Prussian Knappschaften into one pension fund. [...] It has to be emphasized that the surcharge on top of the net payment of contribution to even out fluctuations has to be larger, the smaller the collective of insurants is."[23]

It should be noted here that Caron tackled an issue that is one of the cornerstones of insurance economics: adequate pricing of an insurance contract.[24]

---

21 Hiltrop (1869), p. 223.

22 See Caron (1882), pp. 10–35, Karwehl (1907), pp. 61–71, Imbusch (1910), p. 61, and Bertrams (1912), p. 1 413.

23 Caron (1882), p. 20. As Caron had pointed out in a publication one year earlier, the risk surcharge was commonly considered part of "voluntary, market-based" insurance and was seen as a quantity that mattered; see Caron (1881), p. 15.

24 Emery and Emery (1999) and Broten (2010) recently estimated the implicit degree of risk loading for U.S. fraternal lodges and English friendly societies and concluded that both types of voluntary organizations charged quite risk-adequate premiums. The obvious implication of Caron's consideration is, again, in line with standard insurance theory saying that the reduction in the variance of

Beyond that, Caron made the case for merging KVs into larger entities as the inevitable prerequisite to reforming the unstable organizational structure of KVs, something that the reform of the *Knappschaft* law had failed to address two decades earlier. He actually seems to have been a kind of "structural optimist" since he continued to think of one empire-wide institution that should be formed to the advantage of the entire sector and all insured workers:

> "If one merged, in contrast, all the state's funds into one organization, this would best guarantee the survivability of the whole institution. For it can be concluded that mining in the whole country would not at some time persistently decline below its current state."[25]

Harry Karwehl more explicitly underlined the design flaw that was identified in the high fragmentation as the "cancerous ulcer" of *Knappschaft* insurance.[26] He tried as well to make clear why one did well to consider small KVs inappropriate to deal with the challenges of insuring miners:

> "A small Knappschaft fund is much too sensitive to particular events occurring for the mine or the few mines belonging to it. Such events may be: firedamp, coal-dust explosion, pit fires, man-shaft accidents, epidemics, but also the depletion of the resource deposit or other technical or economic obstacles to operating a mine. Of these events, massive accidents drive claims costs up extraordinarily and might also result, at the same time, in decreasing payments of contributions."[27]

This quote might remind the reader of what can be called in insurance theory and pratice the "accumulation risk" to which an insurer might be exposed; it basically means that the claim of insurant *y* is not independent of the occurrence of a claim by insurant *x*.[28]

Based on this assessment, the natural consequence for Karwehl, as for his previous speakers, was to advocate a necessary selection process during which unviable, small KVs were to be merged into larger KVs:

> "Further concentration is imperative, for, in addition to some individual disadvantages arising for comrades from fragmentation, the small and smallest funds cannot, from the start, fulfill the social tasks that a KV is basically intended to fulfill."[29]

---

average costs is the necessary precondition for an insurer to lower the risk surcharge. In a competitive environment, an insurer would very probably do so, and the premium would develop toward an actuarially fair one, which is roughly equal to expected individual costs; see Emery and Emery (1999), p. 81, and Broten (2010), pp. 35–40.

25 Caron (1882), p. 21.

26 Karwehl (1907), p. 71. Precsiely, he stated: "The German Knappschaft institution obviously suffers from cancer: It is the fragmentation into various small funds."

27 Karwehl (1907), p. 72.

28 The corresponding German term to "accumulation risk" is "Kumulrisiko."

29 Karwehl (1907), p. 72. Heinrich Imbusch, to bring this famous workers' representative into play, points out the same problems that Hiltrop, Caron, and Karwehl wrote about. With respect to the high fragmentation among KVs, he stated: "One of the basic evils regarding Knappschaft insurance is its high degree of fragmentation. Instead of forming large pools of insurants, it was often

Karwehl further makes the case for separating the KVs' pension insurance section from their sickness insurance section, thus substantiating design flaw (2) (as Hiltrop, for example, had already done in 1869 – without proper explanation, however):

> "All these facts force KVs to separate sickness insurance from pensions. A mere accounting separation would not be enough; rather, an organizational separation is imperative. This is because sickness funds require small collectives of insurants such that members can monitor each other, which is more effective and cheaper than administrative control by physicians; this would help to identify causes of excess expenditures. […]. Pension funds, in contrast, require large collectives of insurants with many members such that the law of large numbers comes into effect, […]. […]. So, only for a large stock of insurants can regularities be identified regarding the process of becoming invalid or dying […]."[30]

In this respect, Imbusch claimed that the rate of mergers was not sufficient, not the least because mine owners might have refused to participate in them in order to maintain autonomy and to retain control of their workforce:

> "The belated amalgamation of smaller funds and the formation of larger area-related funds was usually difficult because the state of assets often diverged, and KVs could not agree on how to compensate each other; besides, employers wanted to have small funds per mine to better discipline their workforce."[31]

Another contemporary, Peter Simons, earlier suggested, in 1890, that larger KV areas, achievable by mergers, would result in a de-coupling of local, or even regional, economic growth or decline and a KV's financial state. This is because growing areas, where deposits were still rich, could cross-subsidize stagnating areas, where deposits were close to economic or technical exhaustion.[32] An industry regulator would very probably like this argument since it implies greater stability of the entire insurance system. However, in a prospering mining area, employers as well as miners would very probably not have liked to see their prosperity being redistributed to stagnating areas. So why should have employers decided to absorb a stagnating KV if not for solidarity or pressure by the regulator?

Ferdinand Bertrams's (1912) assessment points in a similar direction:

> "Where there were larger KVs, one could better compensate for recessions or diminishing numbers of members. However, if the few mines (or the single mine) that were linked to a small KV experienced demand crises or membership shrinkage, they were on the verge of ceasing operations. Such insurance institutions must have their feet planted on solid ground and be willing to make immense sacrifices to withstand the storms of bad times. Such sacrifices – increases in contributions without commensurate increases in benefits – were not always easy to sell to the memberships."[33]

---

the case that KVs came into existence for very small areas, if not for single mines;" Imbusch (1910), p. 61.

30 Karwehl (1907), pp. 61–62.

31 Imbusch (1910), p. 61.

32 See Peter Simons (1890), p. 11.

33 Bertrams (1912), p. 1 413.

So far, presented quotes give an impression of how the nexus of the appropriate KV size, mergers and the unbundling of pension and sickness insurance was assessed. Contemporaries obviously believed that running a pension benefit scheme requires a minimum efficient size that was larger compared to the minimum efficient size of a health insurance scheme; we may define it as the size beyond which there were no economies of scale left. The actuarial ideas behind this are, indeed, well known in economics. Relating to moral hazard in the KVs' health insurance section, Guinnane and Streb (2011) and Guinnane et al. (2011) tested the claim that in small KVs, where everybody knew and observed each other, the misuse of the fund's resources by claiming sick pay without being sick at all (*Simulation* according to contemporaries' diction) could have been effectively prevented – or, at least, minimized – through social sanction mechanisms.[34] Moreover, it is commonly agreed that insurers are exposed to a special sort of risk – the actuarial risk, which means that either each individual insured's actual claim or the aggregate actual claim of the insurants' collective, at the end of a period, exceeds expected claims (or falls short of them). Yet, economic theory suggests that the predictability of numbers of claims and, thus, amounts of claims improves alongside a growing collective of insurants. This is because an increase in the number of insurance contracts reduces the variance of the average individual claim towards zero via the (empirical) law of large numbers. Likewise, a decrease in size is equal to an increase in actuarial risk measured by the variance, and a small collective of insurants is generally associated

---

[34] See Guinnane and Streb (2011) and Guinnane et al. (2011). In this context, contemporary observers assumed that the generosity of the sick pay benefit would provide KV members with a strong incentive to malinger. Small funds where everyone knew each other and had close personal ties to fellow workers, were said to help control costs and monitor the behavior of mine workers better than large, more impersonal funds. Social sanction for abusing the funds, or simply the threat of it, was seen as an efficient disciplinary means. Note, however, that waiting periods, which KVs also specified, and a sick pay benefit of less than 100 percent of daily wage (see Chapter 2), may be seen as other instruments (i.e., forms of percentage excess) to control for simulation that do not have anything to do with KV size. More accurately, contemporaries of the KVs addressed a special sort of moral hazard induced by ex-post hidden information (in contrast to ex-ante hidden information leading to the problem of adverse selection) rather than by ex-ante or ex-post hidden action. Simulation was possible if miners had an information advantage regarding their state of sickness as long as KV physicians or other commissioned controllers (such as the Elders; see Chapter 2) could not successfully prove they were not ill at all. Guinnane and Streb (2011) argue that especially rheumatism was used as a cover for simulation. With Nell and Schiller (2002), Schiller (2004), and Lammers and Schiller (2010), we might call this problem of moral hazard more accurately a problem of insurance fraud. In contrast, then, the usual moral hazard problem in the context of insurance is a problem induced by ex-post hidden action – i.e., the mere fact that an individual has insurance coverage leads to a behavioral change that is equal to a reduction in the personal level of carefulness. For example, the fact that a car driver has car insurance arguably leads to a situation in which she drives less careful because she can be sure that possible damage would be reimbursed by contract. However, the bottom line regarding this problem of moral hazard is that the insurant does not intentionally betray the insurer and the collective of insurants, but does simply respond to incentives set by the existence of the insurance contract.

with a higher variance. Hence, from a theoretical viewpoint, growth in contributors would have been good for KVs, and mergers were definitely a measure to jump immediately to a larger size level.[35]

Contemporaries such as Caron or Karwehl argued explicitly on the basis of the law of large numbers. Smaller KVs were said to be more vulnerable than larger KVs to unpredictable events such as accidents or fluctuations in the number of contributors and pensioners. Thus, they had a higher ex ante variance of the average claim, which is equal to a larger bandwidth of possible positive or negative outcomes. Just as buying insurance reduces uncertainty about the future income of an individual, expanding the collective of insurants reduces uncertainty about the future state of finances of the insurer himself. Caron also talked about a steady state – or, in his terms: a *Beharrungszustand* –, in which all fundamental actuarial data – the inflow of new members per age group, the outflow of contributors due to death, turnover or invalidity, and the outflow of pensioners due to death, among others – remain unchanged over a very long time period, so that perfect predictability of financial needs was achieved. This steady state, however, was claimed to require a large size and the law of large numbers coming into effect.[36] As a consequence of actuarial considerations, mergers of small KVs into larger ones, or even mergers of all KVs into one fund for Prussia or Germany as a whole, were again and again demanded. Debates finally culminated in the foundation of the *Reichsknappschaft* in 1923.

Up to this point, some important contemporary critics of the organizational environment in which the KVs were embedded had had their say.[37] But did the regulators and policy makers share their opinion? Even if contemporary opinion suggested the formal

35 See, for example, Albrecht (1982).

36 Caron (1882), p. 8. It reads: "It is undeniable that a steady state, [...], implying that the proportion of active members and invalids inflowing from them remains constant, may occur under certain circumstances. If, during an undetermined period of time, the same number of active members of the same age existed every year, the number of those becoming invalid would be same every year given constant probabilities of death and of becoming invalid. If, on the other hand, from some point in time on, a larger, steadily recurring number of same-aged actives became invalid, there would be a state, after the last of those who became invalid first had died, in which the number of simultaneously living invalids will be maximal. The annual inflow of new invalids of the same age would, from then on, be compensated by those invalids of the same age that exit membership. The combination of both, principally undenied propositions, reflects the steady state as it is targeted by KVs. It basically requires that the same number of active members of the same age exist every year, and that their pool is of such a size that the law of large numbers comes into effect."

37 Note that contemporary observers of the KVs did not come up with an estimate of minimum efficient fund size. Not in the context of KVs, but in the context of sickness funds for factory workers and journeymen, the committee of the Chamber of Commerce Leipzig underlined, in a report of 1867, that a size of 400 to 1 000 members would actuarially be appropriate. Unfortunately, we do not get to know from the report on which theoretical foundation this assessment had rested. It might be that contemporary observers of the KVs thought of such a size in the KVs' sickness insurance as well; see Tennstedt et al. (1999), p. 42, for the report.

unbundling of sickness and pension insurance, this was not implemented by law before 1906 (as will be referred to in more detail in Chapter 2). The basic regulation of 1865 had given KVs the opportunity to split their membership into extra sickness funds per mine or steelworks, while keeping one large pension fund; only a few KVs took the opportunity.[38]

Contemporaries identified the unbundling of the sickness and pension insurance benefits as a necessary precondition to implement the optimal KV size for both types of coverage, though this size was never specified in quantitative terms. Since most KVs refused to split their memberships into several sickness funds per mine, while keeping one single pension fund, contemporaries presumably realized that it makes no sense to highlight the advantages of both small and large KVs simultaneously. Instead, they seem to have focused more extensively on advocating large KVs, thus making pension insurance the top priority. Mergers, in particular, were underlined as a means to release small KVs from financial pressure due to high variability and to help larger KVs increase their size immediately.

So what can we deduce from the regulations themselves and related information about the regulator's intentions? The mining administration as the industry regulator was very probably of the same opinion as the aforementioned experts: namely, that increasing the KVs' size by mergers would be a useful strategy. In fact, the regulations of 1854 and 1865 did not say much on this topic. They stated only that a KV should be neither too small nor too large. More precisely:

> "Regarding the demarcation of KV areas [...] it must have top priority that areas are neither too small, such that the survivability of the fund would be endangered, nor too large, such that the fund's administration and the KV board's effectiveness would be complicated."[39]

In contrast to Hiltrop, for example, who would have surely advocated the slogan "the larger the KV, the better," the legislature seemed to have adopted *de facto* a different view because it implied – perhaps not even intending to – the existence of a kind of optimal scale as the inevitable consequence of a trade-off between actuarial and administrative efficiency. Practically speaking, this would be a scale of operation at which the marginal costs and benefits of KV size, with respect to actuarial and administrative considerations, would be equal.

Towards the reform of the basic regulations in 1906, however, the administration exerted some verbal pressure in favor of larger pools of insurants, but did not force mergers or liquidations upon KVs – at least not in such an explicit way that contemporary observers were in the position to provide us with a clear account of such efforts. Hilt, for example, wrote about an attempt in 1870, initialized by mining official Hermann Brassert, to merge all KVs in the mining administration region of Bonn[40] into one large

---

38 See, for example, Guinnane and Streb (2011), p. 74.

39 Ministerium für Handel, Gewerbe und öffentliche Arbeiten (1856), p. 25.

40 This Prussian mining administration region was the comparatively most fragmented one.

pension fund. This attempt essentially failed because of most firms' opposition to the proposal.[41] Another attempt of the administration to clarify its position regarding size and mergers came in the form of the ministerial decree of 1883, which explained the implications of *Reich* health insurance for the KVs' operations and reads with regard to the issues of size, bundling and mergers as follows:

> "To ensure success in this direction [i.e., counteract rising health expenditures], it is once more recommended to separate sickness from pension insurance […], which has so far been done – except for the former Duchy of Hesse – only sporadically."[42]

It further reads:

> "It has been repeatedly stressed that the fusion of small KVs among themselves or with larger KVs is the appropriate means to improve their efficiency. […]. Some mergers were indeed conducted in the near past, but not to a desirable extent. It must be emphasized from now on that mergers should be conducted all over, where circumstances demand it – not only regarding the one or other small KV, but under systematic consideration of the larger KVs."[43]

The administration concluded that, indeed, some mergers were done, but that this was, by far, insufficient. Yet, before 1906, there is no evidence that the administration actually forced mergers or liquidations upon KVs. There is only evidence that the verbal pressure increased.

The amending law of 1906, then, made it possible for regulators to formally force mergers or closures upon KVs (§177a and b).[44] Bertrams informed us that the liquidation of the *St. Wendeler KV* in 1906, for example, was a consequence of this new regulatory instrument.[45] On the whole, evidence does not suggest that mergers were primarily a consequence of regulatory policy, but a result of business policy. Yet it remains quite unclear whether KVs that absorbed another one did so to help the absorbed KV out of an actuarial trap, to seek their own growth opportunities, or to please the regulator.

In 1910, Imbusch judged the administration's past efforts as follows:

> "The government was busily engaged in seeking an organizational separation of pension and sickness insurance, which was principally allowed by the law of 1865. However, the government did not generally succeed. It was especially in the Ruhr area, which now hosts the largest of the Prussian KVs, where the separation as envisaged by the government was rejected. It was a mere accounting separation that was done."[46]

All of the quotations presented in this section convey a – hopefully – clear enough impression of how the issue of "optimal" KV size was commonly perceived. Contemporaries had developed an understanding of insurance in which the advantages and disad-

41 See Hilt (1877).
42 Ministerium für öffentliche Arbeiten (1883), pp. 79–80.
43 Ministerium für öffentliche Arbeiten (1883), p. 80.
44 See Steinbrinck (1908), pp. 152–158.
45 See Bertrams (1912), p. 1 553.
46 Imbusch (1910), pp. 63–64.

vantages of pool size mattered for the design and efficiency of the mineworkers' insurance scheme. They identified a kind of "design flaw" in the bundling of the sickness and pension insurance benefits, which, they claimed, demanded different minimum efficient sizes of the pool; and they deduced and advocated the necessity to create large funds – if necessary, by conducting mergers.[47] In fact, they approached these questions in a theoretical fashion. However, these questions are, after all, empirical ones.[48]

1.3. Hypotheses

After having pinpointed contemporaries' perceptions of general problems associated with KVs, this subchapter derives three hypotheses that operationalize the central research question and that directly link up with the historical discussion.

To begin with, consider that the nineteenth century actually saw a variety of approaches toward mutual worker insurance, not only the KVs, of course. On the one hand, there were funds with voluntary membership: the friendly societies in Great Britain and elsewhere;[49] the fraternal lodges in the U.S. and Canada;[50] and the non-fraternal industrial sickness funds in the U.S. are examples.[51] On the other hand, English pit clubs and U.S. labor union funds are examples of schemes with compulsory membership. All these funds insured their members primarily against the economic conse-

---

47 With regard to the usefulness of mergers given market imperfectness, Dewey (1961), p. 258, provides us with an interesting conclusion: "Mergers can only make an imperfect market more imperfect." What is meant is that typical adjustments due to suboptimal scales in the market will automatically drive out inefficient firms and bring out those that operate at the optimal scale. In this respect, mergers do not play an important role in this process. To me, it seems justifiable – especially with respect to empirical findings in the subsequent chapters – to understand KVs as reflecting a sort of inefficient institutional design, due to fragmentation and a highly unequal size distribution, where only a few, if any, KVs operated at an optimal scale. So, with Dewey, we should doubt from the start that mergers among KVs pushed the system towards perfection. However, he also concedes that there certainly are sorts of economies of scale that are realizable only through merger (e.g., rationalization based on non-reproducible knowledge held by a particular firm); see Dewey (1961), p. 259.

48 As stated, this overview is selective, as it does not include all available contemporary literature on the KVs. To be mentioned is the subset of studies that write a history of a particular KV; see, for this, Simons (1890) on the *Wurm KV*, Pitschke (1892) on the *Mansfelder KV*, Schneider (1901) on the *Allgemeiner KV Bochum,* Wolff (1907) on the *Wetzlarer KV*, Bülow (1910) on the *Allgemeiner KV Bochum*, Reisel (1924) on the *Saarbrücker KV*, Schlick (1927) on the *Saarbrücker KV*, and Hollborn (1942) on the *Halberger KV*.

49 See Alter and Riley (1989), Van der Linden (1996), Riley (1997), Harris and Bridgen (2007), and Broten (2010).

50 See the seminal contributions of Emery (1994, 1996, 2006) and Emery and Emery (1999), and Beito (2000).

51 See Murray (2003, 2007).

quences of illness, but – as the pit clubs did – also against accidents.[52] Beyond that, it seems as if those funds' histories have two important aspects in common: First, their design, development and potential financial weaknesses, as compared to commercial or government-based solutions, were the subject of – sometimes intense – debate at the time; and second, contrary to the usual assessment that they were rather weak financially, recent empirical research indicates that their operations were financially sound. This finding is striking because they did not use actuarial technology, as commercial insurers had begun to do, and seem not to have been engaged in forming large pools of insurants in order to exploit economies of scale. If the mutual sickness funds' importance declined over time, this was not, indeed, because of financial failure, but because of comparative actuarial advantage of commercial insurers, as in the case of U.S. industrial sickness funds, or because of changing members' preferences, as in the case of U.S. fraternal sickness insurers.[53] In contrast to those schemes, however, it has to be underlined that the size of a fund's pool of insurants was definitely the centerpiece of the historical German discussion.

The empirical part of this study is intended to test for the possible costs and benefits of KV size. The test has been created to accommodate contemporary observers' ideas. In my argument, these ideas can be linked with present-day economics, and doing so is worth a try simply because they reflect timeless economic concerns. Imagine not only insurance providers as the most relevant form of business in our case, but any firm: It is imperative to have some idea about which scale of production is efficient or even optimal; otherwise, inefficiencies might drive the business out of the market. And though firms might not care, the government agency interested in regulating the industry certainly will. Imagine, for example, a typical country that faces decisions regarding the kind of social insurance system to introduce. Once the decision has been made on whether the system should cover a particular fraction of the population (as, for example, industrial workers in the context of Bismarckian insurance 1883–1889) or should be (nearly) universal (as, for example, the recent German social insurance system), another decision must be made: Should covered individuals be organized into one large pool or fund, or should they be organized into many much smaller pools? If the latter alternative is chosen, how should they be divided up? By territory? Occupation? Income? Sex? Or according to a combination of these traits? What the quoted observations on the KVs' history addressed is the actors' search for the unknown nature of that optimal design.

In order to operationalize the central research question and to elaborate on the KVs' historical experience, a number of sub-questions or hypotheses must be derived. This is done against the background of one fundamental presumption: KVs cannot be analyzed in isolation from underlying economic relationships nor from some basic modeling of

52 See Benson (2002) and Murray (2007).

53 See Murray (2007), pp. 218–236, and Emery (2006), pp. 480–481. Moreover, Fishback (1992) makes a related case for state intervention in accident insurance in the U.S.

these relationships. In this regard, economists commonly agree upon two facts: First, an insurer is naturally exposed to some sector-specific risk, which is commonly called "actuarial risk" and which can be influenced to a limited extent by risk policy, but not eliminated to zero; this is, of course, due to the stochastic nature of insurance production. Second, the financial-services sector, including insurance, is one in which economies of scale are claimed to play an important role. Altogether, size is claimed to be generally important for insurance organizations, and – from a theoretical viewpoint – they not only gain from growth in size in terms of reductions in their exposure to actuarial risk, but also have to bear the costs of potential diseconomies of scale. Thus, it seems appropriate to speak of the costs and benefits of an insurer's size in general or, in the context of this study, of KV size.[54] In the end, it is an empirical question of whether economies of scale (and often mentioned in the same breath: economies of scope) exist in a particular industry up to a certain level of production and what this means for the optimal scale and, thus, the preferable design of the market and necessary, or reasonable, regulatory intervention. For example, with respect to the recent European insurance market and the likely effect of firm mergers on that market, Eisen (1991) points out that economies of scale in insurance production may be assumed, but that only an empirical investigation can reveal their nature. Assuming that they exist, three possible scenarios should be distinguished from each other:

(i) Increasing returns to scale exist over the full range of firm size;
(ii) Increasing returns to scale exist up to a certain minimum efficient firm size and vanish beyond that size;
(iii) There exists an optimal firm size implying a U-shaped pattern implying, in turn, the existence of increasing returns to scale up to an optimal scale and decreasing returns to scale afterwards.[55]

Based on the assertion that size was not irrelevant for the performance of KVs, I assess the costs and benefits of KV size in terms of the empirical relationship between KV size and (i) the actuarial risk that KVs were exposed to, and (ii) the operating costs that were necessary to transform insurance claims into insurance benefits. What is added to the analysis of these problem areas is an analysis of external concentration exploring, statistically, the motives of absorptions and liquidations.[56] In what follows, I present the baseline working hypotheses and link them with these categories of analysis in order to establish their relevance.[57]

---

54 In a more general context, Steenbeek and van der Lecq (2007) address the costs and benefits of collective pension systems.

55 See Eisen (1991), pp. 268–270.

56 There are two additional problem areas on that I focus in the study's long version: the empirical relationship between KV size and sustainability of the pension system, and between KV size and pricing; see Jopp (2012b).

57 The question of whether economies of scale might exist in the production of whatever good or service has attracted interest in very different economic contexts, in fact. To mention only few,

(A) Size and actuarial risk: As the brief overview of contemporaries' views has established, the importance of installing large KVs was deduced from the "law of large numbers." According to Albrecht (1982), for example, with more observations available, the relative frequency of an event approaches its true, but unknown, probability of occurrence.[58] Contemporaries apparently believed that the law's impact would help to stabilize the KVs' pension benefit scheme by reducing the variability and, thus, the negative financial consequences of the events mentioned by, for example, Harry Karwehl. By reducing the variability of average claims costs in the end, the predictability of aggregate claims costs was considered improved. This consideration, of course, matches with standard insurance theory, not the least because the "law of large numbers" implies that the variance of the average claim tends toward zero over time if the size of the pool of insurants increases.[59]

Against this background, it seems reasonable to test for the following hypothesis:

H0: The actuarial risk to which a KV was arguably exposed decreased steadily alongside a growing collective of insurants.

H1: There was a minimum efficient or even optimal size of KVs, beyond which the exposure to actuarial risk could not be further systematically reduced or, respectively, beyond which exposure to actuarial risk increased again.

The relevance of addressing actuarial risk stems from a fact with which every economist would certainly agree:[60] It lies in the stochastic nature of insurance production that an insurer is naturally exposed to it. Even though contemporary observers did not use the term "actuarial risk" in their discussion, they definitely addressed its substance.

For an insurance company, production of insurance coverage means giving the insurant a conditional promise to grant benefits in case certain pre-defined events occur – e.g., sickness or invalidity. In the context of KVs, insurance benefits meant, first and foremost, monetary payments to compensate the miner and his family for temporary or permanent income losses. The miner would, in exchange, pay a premium or, more accurately, a social insurance contribution. To meet the principle of collective equivalence, a

---

Stonebraker (1993), for example, assesses optimal church size with an economic model. Pope (1986) assesses economies of scale in state-wide pension systems. Kellermann (2008) concentrates on economies of scale and federalism.

58 See Albrecht (1982), pp. 504–507. Besides the empirical formulation, Albrecht also discusses the more sophisticated mathematical formulations of the "law of large numbers."

59 The literature on economies of scale in financial services production seems to be quite rich. However, economies of scale with respect to actuarial risk seem not to be on the empirical agenda, especially not when turning to history; see Nekby et al. (2001) and Nekby (2004) for studies that use historical data. Empirical studies that focus on the extent of economies of scale in recent banking and insurance include, among others, Hardwick (1994) , Horgby and Söderström (1998), and Amel et al. (2003).

60 For more detailed overviews, see Schwake (1988), Albrecht (1991, 1992), and Bittl and Müller (1998).

KV had to ensure in every period that aggregate revenues would match aggregate expenditures. That, in turn, required the KV to adequately predict future costs so that the average contribution payment could be properly specified. A KV's principal challenge was to predict the number of claims to be reasonably expected, as well as the average claim amount.[61]

From the perspective of insurance economics, ex-post aggregate costs ultimately fall short of ex-ante expected costs due to happenstance and due to an average contribution payment that was possibly falsely specified in retrospect. We do not know precisely how KVs formed their expectations. But we know that contemporaries understood a basic insurance principle: Every insurant is likely to represent a variance greater than zero in making a claim of whatever size. That is to say, there existed a certain range within which the average insurant might have claimed over the year. The larger the range was, the more difficult was it, of course, to predict financial needs.

Let the actuarial risk that a KV, or any insurance provider, was exposed to be defined in three alternative ways: First, it may be the risk that the average ex-post insurance claim deviates from the ex-ante expected; "average" refers to the average insured individual. Second, it may the risk that the aggregate ex-post insurance claim deviates from the ex-ante expected claim. Third, it may be the risk that total actual premium income does not match total costs.[62] For reasons elaborated later, in Section 4.2, I consider actuarial risk according to dimension one. That is, I focus on the deviation of the actual average claim size to be financed by one contributor from the expected average claim size. As the literature points out, an insurance provider is definitely subject to both a stochastic claims process producing the costs of providing insurance coverage and a stochastic revenue process producing the means to finance those costs and to maintain budgetary balance. The implicit assumption behind focusing on the deviation of actual average costs from expected average costs – i.e., in my framework, the variance of the average claim – is that a KV tended to specify social insurance contributions in such a way that expected costs were covered. Thus, if actual costs exceeded expected costs, KVs had underestimated truly required revenues. Put differently, if KVs adjusted their specification of the average contribution payment to expected costs, an assumed reduction in the variance of the average claim through growth in KV size would have led to a situation in which one could be sure that ex-ante specified contributions would match ex-post expenditures.

(B) Size and operating costs: In fact, the historical discussion of the appropriate KV size *did not* focus on whether there was potential to reduce average operating costs by growing larger. Against the background, however, that operating costs nowadays are said to represent a substantive amount of a social insurance system's overall costs, it seems worthwhile to test for the following hypothesis on the miners' historical social insurance system:

---

61 See Murray (2007), p. 221, for a theoretical perspective.

62 See Zweifel and Eisen (2003), pp. 240–243, for this assessment.

H0: A KV's average operating costs reduced steadily alongside a growing collective of insurants.

H1: There was a minimum efficient or even optimal size among KVs, beyond which the operating costs could not be further systematically reduced or, respectively, beyond which operating costs increased again.

As with exposure to actuarial risk, if the null held, there would be room to argue that mergers, and especially the absorption of small KVs, were an appropriate means to both improve the KVs' efficiency and to stabilize the entire insurance system by working towards the ideal – but unknown – institutional setting.[63]

(C) Size and external concentration: Starting in the 1870s, absolute and relative concentration among KVs steadily increased, and the underlying concentration process was not driven just by unequal internal growth, but by liquidations and mergers in particular. Aside from the fact that we can observe this wave of external growth among KVs and link it up with the contemporaries' comments, the decision-making process itself, involving miners, their employers, the state, as well as contemporary observers as external opinion makers, is still pretty much a black box.

In light of the aforementioned basic hypotheses, the third hypothesis to be tested arguably stands out as an even more general extension to the empirical framework hitherto addressed. But it is an extension, nonetheless, of which the relevance almost necessarily follows from the hypotheses addressed above and the historical discussion on the appropriate KV size: Regardless of what the empirical evidence for hypotheses (A) and (B) may reveal about economies of scale and, consequently, the usefulness or efficiency- and survival-improving effect of mergers, how can we explain the observable concentration process among KVs characterized by many exits of KV?

The recent economic literatures, as well as the contemporary critics, suggest two well-known motives for insurance mergers that may also apply to the period this paper studies. On the one hand, mergers might have been conducted to take financial pressure off KVs. Some KVs were struggling due to the structural decline of the mining area to which they were tied and due to business-cycle fluctuations or actuarial disadvantages, such as being too small or lacking actuarial expertise. Both facts, it is surmised, eventually drove the KVs and their insurants to financial disaster. I call this the "rescue hypothesis." From this perspective, mergers were conducted in order to stabilize the job-related insurance scheme in Prussian mining as a whole. This could have worked only if

63 See on this issue, for example, Caswell (1976), Estrin (1988), Mühlenkamp (1995a, 1995b), Mitchell (1998), Whitehouse (2000), Bateman and Mitchell (2005), Dobrogonov and Murthi (2005), and Bikker and de Dreu (2009). Focusing on economies of scale with respect to administrative overhead of modern German sickness funds, Mühlenkamp's (1995a) findings, for example, suggest an optimal size of local sickness funds (*Ortskrankenkassen*) of between 80 000 and 130 000 members. Furthermore, regarding modern Dutch pension funds, Bikker and de Dreu (2009) find evidence for an optimal size of around 50 000 members. Against this background, we do not have any idea of whether this held for the nineteenth century as well and, if not, why.

at least one of the following conditions was fulfilled: First, miners and mine owners, who self-managed their KVs together, felt highly responsible for their fellow members and unanimously considered it an act of solidarity to help financially distressed KVs; and/or second, the mining administration – hence, the industry regulator – forced mergers on KVs according to its own aims and regardless of whether or not miners and mine owners agreed with the strategy of centralizing decisions. On the other hand, mergers might have been conducted simply because the absorbing KV aimed to seek growth opportunities and advantages – such as achieving cost reductions by improving one's own actuarial fundamentals. I call this the "self-interest" hypothesis. Here, solidarity is not required. Rather, the regulator would have been required to tolerate any merger conducted, implying a quite passive position.[64] Since both the rescue hypothesis (H0) and the self-interest-hypothesis (H1) attach importance to the characteristics of the KVs that factually ceased operation, either by absorption or liquidation, they will be tested from the perspective of those exiting funds.

A third hypothesis, however, might be that mergers, and perhaps even liquidations, were the consequence of accidental, ad-hoc decisions rather than planned and conscious action. If the evidence is not conclusive for either of the first two hypotheses, this third hypothesis might be a way of dealing with what happened.[65]

1.4. Placing the study

The following brief overview serves the purpose of placing this study within the body of literature on the German *Knappschaft* institution since 1947, when Varnas's study emerged, apparently the first major post-war publication. In Table 1.1, studies are or-

---

64 Recent literature on the consolidation in the financial-services sector including insurance, combining questions on the effects and determinants of mergers with motives to merge, include Cummins et al. (1999), Fluck and Lynch (1999), Wheelock and Wilson (2000), Ralston et al. (2001), Cabo and Rebelo (2005), Bühler et al. (2006), Cummins and Rubio-Misas (2006), Cummins and Xie (2008), Kötter (2008), and Bauer et al. (2009).

65 In seems to be in order here to point out the approach that I *do not* take to assess the usefulness of mergers and their effects on indicators of performance – their synergy effects, so to speak – but that the literature may suggest. The approach I think of is to conduct an event analysis of a different form than is actually done in Chapter 5, namely in the form of comparing the performance of the pre-merger firms – the "mother firm" and the target(s) – with immediate post-merger performance. This seems to be done often with reference to in-year stock market data or other, preferably in-year, data. The basic logic behind this is to identify whether performance is significantly different – for the new entity in the immediate post-merger period – from the involved firms' pre-merger experience, such that it can be concluded that the merger was a breaking event in one direction or another; see, for example, Koetter (2008). In light of such an approach, I constructed a simple empirical model regarding actuarial risk in Jopp (2010a). However, I will not pursue it further in this study.

dered chronologically and are reported together with the time period and the territory they cover and their main subject. In contrast to what this overview may imply, the studies rarely indicate time and regional focus explicitly, but these usually follow from the content. In this sense, columns two and three, first and foremost, display my personal perception of the scope.[66]

A set of publications that are part of a larger project – *Vergangenheit und Zukunft sozialer Sicherungssysteme am Beispiel der Bundesknappschaft und ihrer Nachfolger* – must be mentioned here in particular. This dissertation grew out of my research as assistant on that project, which studies the origins and development of the occupational social insurance systems developed in the German mining, railroad, and shipping sectors. The project, which takes an interdisciplinary approach combining historical and economic methods, was funded by the Leibniz-Gemeinschaft from 2007 to 2009.[67] This study links up with the sub-project on economic aspects led by the Rheinisch-Westfälisches Institut für Wirtschaftsforschung (located in Essen, Germany) in cooperation with the chairs of economic history at the University of Hohenheim and Yale University. The analysis was intended primarily to derive valuable empirical findings on a topic of ongoing debate in economics and economic history: What are the costs and benefits of the size of an insurance scheme – in particular, with regard to moral hazard in health insurance, intergenerational redistribution in pension insurance, and the actuarial fundament as a whole? The question of what role concentration or, respectively, consolidation played, as well as the effects it had, is one important piece of the analysis. Recent publications that emerged from the larger project include all studies since Bartels (2009) and except for Companje (2009) and Bluma (2011). Bartels et al. (2009) provide an overview of the project's central objective, its interdisciplinary character and the basic questions addressed.

Among the studies published prior to the above-mentioned project, those by Martin H. Geyer and Ulrich Lauf stand out as seminal in the field. Most studies are intended to deal with aspects of the KVs' history in a qualitative way, rather than a quantitative. They are most valuable in providing the reader with historical information on and an historical explanation of the emergence and the development of the institutional setting, and especially the essence of the legal changes. In sharp contrast, Brettschneider (1991) and Rieger (1999) provide two investigations of the *Bundesknappschaft* after 1969 that stem explicitly from the field of economics.

This study is different in two ways: First, it offers a new account of the economic history of KVs based on economic rationale and modelling; in this respect, this study's approach may be placed between the two distinct sets of investigations based funda-

66 The overview excludes, in particular, monographs on the emergence of the German welfare state that might mention KVs in that context, but are not intended to deal with them in more detail; see, for example, Tampke (1981) and Hennock (2007).

67 See Metzger (2007), Bartels et al. (2009), and Bartels (2010). Montz (2010) and Sulzer (2010) provide studies on the railroad and shipping sectors.

mentally on historical science, on the one hand, and on economics, on the other hand. Second, this study explicitly focuses on a central question, and a set of derived sub-questions, that have been hitherto neglected essentially. Historical investigations such that of Geyer (1987) and others provide an historical discussion of the appropriate KV size and, consequently, explain the contemporaries' perception of merging as a way to secure benefits (especially pensions) and to ensure the survival of the institution per se. However, they do not address the underlying economic relationships between KV size and performance. This study shows that, when confronted by empirical analysis, some of the major presumptions behind the idea of "growing" the financially distressed KVs in order to rescue them may no longer be viable; these major presumptions are: (i) the larger a KV, the better; (ii) growth in size serves the "securing of payments" (see Chapter 2); or (iii) mergers would save the KVs from going under.

Finally, the German miners' *Knappschaft* as a quite old, long-standing institution that contributed to the formation of the German and, principally, Bismarckian-type welfare state has essentially been neglected in international research in business and economic history. Thus, telling a story about German KVs may be a valuable undertaking simply because they are, more or less, an unknown piece of German insurance history in international academic circles, but are worth examining due to their broader implications. With the exceptions of the most recent studies from the major research project on the *Knappschaft* institution, as well as Geyer (1992) and Companje (2009), the KVs are quite absent from the English-speaking literature. To name only few examples, in his comprehensive book *The Origin of the Welfare State in England and Germany, 1850-1914*, Ernest Peter Hennock (2007) mentions the KVs, but only briefly.[68] Furthermore, in its introduction to the historical evolution of pension systems, *The Oxford Handbook of Pension and Retirement Income* does emphasize the important role of Bismarckian social legislation for today's benefit systems, but treats the mining sector as some unspecified forerunner. Van der Linden (1996) deals with a wide range of mutual aid organizations, such as the German *Hilfskassen* and, in particular, all sorts of Friendly Societies in Britain and abroad, but offers no information on KVs.[69]

68 This holds for Hennock (1987) as well.

69 See Van der Linden (1996), Thane (2006), Arza and Johnson (2006), Hennock (2007). Karant-Nunn (1993), pp. 79–80, in her investigation of early Saxon silver mining, briefly mentions KVs as well.

Table 1.1: Summary of recent studies on the *Knappschaft* institution (1947–2012)

Author	Time	Focus	Approach
Varnas (1947)	Middle Ages to 1900	Entire economy	Explores the KVs' role as the pioneers of Bismarckian social insurance from a jurisprudential perspective (qualitative)
Kammertöns (1952)	Present	Entire economy	Provides an institutional overview of the current state of the Knappschaft
Thielmann (1958)	1924–1957	Entire Economy	Review of the legal fundament of Knappschaft insurance with special focus on the post-war period
Greuer (1961)	16th to 19th centuries	Harz mountains	Reviews the historical development of the KVs in the Harz mountains
Klein (1965)	1500–1950	Saar	In particular historical overview of the Saar KV (qualitative)
Lingnau (1965)	1767–1961	Prussia (Ruhr)	Social science study of the expansion of benefit categories in Knappschaft insurance (qualitative)
Brinkmann (1967)	1744–1949	Prussia (Ruhr)	Provides an overview of the stages of the Knappschaft institution's development from Frederick the Great's initial legislation until the re-organization after World War II (qualitative)
Gellhorn (1967)	1923–1966	Entire economy	Review of the legal fundament of the Knappschaft with special focus on the period 1956-1966, the period prior to the Bundesknappschaft
Todeskino (1974)	Middle Ages to the present	Entire Economy	Outlines the development of the Knappschaft from its mediaeval origins to the foundation of the Bundesknappschaft (qualitative)
Dapprich (1979)	Middle Ages to the present	Entire Economy	Outlines the development of the Knappschaft from its mediaeval origins to the foundation of the Bundesknappschaft (qualitative)
Bayer (1981)	1854 to the Present	Entire Economy	Sheds light on the role of the Elders as seen from the relevant laws' paragraphs; emphasis is on present law (qualitative)
Dapprich (1981)	Middle Ages	Entire Economy	Discusses mediaeval mining regulations relevant for risk provision through the KVs (qualitative)
Dapprich (1982)	Middle Ages to 19th century	Entire Economy	Sketches the evolution of benefit categories within the miners' risk provision scheme (qualitative)
Verweyst (1983)	1923–1983	Entire Economy	Outlines the Reichsknappschaft law's implications for the institutional design of the Knappschaft institution and the revisions in the sixty years following its implementation (qualitative)

Table 1.1 continued

Author	Time	Focus	Approach
Geyer (1985)	1880–1910	Prussia	Overview of the debate on potential organizational failures of KVs, discussed potential solutions and reforms (qualitative)
Lauf (1985)	1923–1945	Entire Economy	Brief description of the Reichsknappschaft (qualitative)
Lauf (1986)	1853–1927	Entire economy	Role of Viktor Weidtman for the development of KVs (qualitative)
Geyer (1987)	1900–1945	Entire economy	Encompassing history of the Reichsknappschaft's operation (qualitative, some quantitative data presented)
Hermann (1987)	1929–1933	Entire economy	Asks for the situation of the Reichsknappschaft during the Great Depression
Lauf (1987)	Middle Ages to 1947	Saxony	Tells an illustrative story of the Saxon KVs from their very beginnings until the abolishment of the Knappschaft institution by the Soviet occupying force
Graf (1989)	Middle Ages to the present	Entire economy	Tells an illustrative story of the Elders' role in operating KVs (qualitative)
Geyer (1990)	1770–1870	Prussia (Ruhr)	Reviews the KVs' invalidity pension provision and compares it briefly to Bismarck's social legislation (qualitative)
Lauf (1990)	16th century	Entire economy	Sheds light on the effects that Reformation had on the way miners' associations were operated
Brettschneider	1991 (1991)	Entire economy	Detailed description of the legal configuration of the miners' pension insurance on the occasion of the 1992 pension reform
Lauf (1991)	1957–1960	Entire economy	Outlines the process of inclusion of the Saar Knappschaft into the Bundes-knappschaft (qualitative)
Geyer (1992)	1854–1979	Entire economy	Overview of the KVs' history since the mining reform; places KVs into the emergence of social insurance (qualitative)
Geyer (1994)	1923–1933	Entire economy	Overview of the Reichsknappschaft during the Great Depression (qualitative)
Lauf (1994)	Middle Ages to the present	Entire economy	Tells a vivid story of the Knappschaft institution since medieval times
Boyer (1995)	1885–1945	Entire Economy	Encompassing history of Bismarckian accident insurance for miners
Lauf (1996a)	1914–1918	Entire economy	Overview of the KVs' role in World War I (qualitative)
Lauf (1996b)	1906	Prussia	Review of the causes and consequences of the amending law of 1906 (qualitative)

Table 1.1 continued

Author	Time	Focus	Approach
Horn and Buschfort (1997)	1945–1948	Soviet occupation zone	Illustrates how the Knappschaft was reorganized towards the socialist ideal (qualitative)
Graf (1998)	1923	Entire economy	Reports on the meeting that constituted the Reichsknappschaft
Lauf (1998)	1919–1923	Entire economy	Reviews conditions shortly before the installment of the Reichsknappschaft (qualitative)
Verweyst (1998)	1923–1992	Entire economy	Overview of changes in the legal framework (qualitative)
Lauf (1999)	1933–1945	Entire economy	Description of the way in which the National Socialists adjusted the Reichsknappschaft according to their economic and political interest (qualitative)
Rieger (1999)	1994–1996	Entire economy	Economic analysis of the rationale behind subsidization of the miners' pension insurance (qualitative with some quantitative exercises)
Lauf (2002)	12th–16th centuries	Harz mountains	Historical investigation on the geographical origins of KVs in the Middle Ages (qualitative)
Lauf (2003a)	See Lauf (2002)	Harz mountains	See Lauf (2002)
Lauf (2003b)	See Lauf (2002)	Harz mountains	See Lauf (2002)
Lauf (2004b)	See Lauf (2002)	Harz mountains	See Lauf (2002)
Tenfelde (2004)	1854	Prussia	Evaluation of the new Knappschaft law (qualitative)
Lauf (2006a)	1854–1923	Entire economy	Descriptive overview of KV operation during the formative period, e.g. mining reform, fragmentation, self management (qualitative)
Lauf (2006b)	1915–1916	Prussia (Ruhr)	Outlines the conflict at the Ruhr between employers and miners on how to pay physicians
Lauf (2006c)	1918	Prussia (Ruhr)	Illustrates the interesting case of the influenza pandemic at the Ruhr that constituted a risk to many miners
Lauf (2008a)	1892–1913	Prussia (Ruhr)	Descriptive overview of the Allgemeine KV's pension provision (mainly qualitative, some data are presented)
Bartels et al. (2009)	–	–	Overview article outlining the Knappschaft project that this investigation is also part of (qualitative)
Bingener (2009)	13th–16th centuries	Harz	Reviews the development of the Knappschaft institution around the town of Goslar before and after Reformation (qualitative)

Table 1.1 continued

Author	Time	Focus	Approach
Bluma (2009a)	1871–1913	Prussia (Ruhr)	Sheds light on the importance of the Allgemeine KV Bochum in fighting the hookworm epidemics
Bluma (2009b)	1871–1913	Prussia (Ruhr)	See the preceding record.
Companje (2009)	1770–2008	Entire Economy	Compares German KVs with related funds in the Netherlands and Belgium (qualitative)
Lauf (2009a)	1890–1923	Prussia (Ruhr)	Tells a vivid story of the largest and, thus, single most important KV at the time, the Allgemeine KV Bochum
Lauf (2009b)	1871–1913	Prussia (Ruhr)	Addresses the practice of KVs, especially of the large ones at the Ruhr, to allocate part of their assets to housing loans (qualitative)
Lauf (2009c)	1850s	Prussia (Ruhr)	Presents the impliciations of an early contract specifying how KV physicians were paid
Bartels (2010)	–	–	See Bartels (2009)
Bingener (2010)	1744–1854	Prussia	Discussion of the conditions under which the Knappschaft law of 1854 was implemented
Bluma (2010)	1900–1930	Prussia (Ruhr)	Medical-historical study focusing on hospital architecture
Bluma et al. (2010)	1871–1913	Prussia	Economic historical study of how contemporaries perceived moral hazard and related solutions (qualitative)
Jopp (2010a)	1854–1923	Prussia	Asks for the relevance of the KVs' exposure to actuarial risk regarding the concentration process they underwent (economic and quantitative approach)
Jopp (2010b)	1854–1923	Prussia	Description of the Prussian KVs' insurance operations between the Prussian mining reform and the foundation of the Reichsknappschaft (quantitative)
Lauf (2010)	1969–2005	Entire economy	Presents an overview of the Bundesknappschaft's development as the nation-wide carrier of mining-specific social insurance
Bluma (2011)	Middle Ages to early 19th cent.	Prussa (Ruhr)	Concentrating on the late 19th century, elaborates on the biopolitics at the Ruhr which concerned the miner's health as well as working safety
Jopp (2011a)	1861–1920	Prussia	Explains external concentration among KVs using survival analysis (economic and quantitative approach)

Table 1.1 continued

Author	Time	Focus	Approach
Jopp (2011b)	1854–1913	Prussia	Calculates implicit rates of return for a subset of KVs to assess the sustainability of their operations (economic and quantitative approach)
Lauf (2011)	1969–2005	Entire economy	See Lauf (2010)
Guinnane and Streb (2011)	1867–1914	Prussia and Bavaria	Econometric modeling of moral hazard regarding the KVs' sickness insurance section (economic and quantitative approach)
Guinnane et al. (2011)	1867–1907	Prussia and Bavaria	Discusses the usefulness of the KVs' historical experience for examining moral hazard in modern sickness insurance
Guinnane et al. (2012a)	1861–1920	Prussia and Bavaria	Provides a quantitative overview oft he KVs
Guinnane et al. (2012b)	1861–1920	Prussia and Bavaria	See Guinnane et al. (2012a)
Jopp (2012a)	1867–1913	Prussia	Assesses the empirical relationship between KV size and actuarial risk (economic and quantitative approach)
Jopp (2012b)	1867–1913	Prussia	This study's long version
Jopp (2013)	1861–1920	Prussia, Bavaria Saxony	Introduces the KV statistics of 1861–1920 as a source for historical economic and social research and

Source: My own depiction.

1.5. Structure of the study

In order to answer the research question posed, this study is organized into four main chapters. Chapter 2 provides a baseline institutional overview of the KVs' main structural features. It begins with a brief description of the historical origins of the miners' *Knappschaft* institution in the mediaeval period and its development until the mining reform in the middle of the nineteenth century, which marks the formal starting point of this study. After that, I lay a foundation for understanding the subsequent chapters by describing the basic regulatory framework that the Prussian KVs were principally bound to within the observation period of 1854 to 1923. This implies raising basic issues including who was covered, which contingencies were addressed, how the finances were organized and administered, and what the contents of the individual insurance contract were. From the basic regulations that formed the essence of the miners' social insurance scheme – the *Knappschaft* Law (*Knappschaftsgesetz*) of 1854, the Prussian *Knappschaft* Law (*Neues Knappschaftsgesetz*) of 1906, the Bismarckian social insurance laws of 1883–1889, and a few others – I infer the necessary knowledge of both the basic regulatory framework set by the state and the extent to which KVs were free to maneuver regarding business policy. It will become clear that KVs definitely had a free hand in specifying some parameters of their own – e.g., the concrete amount of a pension to be paid in the event of invalidity or survivorship. In fact, these operational margins established notable heterogeneity among the universe of KVs.

Chapter 3 quantifies, in very basic terms indeed, the KVs' operations. A look into the literature (see Table 1.1) has shown that, to date, there is no comprehensive quantitative descriptive analysis of the German miners' insurance that would use available mass data on *all* Prussian, but also Bavarian and Saxon, KVs to generate the basis for any empirical analysis. Therefore, this chapter starts with a section on data issues, in which the main data sources – the Prussian, Bavarian, and Saxon *Knappschaft* statistics, as officially compiled and published by the states' governments – are introduced and briefly discussed. Then, I present the collected quantitative data that are central to a deeper understanding of the KVs' operations and the analysis of size, thereby making a self-standing contribution to German historical insurance statistics. Lines of developments regarding membership size and structure, as well as regarding the revenue and expenditure side of the KVs, are worked out. Where appropriate, the KVs are compared with other insurance institutions, especially with Bismarckian social insurance.

Then, having now some idea of how the KVs worked, and based on one of the contemporary observers' major arguments, Chapter 4 sets out to address the two basic hypotheses. First, KVs arguably reduced their exposure to actuarial risk considerably by growing large; and, second, larger KVs were more cost-efficient than smaller ones in terms of operating costs. If both hypotheses on the relationship between fund size and fund performance were empirically true, KVs would have done well to engage in mergers to improve survivability or, depending on the focus, to rescue inefficient funds to the benefit of the whole insurance system's stability. Some basic econometric models are specified to test both hypotheses and shed light on the question of whether there was a minimum efficient scale or, respectively, an optimal scale of KV size. Findings are then used to generalize on the appropriateness of the KVs' organizational system at the time.

To open the black box of KV survival, Chapter 5 combines the qualitative analysis of contemporary economic arguments for mergers and liquidations with statistical evidence on the determinants of the concentration process, partly derived from the previous chapter. The basic assumption of this study that external growth was due mainly to an economic rationale seems to be justified for two reasons. On the one hand, this view dominates the contemporary literature on KVs. On the other hand, source material on board meetings, which would indicate otherwise – i.e., whether decisions to merge or to liquidate were done for non-economic reasons – is currently unavailable. After having rearranged the data from the KV statistics so as to get an impression of the quantitative characteristics of target KVs, absorber KVs, liquidated KVs, and the uninvolved rest, a straightforward duration analysis is conducted, explaining the conditional probability with which a KV was absorbed or, respectively, liquidated.

Finally, the concluding argument of the study is developed in Chapter 6 by summarizing evidence from the main chapters and using the evidence to provide an answer to the central research question.

2. Institutional history of the *Knappschaften*: Regulatory framework and room for organizational maneuver

I begin this chapter with a brief discussion of the *Knappschaft* institution's historical origins. After demonstrating that "social security mutualism"[1] in German mining apparently has a very long tradition, I concentrate on the basic regulatory framework of the *Prussian* KVs during the observation period. Since, as mentioned in Chapter 1, the available literature on the miners' funds tends to concentrate on such descriptions of the KVs' institutional history and its complexity, I keep my overview as brief as possible.

2.1. The historical origin of miners' solidarity[2]

The mid-nineteenth-century KVs trace their roots to associations privately and voluntarily created by miners to serve religious purposes, maintain traditions, and to support one another in case of both temporary and permanent income losses due to sickness, injuries and invalidity. Most accounts, such as those of Lauf (2003a, 2003b, and 2004) and Bingener (2009), locate their origins in the medieval ore mining regions in the Harz Mountains (around Goslar) and the Erz Mountains (Saxony). Their existence was documented the first time for 1260.[3] Nineteenth-century observers usually assumed that the formation of those voluntary associations reflected the uniqueness of mining at the time and the greater need for financial security because of mining's unprecedented hazards to health and the ability to work.[4] Recent research, in contrast, stresses that association activities of this sort were a widespread phenomenon at the time and not limited to this particular occupation – as the formation of craft guilds shows. In fact, these early asso-

---

1 This term refers to Van der Linden (1996).

2 An earlier version of this chapter is published as part of the article "Old times, better times? German miners' Knappschaften, Pay-as-you-go Pensions, and Implicit Rates of Return, 1854–1913;" see Jopp (2011b). The subchapter also contains parts (i) published in the article "The hazard of merger by absorption: Why some Knappschaften merged and others did not, 1861 to 1920" and (ii) published in the article "The costs and benefits of size in a mutual insurance system: The German miners' Knappschaften, 1854–1923;" see Jopp (2011a), pp. 83–84, and Guinnane et al. (2012).

3 See Lauf (2003a, 2003b, 2004) and Bingener (2009), p. 294.

4 See, for example, Simons (1890), pp. 13–15, Simons (1895), p. 3, Imbusch (190), p. 10, and Köhne (1915), p. 5.

ciations also fit well into a more international picture of emerging collective action, as drawn for Western Europe, for example.[5] Nevertheless, there is no doubt that miners felt the need to jointly shoulder financial uncertainty due to erratic, immediate income losses resulting from sickness or injuries long before the beginning of this study's observation period.

The *Knappschaft* institution's history, from its origins around 1260 to the beginning of my study period, can be divided into three "developmental stages." At first, miners' contributions were voluntary and paid when they were needed, usually after an accident. A fund in the sense of a permanently existing and replenished financial reserve had not yet been established. Accordingly, benefits were needs-tested and not high, and the organizing principle was charity. Thus, a legal right to be supported by one's own KV did not initially exist.

In the second stage, which began roughly with the first enactments of local mining codes in 1300 (the *Kuttenberger Bergordnung*), the KVs' ability to provide protection against life risks (including survivorship of miners' dependants) improved insofar as associations now demanded regular and obligatory contributions from their members in the form of the so-called *Büchsenpfennig*. These contributions were used to create and maintain permanent funds. Nonetheless, members were still not formally entitled to a minimum level of payment, not to mention a guaranteed payment commensurate with contributions paid. Benefits were provided in the case of need, simply so long as the cash box allowed. Various mining codes actually strengthened compulsory membership among the local workforce and also often required the proprietors of mines (the *Gewerken*) to share in the costs of sickness and invalidity claims. In this context, Hiltrop (1865) explains that owners often had to provide replacement pay for sick miners for four to eight weeks, while support for those permanently disabled had to come directly from the KVs.[6] Obviously, the principles of both compulsory membership and joint financing by miners and mine owners have quite a long history. Still, during this second stage, the Reformation removed much of the KVs' religious function.[7]

In the third stage, KVs were integrated into the absolutist-mercantilist regime that emerged around the middle of the sixteenth century. Patronage by the sovereign became a substitute for miners' fraternalism. It was around the middle of the eighteenth century that the geographical focus of mining within German states sustainably shifted towards Prussia with its hard-coal deposits; Prussia emerged as the new mining core region.[8] Since it was a characteristic of the absolutist-mercantilist state to oversee the exploitation of natural resources in every respect, Frederick the Great and his followers ensured

---

5 See Wagner-Braun (2002), pp. 28–32, Ogilvie (2004), Kluge (2007), and De Moor (2008).

6 See, for example, Hiltrop (1865), pp. 216–217, Karwehl (1907), pp. 15–17, and Klein (1965), p.13. Brassert (1858) has introduced a collection of early mining codes into the literature. Bülow (1905), in particular, tells the institutional history of the Ruhr coal area until the mining reform of 1865.

7 Bülow (1905), pp. 1–34, Simons (1895), p. 3, and Lauf (1990).

8 See Bülow (1905), pp. 34–62, Bartels et al. (2009), p. 198, and Tenfelde (2004), p. 21.

that miners and mine owners became subject to the royal administration's mercy. According to the *Revidirte Cleve-Märkische Bergordnung* of April 29, 1766, miners had to swear fidelity to the sovereign and the mining authorities. Their complete subordination is reflected, for example, in the term *Gnadenlohn* ("wage by mercy"), as the benefit to incapacitated miners and survivors was called. As instruments of the mercantilist state's resource policy, miners and mine owners saw the Prussian administration establish direct control over mineral production and every related operational aspect, such as decisions on buying inputs. If there was something good about full subordination for mineworkers, it might have been guaranteed workplace security and exemption from military service. The KVs, however, lost their role as self-managed risk-sharing communities. The bottom line: Charity was replaced by government's paternalism.[9]

This third stage ended first in Prussia and then in all the German territorial states with the mining reform that took place between 1851 and 1865. This reform reorganized mining and miners' mutual insurance mechanisms as part of a shift towards a more liberal economic system. The absolutist-mercantilist economic order was overthrown in favor of a regulatory framework that again allowed mine owners a free hand in all business-related decisions. KVs became real insurance providers. The state, nonetheless, remained a passive supervisor that continued to create the regulations that established the basic behavioral norms for KVs.[10]

---

9 See Hiltrop (1869), p. 217–218, Achenbach (1871), Nöggerath (1873), Simons (1895), p. 3–4, Bülow (1905), pp. 1–34, Ptok (1922), pp. 3–27, Varnas (1947), pp. 3–25, Kaufhold (1994), p. 35–40, and Springer (2010). The "direction principle" (*Direktionsprinzip*), as the subjection of miners and mine owners to full state control was called, was established through a number of mining codes, of which the most important for Prussia were: the *Revidierte Cleve-Märkische Bergordnung für das Herzogtum Cleve, das Fürstentum Moers und die Grafschaft Marck vom 29. April 1766;* the *Revidierte Schlesische Bergordnung vom 5. Juni 1769;* the *Revidierte Magdeburg-Halberstädtische Bergordnung vom 7. Dezember 1772*; the *General-Privilegium für die Bergleute im Herzogtum Cleve, im Fürstentum Moers und in der Grafschaft Marck vom 16. Mai 1767;* and, finally, the *Instruction zur Einrichtung und Führung der Knappschafts-Casse für die Bergwerke im Herzogtum Cleve,Fürstentum Moers und Grafschaft Marck vom 16. Mai 1767*. Bingener (2010), for example, has recently provided an overview of those regulations and the immediate pre-reform period. Krampe (1961) focuses on the period 1800–1865 and Wixforth (1995) on the period 1826–1846.

10 Evaluations of the reform are provided by Fischer (1961a, 196b, 1961c), Tenfelde (1977), pp. 164–190, and Brown (1995), among others. The *Gesetz über die Verhältnisse der Miteigenthümer eines Bergwerks* of 12 May 1851 (Ownership Law), for example, introduced economical self-determination of mines; employers were, at first, still not allowed to bargain on their own with employees over wages. The *Gesetz über die Beaufsichtigung des Bergbaus durch die Bergbehörden und das Verhältnis der Berg- und Hüttenarbeiter* of 21 May, 1860, commonly referred to as the *Freizügigkeitsgesetz* (Freedom of Movement Law), then formally deregulated the labor market by granting freedom of contract and, thereby, mobility.

2.2. The *Knappschaften* and their economic niche: industry-related insurance[11]

With the mining reform of 1851–1865, the various German KVs became non-profit *insurance* providers for all those workers employed in mining. While, at first, only KVs in Prussia were targeted, Prussian mining legislation was soon imitated by most of the remaining German states. In practice, the miners' funds developed into a self-standing social insurance system that had to fulfill numerous insurance functions at once.[12]

The Prussian mining reform established the government regulations to which the KVs were subjected for the next 40 years, and beyond. They gave *Knappschaft* insurance its general institutional form, but also left "room for organizational maneuver," mainly with regard to the monetary level of payments to and benefits from the KV, as well as the way payments were scaled to the insured's characteristics. In short, which were the most important regulations enacted between 1854 and 1923? To begin with, the legal basis was formed by the path-breaking *Knappschaft* Law (*Knappschaftsgesetz* in shorthand or, to be precise, *Gesetz betreffend die Vereinigung von Berg-, Hütten- und Salinenarbeitern in Knappschaften*) of 1854[13], the related Instruction Manual of 1855, and the Prussian General Mining Law (*Allgemeines Berggesetz für die preussischen Staaten*) of 1865, through which the miners' benefit scheme, formerly run by the royal administration, was converted into workplace-based social insurance.[14] According to

---

11 Earlier versions of this section are published in my articles "Old times, better times? German miners' Knappschaften, Pay-as-you-go Pensions, and Implicit Rates of Return, 1854–1913" and "Insurance, Size, and Exposure to Actuarial Risk: Empirical Evidence from Nineteenth- and Early Twentieth-Century German Knappschaften;" see Jopp (2011b, 2012a).

12 Compare, for example, Karwehl (1907), pp. 20–23, for the stages of the empire-wide diffusion process. He reports the following dates of adoption: Brunswick (1867); Saxe-Meiningen and Saxe-Gotha (1868); Waldeck and Bavaria (1869); Reuß jüngere Linie (1870); Saxe-Altenburg (1872); Alsace-Lorraine (1873); Wurttemberg (1874); Anhalt (1875); Hesse (1876); Baden (1890); Birkenfeld (1891); Schwarzburg-Rudolstadt (1894); Lübeck (1895); and, finally, Saxe-Coburg (1899). Although the reasons for having compulsory social insurance are not of primary interest here, they should at least be mentioned. Diamond (1977), pp. 278–282, points out three such reasons: income redistribution, market failure, and paternalism.

13 See Friedrich Wilhelm IV. (1854). In this study, it is taken for granted that the *Knappschaft* institution adopted the character of insurance in a more narrow sense of the word in 1854. This transformation of a system that had been under strict control of the sovereign shortly ago was not necessarily welcomed by miners. Serlo (1859) exemplarily elaborates on this point regarding miners at the Ruhr.

14 See Klostermann (1866) and Ministerium für Handel, Gewerbe und öffentliche Arbeiten (1856). In case you wonder how the *Knappschaft* Law and the Prussian General Mining Law relate to each other, here is the answer: Between 1851 and 1863, 14 self-standing laws – of which the *Knappschaft* Law was one – were enacted to regulate all sorts of problems related to mining, all of which were merged together into one single legal text, the aforesaid Prussian General Mining Law of 1865. As we will see, only some minor changes were made compared to the original *Knappschaft*

Tenfelde (2004), miners were now legally entitled to benefits broadly equivalent to their mandatory payments. In 1906 and 1912, two amending laws were enacted. The former can be seen as a direct consequence of the ongoing discussion of organizational failures and related solutions discussed in Section 1.2. The latter was intended to adjust the KVs' scheme to the *Reich* insurance regulation of 1911 (*Reichsversicherungsordnung*), introducing white-collar and survivorship insurance on the national level. It was, however, not the first *Reich* law that required adjustments on the KVs' side. Actually, the introduction of Bismarckian social insurance through the Health Insurance Law of 1883, the Accident Insurance Law of 1884, and the Disability and Old-age Insurance Law of 1889 influenced the way KVs operated – precisely how is also addressed below.[15]

With regard to the aforementioned regulations, the following subsections answer three interrelated questions: i) How did the KVs work? ii) What was the typical KV insurance contract like? and iii) How did Bismarckian social insurance affect KVs? To be precise, I will focus on the institutional bottom line, and not on the exceptions. The bottom line is derived from the relevant laws. Since KVs were obliged not only to send their membership and financial statements to the mining administration for publication, but also to publish statutes stating their "terms of business," we principally have an additional source at hand to illuminate how KVs responded to the lack of certain rules.[16] However, I will not refer too much to these statutes below because they would bring us away from the institutional bottom line. Wherever it reads "KVs could decide," this means that the regulator left an issue unregulated.[17]

2.2.1. How did the Knappschaften work?

As of 1854, KVs had to be re-established in every area where mining and complementary economic activity took place.[18] Two basic types of carriers have to be distinguished from each other. The first is quite similar to what Murray (2007) calls "establishment

Law. A noteworthy prominent exception to Prussian mining legislation is the Kingdom of Saxony, which had its own tradition and was organized in a different way. Traditionally, it was separated strictly between the subsectors "ores" and "coal" and, within coal, between "hard-coal" and "brown-coal;" besides, it was also strictly separated between pension funds, on the one hand, and sickness funds, on the other hand; see Böhmert (1879, 1885), Wächter (1893), Dannenberg (1901), and Elsholz (1910) for the Saxon KVs' insurance tradition, and Lauf (1987) for a recent account.

15 See Steinbrink (1908), Gottschalk (1912), and Tenfelde (2004), p. 25. On the *Reichsknappschaft* Law, see Wissmann (1930), who provides a contemporary overview.

16 The Instruction Manual of 1855 basically includes suggestions on how to shape statutes.

17 I do not start from a formal taxonomy. However, the KV features discussed below are ordered according to certain "problem areas:" coverage in terms of addressed people and contingencies, ways of financing, specification of contributions, specification of benefits, and administration. Homburg (1997), Mesa-Lago (2006), and Whitehouse (2006) provide formal taxonomies.

18 See Friedrich Wilhelm IV. (1854), § 1, § 2 and § 10, and Klostermann (1866), § 165 and § 166.

funds" when addressing U.S. industrial sickness funds in the late nineteenth century.[19] In the context of KVs, I call them firm- or company-related funds (*Werksknappschaftsvereine*). Their pools of insurants were naturally much smaller than those of the second type – the area- or territory-related KVs (*Bezirksknappschaftsvereine*). A KV area was, by and large, equal to a mining area (*Bergbaurevier*), which could have been quite small or quite large and could have contained different resource deposits (hard-coal, brown-coal, iron-ore, miscellaneous ores, salts and stone) at once. Which area a KV took for itself depended on two things: It had to claim the area in its statute before another KV did, and it had to receive the "ok" from the mining administration. In its area, a KV was the exclusive supplier of insurance, meaning that no other KV could have a subsidiary there.

It is straightforward to identify the size of the area or the company as the principal limiting factor to growth or, alternatively, to finances and actuarial underpinnings.[20] As a consequence of running locally-focused social insurance schemes, the KVs' economic fate was directly linked with the development of mining activity in their operating areas. If mining there was subject to long-term stagnation, so were the KVs. If mining areas thrived, so did the KVs. Apparently, membership growth of KVs was bounded in two ways: on the one hand, by their limited space and, on the other hand, by their focus on one particular occupational group or sector. Moreover, that KVs could not have had subsidiaries in another fund's area tells us something about competition among KVs: At first glance, KVs were not forced into it. However, if KVs – reflecting the supply-side – could not compete by penetrating others' areas, miners – reflecting the demand side – might have forced them into competition by switching to the KVs that supplied, in their eyes, superior insurance coverage.[21] Though, the decision to move to another KV area was probably much more dependent on the employment prospects than on the local quality of insurance coverage.

The initial regulatory framework did strengthen the principle of compulsory membership, but did, indeed, not invent it. As the overview in Subsection 2.1 explained, compulsion had long since been a key characteristic of the miners' social security mutualism. Beyond all question, KVs were obligated to provide insurance for all Prussian miners.[22] Corresponding to the fact, however, that KVs had also existed for what I call, in shorthand, the "related industries" (steelworks and other foundries), steelworkers and the like were not excluded from membership once the firm owners voluntarily decided

---

19 See Murray (2007), p. 98.

20 As the overview of the Prussian KV statistics in Subsection 3.1 reveals, data do not allow doubtlessly distinguishing firm-related from area-related funds.

21 Linking particular KVs to certain areas is somehow reminiscent of assigning territorial supply monopolies (such as the large German energy suppliers in the second half of the twentieth century).

22 According to, for example, Simons (1895), p. 8, these also included the so-called technical civil servants (*Werks- und Revierbeamte*).

to continue being part of *Knappschaft* insurance. The set of requirements they had to meet – (i) firms had already existed before 1865, (ii) firms had insured their employees in KVs already installed before 1865, and (iii) firms had decided to continue their membership in *Knappschaft* insurance beyond 1865 – implies that firms founded after 1865 would not gain access to KVs.[23]

Membership in KVs was linked with the workplace. Principally, if a miner lost his work, he lost insurance coverage and accumulated entitlements.[24] Contemporary observers, as well as the workers themselves, identified this feature as a major design flaw, as it constituted a *de facto* obstacle to *de jure* freedom of movement. Miners willing to take employment in another mining area would be prohibited from relocating unless acquired entitlements to insurance benefits – pensions, in particular – were portable.[25] Formally, the issue of portability was not addressed by the regulator until 1906, when the amending law certified the right to retain acquired entitlements when switching KVs. Informally, however, many KVs had been reacting since the late 1870s to the rising discontent with their practice of withholding payments from exiting members by concluding bilateral contracts.[26] By way of these "reciprocity contracts" (*Gegenseitigkeitsverträge*), KVs bound themselves to pay for the time that they had been responsible

---

23 See Klostermann (1866), pp. 302–303, and Wirtz (1911), pp. 80–81. It is paragraph § 166 of the Prussian General Mining Law that contains the relevant information. As of 1854, steelworkers were an integral part of *Knappschaft* insurance. As of 1861, however, the *Gesetz über die Kompetenz der Oberbergämter* of 10 June, deprived steelworks and other smelting works of the mining administration's control and principally assigned them to trade control (*Gewerbeaufsicht*). This meant that steelworks and other smelting works lost access to *Knappschaft* insurance. However, as of 1865, steelworkers were allowed to stay in if they met the outlined conditions; for the law of 1861, see Fischer (1961a), p. 182. Steelworks and salt mines, in particular, had their own firm-related KVs; see the first paragraph of this subsection.

24 The reader familiar with prominent institutions from the Anglo-Saxon area, will recognize that nineteenth-century KVs were much more similar to, for example, U.S. industrial sickness funds, as described by Murray (2007), than to English friendly societies or U.S. and Canadian fraternal lodges, where membership was not linked to the workplace; see Murray (2007), p. 92–98. Gilbert (1965), for example, provides an overview of English friendly societies. Emery and Emery (1999) do so for the American fraternal lodges. Van der Linden (1996), furthermore, contains examples of non-English friendly societies that also do not reflect the workplace principle.

25 One might wonder why miners, as they self-managed their funds (see below), were apparently eager to restrict themselves by way of such practices. The answer arguably lies in the fact that they self-managed their funds together with their employers, who are generally said to have quite outmaneuvered their employees for the sake of their own interest. Their own interest was to be able to command a settled stock of workers into whose "training on the job" they invested; see Bertrams (1912), p. 1 413, and Lauf (2006a), p. 265. The problem of how to secure the portability of acquired entitlements recently played a role in the context of the German health reform of 2006; see, for example, Reimers (2006), p. 594.

26 As a consequence, KVs engaged in contracting on portability of entitlements more rigorously, as the *Darmstädter Freizügigkeitsvertrag* of 1908 and the *Freizügigkeitsvertrag* of 1917 show; for the former, see Lauf (2006a), pp. 279–280, and for the latter, see Reuss (1917).

for insuring a member, that is, for the time a member paid regular contributions to the KV.[27] A related case occurred if a miner was formally "on leave," a status of membership implying that he did not regularly contribute to the KV, as he would have surely done if classified as "active."[28] Contemporaries tell that a member classified "on leave" usually had the opportunity to maintain entitlements acquired in the past by continuing to pay some fee to its KV; the Prussian KV statistics report those payments as "contributions by the ones on leave" or "acknowledgement fees." In any case, paying those fees did not constitute new entitlements, but only increased the price of insurance coverage retroactively.[29]

Regarding membership, another major feature has to be mentioned: KVs discriminated between two categories of working membership and invalids, the so-called "established" and "unestablished" members (or, in German, the *ständige* and *unständige Mitglieder*). Originally, this distinction had been developed to identify full-time, or professional, miners (the established ones), on the one hand, and day-laborers, who were not permanently employed in the mining sector and used to satisfy short-run labor demand peaks, on the other hand.[30] With the new 1854 and 1865 regulations, all miners – i.e., full-time workers in particular – who did not meet certain criteria of admission as defined by the KVs themselves, were classified unestablished. To achieve the status of an established insurant, a miner usually had to be between ages 20 and 45, had to be of good certified health and of moral integrity, and had to have undergone a waiting period of some years.[31] So, in the miners' compulsory system, every individual insurant, regardless of the "risk" he represented, had to be insured by a KV (as long as the miner was employed within the KV's area and no company-related KV was responsible for him). From an economic viewpoint, indeed, the discrimination between established and unestablished members might be interpreted as an opportunity to give KVs some leeway in shaping their memberships' risk structure. That is to say, unless it were not, in general, the relatively more unhealthy men that chose to become miners, compulsory membership implies that KVs were not subject to one of the major economic problems associated with insurance, namely adverse selection. However, by being able to refuse miners that did not meet the mentioned criteria full rights – or, put differently, a relatively better actuarial equivalence of contributions and benefits – they could be used to cross-subsidize the established ones. Such a redistributive scheme, which treats the ones

---

27 See Lauf (2006a), p. 265.

28 Is seems likely to me that a miner that had fallen into unemployment was classified "on leave."

29 There is no evidence that those miners switching membership could pay such fees to keep their entitlements alive in the KV they left.

30 See, for example, Imbusch (1910) pp. 65–68. Originally, the law of 1854 only made health care benefits and sick pay obligatory benefits for the unestablished miners, while pensions were excluded. One of the few alterations that came with the law of 1865 was that unestablished miners were now eligible for an invalidity pension as well; see Klostermann (1866), p. 305.

31 See Bertrams (1912), pp. 1 459–1 461. As Murray (2007) tells us, the industrial sickness funds in the U.S. had similar criteria for joining; see Murray (2007), pp. 99–100.

that were generally better off even better and the ones that were generally worse off even worse, is probably not to everybody's taste.[32] Eventually, the law of 1906 abolished that opportunity.[33]

Besides abolishing established and unestablished membership, the reform of 1906 included two other improvements affecting membership and "market structure," both relating to one of the major design flaws on which this study focuses. First, while the KVs had begun by running, what I call, a compound insurance scheme, they finally had to unbundle pension from sickness insurance; and second, the mining administration now had the formal capacity to merge unviable KVs into other funds or to shut them down entirely. Regarding the unbundling of benefits, one might think that this made two self-standing KVs out of one. However, that was *not* the case. In fact, the targeted institutional separation was reflected in the accounting framework (see Subsection 3.1). The KVs' financial operations and their memberships were to be separately reported by the pension insurance and the sickness insurance sections, which forced KVs to make source-specific allocation of expenditures and revenues explicit. Contemporaries, indeed, believed that this form of unbundling was, at very least, a prerequisite for regaining financial stability. Unfortunately, we still do not know whether the unbundling eliminated diseconomies of scope, possibly inherent in the compound scheme of the pre-reform period, or only created such. What we *can* say is that this separation was definitely a prerequisite for being able to establish different KV sizes in the two, assumingly different insurance sections. As a matter of fact, the KVs' sickness insurance sections were reported as having slightly more members than the pension insurance sections. This was because not every formerly unestablished member gained access to both sections; if he did not meet certain health or age requirements, he was not eligible for pensions.[34]

Regarding the ability to merge or shut down KVs, contemporaries wanted the mining administration to have an instrument on hand, the use of which would definitely improve the financial situation or stability of the whole insurance scheme. Subsection 1.2 presented evidence that the regulator definitely exerted some verbal pressure and made clear that he also favored the idea that financial stability might be connected with a larger KV size.[35] However, I could not locate evidence that the mergers and liquidations that occurred before 1906 were a direct consequence of "regulatory persuasion." Ber-

---

32 Contemporaries felt it unjust that established members gained full rights from their KVs, whereas unestablished members did not gain full access to benefits and were usually said to receive lower insurance benefits per unit of contribution paid than established ones.

33 See Steinbrink (1908), Reif (1908), and Bertrams (1912).

34 Another alteration of the 1865 law – as compared to the 1854 law – was to give a KV the opportunity to create different self-standing sickness funds, each organized to cover a single mine, for example, while still keeping one collective of insurants regarding pension provision. Actually, only a few KVs took the opportunity.

35 See Steinbrink (1908), pp. 29–31, for the relevant paragraphs, § 177a and b, and Reif (1908) for an account.

trams (1912) wrote that the liquidation of the *St. Wendeler KV* in 1906, for example, was the first liquidation to be based on the regulator's new capacity.[36] Unfortunately, the law did not suggest a set of indicators for financial instability. If the regulator concluded, as, for example, in the case of the *St. Wendeler KV*, that a liquidation was necessary, the decision making process behind is a black box to us (as with the decision process on business matters in the KVs).

Each KV offered its members a legally pre-determined bundle of insurance benefits that could not be purchased separately;[37] even though a miner might have preferred one benefit over another, he was obligated to pay for the bundle. KVs had to insure against income losses due to temporary sickness, permanent invalidity, and survivorship of a miner's wife and children; this includes the consequences of job-related accidents, as they were often linked with inability to work, either temporarily or permanently. Because an invalidity pension was guaranteed until the recipient's death, it is reasonable to add the risk of longevity to the set of life risks explicitly addressed in the regulations. This holds likewise for a widow's pension, as long as she did not remarry; an orphan's pension had to be granted until the age of 14.

So the KVs evidently provided old-age coverage, too, even though the payment-triggering event was certified permanent incapacitation, and not, as in today's matured welfare states, "having reached a threshold age" of 65, for example, separating an "employment phase" from a "retirement phase."[38] Daily sick pay, medical treatment, as an in-kind benefit, and funeral costs were also part of the obligatory package. Beyond these benefits, each KV could decide on its own about offering further case-specific payments. These often included needs-tested support (the statistics call this "extraordinary" support) and financing for schooling of the miners' children. KVs even offered loans for house building and operated their own hospitals and sanatoriums.[39] However, the five most costly benefit categories were the three types of pensions (for invalids, for widows, and for orphans), daily sick pay, and medical benefits, including physicians' care, drugs and rehabilitation. The expenditure data presented below generally draw a picture of expansion and, indeed, highlight the weight of those five benefit categories in

36 Bertrams (1912), p. 1 553.

37 See Friedrich Wilhelm IV. (1854), § 3, and Klostermann (1866), § 171, for the prescribed benefits.

38 In the context of Bismarckian old-age insurance, Conrad speaks of the "profiling of old-age as a special phase in life." Hardach (2003 and 2011) goes in the same direction; see Conrad (1994), p. 14, and Hardach (2003 and 2011). Göckenjan argues that Bismarckian invalidity and old-age insurance was a tool to maintain remaining capacitation for work on the side of the insured such that a true mental and material separation between working life and work-free old-age had still not existed; see Göckenjan (1993), p. 731.

39 From the recent literature, see Lauf (2005) concerning the KVs' hospitals and Lauf (2009b) concerning examples of loan policy in the Ruhr area. He mentions two rationales for offering house loans. The first rationale has already been mentioned in a different context: forming a stock of settled, experienced – that is, productive – miners; in this sense, such loans were a complement to corporate social policies. The other rationale was to improve unhygienic living conditions.

the payment mix. It has to be pointed out that none of the regulations, either the initial or the follow-up ones, stated how generous benefits should be – that is, any upper or lower limits for how many thalers (until 1873) or marks should be spent per capita.

Although membership in KVs was linked with the workplace, it might be erroneous to assume that KVs exclusively protected against job-related risks. While many of the sickness or invalidity claims, perhaps most, were connected with doing the miner's work, they might well have occurred due to non-job-related reasons such as genetic predisposition for bad health or lifestyle (even if KVs made some effort to control the latter; read about the so-called "Elders" below). According to the distinction made by Köhler-Rama (2008), we should bear in mind that KVs insured the "job-related" risk *and* the "general risk of becoming incapacitated for work."[40]

Regarding the First World War, entitlements of those drafted for military service were regulated by the RVO emergency laws, which had no influence on KVs. It was the *Knappschaft* War Law of 1915 that addressed the relationship between those Prussian miners enrolled and their KV in more detail. Originally, the bottom line was that miners participating in war lost their entitlements since membership was linked with working as a miner. However, the law determined that returning soldiers were allowed to re-join their KVs without an additional health check, and their military service was applied to their length of service in the mines and to waiting periods for pension and sickness benefits.[41]

With regard to sources of revenue, the regulations mentioned only premiums (or, as I call them, social insurance contributions). Those were to be paid by the mineworkers themselves and their employers and were to be specified as a fixed-amount per capita or as a percentage of wages.[42] As with compulsory membership, the regulations merely strengthened this existing principle of obligatory contributions for all Prussian KVs and did not introduce something entirely new. Until 1906, the employers' portion of a contribution was simply not to fall below one-third. After 1906, however, when pension and sickness insurance benefits were "institutionally" separated from each other – meaning that each had its own sources of revenues and assets – the miners' and employers' portions of total contributions had to be exactly equal. Beyond contributions, KVs were free to explore other sources of revenue. Common among KVs was to specify a set of case-specific fees, such as marriage fees[43], fees relating to the act of joining a fund, or for punishment if the lifestyle was inappropriate. Further income sources might

---

40 See Köhler-Rama (2008), p. 123. The difficulty certainly is to distinguish between both sorts of risk empirically.

41 See Lauf (1996a), pp. 589–590. Bavaria and Saxony, among others, established similar regulations.

42 See Friedrich Wilhelm IV (1854), § 4, and Klostermann (1866), § 175. Contributions were to be deducted at the source by employers who would transfer the money to the funds.

43 I cannot say with any certainty that KVs tried to regulate their membership's marriage market by introducing such a payment. Instead, KVs very probably played a role in processing administrative aspects linked with the act of marriage.

have been rent and interest on assets. Although KVs were initially not obligated to generate reserve funds, almost all KVs accumulated financial reserves from income not required to settle claims. These reserves were the more important since each KV did carry its actuarial risk on its own.[44] Risk structure compensation between funds still had not been established – unlike today, as the example of German sickness funds shows. The possibility to refinance on the capital market was presumably very limited, although we know from the statistics that some KVs contracted debts on their property. However, KVs operated not only independently of each other, but also independently of the state's budget, which means that they did not receive state subsidies at any time. The Instruction Manual of 1855, intended to substantiate the meaning of the 1854 law and to make some recommendations on business policy, merely suggested that the KVs specify contributions to meet their benefits, but also to generate reserves. Precisely, it reads:

> "The contributions to be charged should be specified according to needs – that is, such that they cover not only benefits prescribed (§ 3) as to the amount experience suggests, but also a proper surplus remains to cover rising expenditures over time, as on invalids and widows, and to have a reserve at hand for times of trouble."[45]

Regarding the choice between a fixed amount per capita or a percentage rate of wage, the regulator apparently favored the former, writing that it

> "[...] would be assumed a fixed amount per capita (per month), calculated on the basis of average wage, to remove, as observable for most existing KVs, complicated calculation principles and injustice, reflecting the fact that the diligent worker must contribute more, without acquiring entitlements to higher benefits."[46]

A fixed amount would certainly unburden the high-income earners among the members and contradicts the principle of paying according to one's own economic capability.

The regulations left room to maneuver not only with respect to the extension of the benefit package with non-prescribed benefits and the exploration of other revenue sources, but also with respect to the determination of monetary benefits or levels of contribution or, respectively, calculation principles (including decisions on waiting periods, on eligibility criteria, and on whether contributions and benefits should be scaled somehow). How KVs approached these issues is part of the next section. A note on whether revenues and expenditures might have been lowered or increased in a discretionary fashion, however, is necessary right here: To the best of my knowledge, the official, non-discretionary way towards adjustments of revenues and benefits was to change the statute visibly for everyone. This process of revision arguably took some time, and available evidence suggests that KVs did not change their statutes from one day to another, but rather one time in a couple of years. So, whether a KV could really

44 See Klostermann (1879) for a discussion of the necessity to accumulate reserves.

45 Ministerium für Handel, Gewerbe und öffentliche Arbeiten (1856), p. 27.

46 Ministerium für Handel, Gewerbe und öffentliche Arbeiten (1856), p. 27.

charge to any insured miner a kind of temporary ex-post assessment or increase contributions permanently without making it official via a change in its statute is not clear. I would not rule out the possibility that this may have happened sometimes. In this respect, we should carefully interpret the various per capita revenue and expenditure figures presented in the subsequent chapters because they certainly, for technical reasons, convey the impression that adjustments were done year-by-year.

The KVs had a free hand in another issue, namely the financing mechanism to be applied. Whether to use full funding which was already known from the example of private insurance companies or the PAYG method was entirely the funds' choice. Contemporaries informed us that KVs overwhelmingly used the latter, thus converted in each year the contributions of currently working members and miscellaneous revenues into sickness benefits and pension benefits for non-working invalids and survivors.[47] However, KVs universally combined the PAYG mechanism with some reserve-building. Originally, reserves simply served as a financial buffer to compensate for unforeseeable deficits that might – and actually did, as we will see below – occur because actual costs did not meet predicted financial needs. If actual revenues exceeded actual costs at the end of a year, reserves were replenished. Data from the Prussian KV statistics, which will be introduced in the next chapter, illustrate the asset structure and the magnitude of the assets in reserve for unpredicted extra costs. So, in what form did KVs hold their reserves if not in cash? In fact, cash reserves made up only a minor fraction, on average. Interest-bearing assets – or interest-bearing "papers," as this item became known in 1908 – accounted for the highest percentage. The term itself implies that KVs were investors on the capital market. Unfortunately, we cannot tell from the statistics in what obligations they invested or whether "interest-bearing assets" also included loans to other KVs. Fortunately, though, Jüngst (1913) gave an account based on what the largest Prussian KVs reported in their original business reports. As productive investments, he enumerates mortgages, Prussian government bonds, German *Reich* bonds, municipal bonds, other loans, and reinsurance funds (see below).[48]

Contemporaries always expressed their worries about the financial reserves' inadequacy to cover implicit liabilities from pensions, which were the actual pensioners' future pensions and the actual working membership's accumulated entitlements on future pensions. We will see below that reserves were, in most cases, sufficiently large to cover the deficits that occurred. This holds under the assumption that "interest-bearing assets" were easy to liquefy at all times. If reserves were not sufficient, KVs were in the position to increase contributions immediately, perhaps request a loan from the mine owners, or borrow against assets not immediately liquefiable. However, regarding loans from fellow mine owners, I do not have sufficient evidence to state this with certainty.

---

47 See Caron (1882), p.7, Bertrams (1912), p. 1 417, and Köhne (1915), pp. 18–19.

48 See Jüngst (1913), pp. 266–269.

The reform law of 1906 also brought about improvements regarding the financing mechanism and partly relieved contemporaries' worries about the PAYG method's inferiority. This was the first time that the regulator prescribed a certain financing method to be used among KVs: the *Rentenwertumlageverfahren*, commonly called the *Kapitaldeckungsverfahren*. Although the best translation of the latter German term might be "capital funding method," it did not match our modern understanding of funding. According to the new method, which had already been used to finance Bismarckian invalidity insurance for the first ten years, KVs were obliged to charge contributions in a particular year (minus miscellaneous revenues) so as to cover, first, the annual expected explicit liabilities from pensions, and, second, the expected implicit liabilities from pensions newly granted in that particular year over their expected pension duration.[49] To understand what the second requirement means, consider the miner Wilhelm who became invalid in January, 1908: Besides covering the pension payment for the remaining year 1908 (according to requirement one), the KV had to put aside, in that same year, an amount equal to the (expected) discounted value of Wilhelm's nominal annual pension times his expected remaining lifetime and, thus, expected pension duration. If he received an annual pension of 200 marks and was expected to live another ten years, from 1909 to 1918, the KV had to accumulate a reserve equal to a discounted 2 000 marks. The first requirement, of course, had already been met before 1906, but it seems that reserves had not been accumulated to explicitly cover a certain percentage of future implicit liabilities. Actually, the difference between "having accumulated reserves to provide for times of trouble" and "having accumulated reserves to prefund a certain amount of future liabilities" might be more semantic than technical.[50]

Nevertheless, KVs continued to operate on a PAYG basis after 1906. Even though they now had to meet a formal prefunding requirement, the currently-working membership's contributions were going to be redistributed to the currently-non-working per period of time, implying an "intergenerational contract" between those factions. Thus, unlike a funded system in the modern sense of the term, funds were still not accumulated per individual or per generation (i.e., a particular birth-cohort) from contributions of exactly that individual or that generation. Furthermore, it might have been a compromise to exclude some portions of liabilities from prefunding not to let adjustment costs explode from one day to the next: Invalids and widows that became eligible for their pensions before the new rule were not to be covered over their expected remaining lifetime;[51]

---

49 There is at least one KV that has been pointed out as an exception: The *Saarbrücker KV* is said to have introduced the *Rentenwertumlageverfahren* from the law of 1906 as early as in the 1880s.

50 Thullen (1980, 1985, and 1986) provides formal insights into the different financing methods used in German social insurance from its beginnings to modern times. Stein (1991) provides another useful overview.

51 This, likewise, held for working members' entitlements to future pension benefits. Imbusch's assessment of the 1906 reform illustrates the effect of the law: "The law of 1906 had a deep-

A social insurance scheme such as that carried by KVs is likely to encounter risks to its financial stability. Principally, systemic risk may be of a macroeconomic, political, structural, demographic or actuarial nature. Depending on the institutional design, one systemic risk might be more significant to the well-being of the system than another.[52] One point clearly follows from the KVs' – and, as of 1906, from the regulator's – decision to apply the PAYG method: Changes in underlying demographic parameters, such as the old-age dependency ratio, gained special importance because of the general sensitivity of the PAYG method to them.[53]

To provide the right incentive to comply with the regulations and to accumulate the appropriate amount of reserves to secure benefits, the installation of a reinsurance institution to collect the required funds came along with the introduction of the *Rentenwertumlageverfahren*. Though, this reinsurance fund, the *Knappschaftliche Rückversicherungsanstalt Charlottenburg a. G*[54], which was created in 1907, was based on voluntary membership initially. According to its statute, the *Rückversicherungsanstalt* fulfilled the following purpose (article one):

> "The purpose of the institution is to provide, according to this statute, reinsurance for participating KVs regarding their provision of pension benefits for invalids, widows and orphans."[55]

In practice, reinsurance covered survivors' pensions in full and half the invalids' pension, provided that pensions had not been granted before the institution's foundation. Nonetheless, as long as two or more KVs shared the costs of a pension (because the miner had changed KVs once), costs of invalidity pensions were covered in full as well.[56] How, exactly, were premiums to the reinsurance fund specified? Article 15 gives an answer:

> "Each participating KV's premiums are to be specified so as to cover the net present value of all payments that would expectedly be taken in reinsurance over the period of contribution [five years first; then another five] plus miscellaneous costs of the institution."[57]

This statement is not overly precise. However, according to how the statute reads, this is the highest degree of precision we can extract.[58] Finally, three other organizational fea-

---

cutting influence on the level of contributions and benefits. Many funds considerably raised contributions and simultaneously reduced benefits. The currently living generation had in many ways to deal with the past's sins;" Imbusch (1910), p. 72.

52 See, for example, Mattil (2006), pp. 27–35, who, in the context of pension provision, refers to systemic risks specifically as "exogenous risk factors."

53 See, for example, Schmähl (1995) and (2001) for a more formal assessment of the PAYG method.

54 Here, "a. G" means *auf Gegenseitigkeit*, so "mutual."

55 Knappschaftliche Rückversicherungsanstalt a. G. (1911), p. 3.

56 See Knappschaftliche Rückversicherungsanstalt a. G. (1911), p. 13.

57 Knappschaftliche Rückversicherungsanstalt a. G. (1911), p. 15.

58 Actually, what follows in the next article is only the remark that KVs should specify their premiums according to their past experience regarding how long pensioners would, on average, live; see Knappschaftliche Rückversicherungsanstalt a. G. (1911), p. 15.

tures might be interesting to know: At first, only Prussian KVs could participate; second, the reinsurance arrangement would not cover an increase in a pension exceeding the original pension's amount when taken into reinsurance; and, third, all incoming premiums were to be invested to generate interest.[59] So, in addition to its voluntary character, we know that under this regulation, a KV was free to specify the scale of its reinsurance premium on its own – this means: so as to cover the present value of *expected* pension payments –, and the money to settle pensions claims (which had to be raised as contributions from KV members with or without this institution existing) was actually re-directed towards the *Rückversicherungsanstalt*. As a consequence of the new funding requirement, many KVs raised the contributions required from mineworkers and employers in order to finance the reinsurance premium, or to accumulate the necessary reserves without using the *Rückversicherungsanstalt*.[60] In fact, not every KV joined the institution at first.[61] Some larger KVs, including the *Oberschlesische*, the *Brandenburger*, and the *Halberstädter* KVs, were said to stay away from the arrangement because they feared losing their autonomy.[62]

In 1916, the *Rückversicherungsanstalt Charlottenburg a. G.* was redefined as the *knappschaftliche Rückversicherungsverband*.[63] Its function seemed to have remained, in all, the same, and the one major change concerned the switch from voluntary to compulsory membership of all Prussian funds. Recently, Lauf (2006) evaluated the newly-founded reinsurance institution as having established "a distribution of risks over a coalition of KVs."[64] I myself am a bit sceptical about this view because, to date, the *Knappschaftliche Rückversicherungsanstalt* has not been analyzed rigorously regarding its functioning. So it is all but clear as to whether it really complied to what we call a "reinsurance" institution that would really have pooled the KVs' individual actuarial risk so as to reach a higher level of balancing-of-risks-in-the-collective. What the *Rückversicherungsanstalt* (or, respectively, the *Rückversicherungsverband*) definitely did,

---

59 See Knappschaftliche Rückversicherungsanstalt a. G. (1911), p. 14 and p. 18. Taking exemplarily the1908 and 1916 business reports of the *Rückversicherungsanstalt* or, respectively, the *Rückversicherungsverband* into account, we can specify the annual asset mix from which that interest was generated more clearly: mortgages, municipal loans, bank deposits, securities (own holdings as well as KVs' holdings which were provided as collateral), and outstanding premiums (default interest!). While, in 1908, the single most-important asset item were municipal loans with 63.2 percent of total assets, the single-most important item in 1916 were mortgages with 57.4 percent (municipal loans declined to 9.6 percent); see Knappschaftschaftliche Rückversicherungsanstalt a. G. (1910), p. 8, and Knappschaftlicher Rückversicherungsverband a. G. (1917), p. 10.

60 See Milde (1907), pp. 124–129, and Lauf (2006a), p. 276.

61 See Reuss (1916), p. 397, who tells of 29 KVs constituting initial membership. According to the report on the constitutive meeting of October 30, 1907, 30 KVs participated: 26 from Prussia, one from Saxony, one from Anhalt, and two from Brunswick; see Allgemeiner Deutscher Knappschaftsverband (1907), pp. 1–2.

62 See Köhne (1915), p. 20.

63 See Knappschaftlicher Rückversicherungsverband a. G. (1916), p. 3, and Reuss (1916).

64 Lauf (2010), p. 46.

was to collect and administer the funds that KVs were obliged to raise to cover granted pensions over their expected duration.[65] More to the point: Once the funds had been raised according to past experience, KVs faced two investment alternatives – i) to directly invest on the external (all kind of securities) or internal (housing loans for KV members) capital market, or ii) to invest in the reinsurance fund, which, in turn, would search for investment objects. As, from some perspective, it was more of an investment club – a financing instrument – than a real insurance institution, membership in the *Rückversicherungsanstalt* might have benefited the KVs since it lowered transaction costs (asset management was partly outsourced) and might have led to a higher average return on investment (as the *Rückversicherungsanstalt* bundled – sometimes small, sometimes larger amounts of money – to an even larger amount).[66] Whatever the matter, we might accept the observations of contemporary observer August Köhne (1915), who stressed the educational value of the whole arrangement by stating:

> "The participating KVs transfer their funds to cover yearly granted pensions to the Knappschaftliche Rückversicherungsanstalt in Berlin, which, in turn, transfers the pensions coming due back to KVs. The security of [working members'] entitlements to future benefits has not been increased much, but certainly of existing invalids' pensions. However, the necessary accumulation of funds educates KVs to be cautious with their financial operations, and this is no less a benefit."[67]

Until now, I have discussed what the KVs did or did not do and what requirements they had to or did not have to meet. This may evoke the picture of an impersonal and monolithic entity that made seemingly consistent decisions and set standards for the ordinary insured. This was not the case, however. Owing to the law of 1854, and fitting into the process of the "self-dismantling" of administrative and business responsibilities, state

---

65 According to Pearson (1996), p. 558, reinsurance has to basic functions: On the one hand, it serves to diversify the first insurer's actuarial risk, provided it is too large to be carried alone, among other companies and geographical locations. On the other hand, reinsurance coverage can extend a first insurer's potential to accept more risks than he actually can.

66 In fact, there seems to be one way that the whole reinsurance arrangement might have really spread the KVs' risk over the aggregate of funds: What still remained was the longevity risk linked with the individual pensioner – that is, the risk that a KV still underestimated the length of time, for which, for example, miner Friedrich would receive pension payments. So, who would have cared about the excess payments occurring if Friedrich had lived longer than expected? If the difference occurring for one KV had been compensated with the funds of another KV, then the whole arrangement would have looked like an arrangement spreading risks. However, it seems likely to me that such occurring differences would have been settled by the funds of the respective KV itself; there certainly were funds left in each year that were not needed since a miner died earlier than expected. Besides, an article was established saying that a KV for which the accumulated funds would turn out insufficient ex-post would be charged with an immediate assessment to fill the gap due to underestimation of pension liabilities. This institutional characteristic does not really match with the idea of spreading risks over the participating KVs; see Knappschaftliche Rückversicherungsanstalt (1908), p. 19 (§ 25).

67 Köhne (1915), p. 20.

officials no longer ran the KVs on their own. Instead, each KV was to form a managing board.[68] Decisions on how to run the KV (e.g., which technical civil servants and *Knappschaft* physicians to employ) and how to specify parameters was the responsibility of that KV board, which had to consist half and half of miners' and employers' representatives; the state, besides its general supervisory tasks, was still involved, in that an administrative official had the decisive vote if a decision had not received a majority vote among the members' representatives. The way the KVs' self-management worked in practice – thus, how the parties involved influenced the decision-making process – is actually kind of a black box. While employers, either mine owners themselves or hired managers, chose their representatives directly, miners chose the so-called Elders (*Älteste*), who, in turn, elected miners' representatives for the board. The Elders, who could have been active workers or retirees, were probably chosen by workers because of their prestige among the local membership. As Graf (1989) points out, the institution of the Elders is as old as the *Knappschaft* institution itself.[69] Since mediaeval times, they had been given some important tasks, such as administering the cash box. As of the reform, the Elders had to manage the membership list, monitor whether members' behavior met the standards, visit the sick and look out for malingering, and present members' concerns to the board.[70] In their capacity as sickness controllers, they were one instrument to fight malingering, or as Germans called it: *Simulation*, or the illegitimate obtaining of sick pay.[71]

Both the 1854 and 1865 laws allowed the Elders to elect ordinary KV members or even mining officials (private or state) to the board. It is a widespread theme of nineteenth-century writings, as well as of recent historical works, to claim that mine owners, via their representatives, had relatively stronger influence in the self-management bodies – if not virtually had dominated them. The miners' chosen representatives were usually said to be friendlier toward their bosses – the mine owners – than to the miners they were elected to represent. The Elders, as intermediaries, are said to have made the difference in favor of a stronger position for employers.[72] According to Lauf (2006), it is not surprising that owners effectively dominated the decision process since they were the intellectuals.[73] Heinrich Imbusch (1910), who became the trade union leader of the *Gewerkverein christlicher Bergarbeiter* in 1918, stressed that many mine owners intended to use KVs to monitor and discipline their workforce:

---

68 See Friedrich Wilhelm IV. (1854), § 6, and Klostermann (1966), § 178. Due to the law of 1906, a second self-management body was introduced, the general assembly (*Generalversammlung*), which was intended to elect the board members and to control the balance sheets; see Lauf, (2006), p. 273.

69 See Graf (1989).

70 See Graf (1989), p. 36.

71 See Guinnane and Streb (2011) for details on the existence of moral hazard in the KVs' sickness insurance and on how they fought it.

72 See, once again, Tenfelde (2004), pp. 26–33.

73 See Lauf (2006a), p. 272.

> "Many employers saw in the KVs an instrument to pursue their own interests. The KVs' and, thus, the workers' interests had to be subordinated to owners' intentions."[74]

Imbusch goes on to say:

> "When the 1854 law put workers in charge to help manage their funds, they were not informed well. Thus, it was not difficult to push mining officials through as workers' representatives. Accordingly, employers' influence on the election of workers' representatives was sustainably ensured. In most funds, the Elders, who would elect the workers' representatives, were elected publicly or by being officially mentioned in the minutes; in others, a number of candidates had to be elected in order that the board would be able to select the appropriates ones. A reckless 'voting terrorism' led to the election of candidates that would please employers. But even in case of secret elections, employers mostly managed to secure their reign. If there a respectable workers' representative was chosen, he was fired, especially in small KVs. Under these conditions, the KV boards usually identified employers' interests as KV interests [...]."[75]

In this context, Ferdinand Bertrams (1912) points once more to the fact that mine owners had an interest in establishing a settled – that is, immobile – stock of experienced miners.[76] This makes sense insofar as becoming a productive hewer at the time, and probably still today, surely required a learning-by-doing process. Taking a young, inexperienced miner to his most productive level was, thus, a costly investment for the mine owner, who, so the argument runs, would not have really been willing to develop a productive worker, only to lose him to a competitor. This would very probably have felt to him like an expropriation of future returns.[77]

I have not yet been able to locate source material on board meetings. Contemporary observers are also not very informative about how decisions on contributions and benefits were made and justified. However, we can be quite sure that KVs ran an expenditure-oriented revenue policy; that is, they specified, for a given time interval, the monetary level of benefits and formed in some (yet unknown) way expectations on aggregate claims costs, which were converted into an average contribution required; they also accounted for expected miscellaneous revenues and room to maneuver because of accumulated reserves. That is to say, we *can* assess how average contributions and average benefits were effectively adjusted from an historical perspective because we have the quantitative data from the KV statistics. What is harder to assess are the underlying ideas and the actuarial knowledge available, as well as the fraction of it actually used.

---

74 Imbusch (1910), p. 63.

75 Imbusch (1910), pp. 64–65.

76 See Bertrams (1912), p. 1 413. Actually, workers' representatives articulated this on several occasions. In the *Bergarbeiter-Zeitung* of 1906 (January 6, first issue), one of the mineworkers' voices, it reads on occasion of the government's draft on *Knappschaft* reforms: "Since fifty years mine owners purposefully operate *Knappschaft* funds to do harm to workers' interest. They took completely over administration; managing boards almost everywhere do no longer comprise workers; mine owners' contributions are reduced as much as possible; miners' contributions increased;" see Henning (1987), p. 13.

77 The learning-by-doing process that a miner had to undergo is, for example, mentioned by Brüggemeier (1983), pp. 99–104.

2.2.2. What was the typical insurance contract like?

We know how the KVs worked in general, but not exactly what the typical KV insurance contract looked like. Until 1906, KVs charged each working member one "composite" contribution (weekly or monthly) covering all the insurance risks at once; as of 1906, however, two separate contributions were collected, one for pension coverage and one to finance sickness benefits. In contrast to, for example, the Odd Fellows described by Emery and Emery (1999), the KVs did usually not price the entire membership equally by using an age- and other factor-independent flat-rate payment.[78] The 1855 Instruction Manual suggested that KVs determine contributions as described here:

> "The level of contributions would be specified as follows: about the fund's benefits, [...], expectations on the yearly amount are commonly formed; this amount is raised by an appropriate percentage to replenish reserves, and this covers the annual financial need; interest on assets or irregular revenues are deducted, and the remaining amount is divided by the number of working members. It follows that the worker and his employer pay the annual per capita contribution of the Knappschaft comrade. How much each pays is to be specified according to the law, whereby an equal share is preferred. Having calculated the average payment this way, the contribution by worker category, according to the statute, is to be specified by taking into account the worker proportions, and to be included in the draft."[79]

KVs, in exchange for promised insurance benefits, did not scale contributions to age in a biological sense, but, instead, to the miner's length of service. A miner employed and insured for twenty years had to pay more than a miner who had worked for only five years in the mining sector. In addition, it might have also been scaled according to occupational classes, which was, in fact, equal to scaling according to wage. Given the same length of service, a hewer underground – the worker who earned most – had to pay more than an assistant worker at the surface, whose wage was lower. Actually, a KV was free to determine the number of classes, in which members were divided due to characteristics like length of service or wage. Usually, the contribution per class was a flat rate rather than a percentage of labor income. In particular, there is no evidence suggesting that KVs would have scaled by, for example, linking contributions or fees with the worker's age at the point in time he joined the KV.[80] Hence, contrary to the Odd Fellows and Friendly Societies, the KVs' pricing techniques seem to have been somewhere in the middle between non-actuarial pricing and commercial insurers' technology.

Pensions were scaled to length of service and, perhaps, income, too. Regarding the level of the invalidity pension benefit, KVs had a free hand in specifying amounts and calculation principles. Regulations did not even set minimum monetary standards or

---

78 See Emery and Emery (1999), p. 87.

79 Ministerium für Handel, Gewerbe und öffentliche Arbeiten (1856), p. 30.

80 Emery and Emery (1999), p. 87, describe that scaling the joining fees according to age was, indeed, the practice of Odd Fellows.

eligibility criteria. The fact that KVs had so much room to maneuver explains most of the heterogeneity among them. Invalidity pensions consisted predominantly of a flat rate, varying in its level over KVs and interpretable as a minimum pension, plus build-up rates for each contribution period during employment (week, month or year).[81] These rates often depended on the miner's classification according to his length of service, wage, or occupation within the mine; build-up rates also varied across insurers. In particular, there is no indication that KVs directly indexed pensions, once granted, to price inflation or productivity; a miner receiving a nominal annual pension of 100 marks in one year would most likely have received those 100 marks also in the subsequent years (except for a non-built-in increase, of course). This rendered KV pensions static.[82] Furthermore, the law of 1906 modified the typical calculation principles insofar as it removed the flat-rate payment. From then on, a miner exclusively collected build-up rates per unit of time of contribution.[83]

When, exactly, was a miner eligible for a pension? According to the KVs' practice, when he was "finished," as the German term for the necessary state of incapacitation, *bergfertig*, most basically implies. The regulations stated only that the state of incapacitation must not be due to the miner's own carelessness. Beyond that, each KV could decide about eligibility rules autonomously. The widespread criterion to qualify for an invalidity pension, however, was simply the inability to work as a miner, meaning that the miner could no longer earn one half of his past wage in his current condition. This criterion was quite clear, and a KV physician had to certify permanent incapacitation.[84] A related distinction that was sometimes made was between so-called full and semi-invalids; the latter simply earned half the standard pension according to length of service and were certified capable of doing some minor work. In contrast to KVs, Bismarckian invalidity insurance (introduced in 1889) granted pensions if the employee was no longer able to earn one sixth of his average wage of the preceding five years and one sixth of the average wage in his particular job. Thus, the KVs' eligibility criteria can be labeled, by and large, as comparatively less strict.[85] Finally, survivors' pensions were usually specified as a proportion of invalidity pensions (often 50 to 60 percent for the widow's pension and ten to 20 percent for the orphan's).

---

81 Contemporaries even tell of KVs that set maximum pensions, as well.

82 For Germany, the change from static to dynamic pensions was, of course, an achievement of the pension reform of 1957; see Schlegel-Voß and Hardach (2003).

83 See Bertrams (1912), p. 1 465.

84 Kleeis, for example, informs us about different practices of certified invalidity. Mostly, KV statutes required that a KV physician had certified permanent incapacitation before a pension was granted. However, there also were KVs that granted an invalidity pension simply in case a working member had reached a certain biological age or had accumulated a certain number of working years (e.g., the *Halberstädter KV*). This automatism, then, did not require certified incapacitation (unless a miner became invalid before the appropriate age); see Kleeis (1910), p. 7.

85 See Geyer (1990), pp. 192–193, and Frerich and Frey (1993), p. 100.

As with invalidity, sickness had to be certified by the KV physician. KVs, moreover, often imposed waiting periods of some days before giving sick pay.[86] Usually, a replacement rate of 50 percent of daily shift earnings was targeted[87], although the data suggest that the average income replacement per day was, in reality, roughly between one third and 39 percent from 1867 and 1907, and increased to slightly more than 50 percent only therafter.[88]

2.2.3. Double your coverage: Consequences of introducing Bismarckian insurance

Workmen's insurance in the form of sickness, accident, invalidity and old-age benefits was introduced on the *Reich* level between 1883 and 1889.[89] But, obviously, KVs were neither abolished as carriers of a self-standing occupational insurance scheme nor turned into organizations with voluntary membership. Instead, mineworkers were insured in both compulsory systems simultaneously and, thus, were privileged in that they gained access to more benefits than the "ordinary" worker. However, where there is light, there is shadow as well. While having been entitled to benefits of two insurance institutions was the bright side, the dark side was, of course, their obligation to pay more contributions. Thus, the mineworkers were required to pay in a larger fraction of their wages than other workers did. Whether this was a truly uncomfortable situation for miners is a question of miners' preferences. If a miner valued the protection of future consumption possibilities higher than he valued his present consumption (or, respectively, the restrictions on it), he might well have accepted an additional obligatory draw upon his disposable income. What seems clear, indeed, is that employers might not have liked to see their employees be double-insured since this meant that they, too, had to pay more.

Besides creating cross-over membership, did the presence of Bismarckian insurance directly influence KV operations? Regarding monetary benefit levels, it did indeed.

---

86 See Guinnane and Streb (2011), pp. 74–75.

87 See Lauf (2006a), pp. 277–281.

88 There is a concise treatise of 1876 (the author of which is unknown) entitled *Die statutarischen Bestimmungen der Knappschaftsvereine im Oberbergamtsbezirke Bonn* that conducts a kind of comparative statute synopsis focusing on the 46 KVs in the mining administration region of Bonn at the time.

89 It is beyond the study's scope here to provide a comprehensive picture of Bismarckian social insurance. This is especially the case since the literature on its emergence and shape is quite rich; see, for example, Ritter (2010), Companje et al. (2009), Hennock (1990, 2006), Haerendel (2001), Eghigian (200), Steinmetz (1993), Mommsen (1981), Saul (1980), Stolleis and (1980). Manow (2000), in particular, focuses on the historical discussion about how to finance social insurance. For an international perspective on the emergence of different sorts of welfare states, see Kuhnle and Sander (2010), Cutler and Johnson (2004), Lindert (1994, 1998, 2004), Flora and Heidenheimer (2003), and Köhler and Zacher (1982).

According to § 74, the Health Insurance Law of 1883 required KVs to raise their sick-pay benefit to the monetary standard of factory sickness funds (*Betriebskrankenkassen*), which was apparently higher.[90] Beyond that, the "decree concerning the modification of the *Knappschaften*'s statutes due the Health Insurance Law" (*Ministerial-Erlass vom 1. October 1883 an die sämmtlichen Königl. Oberbergämter, betreffend Aenderung der Statuten der Knappschaftskassen auf Grund des Reichsgesetzes vom 15. Juni 1883 über die Krankenversicherung der Arbeiter*) made clear that sick pay was uniformly to be paid after a waiting period of three days, and then for 13 weeks.[91] Up to this point, KVs had generally granted sick pay for up to eight weeks. In particular, the decree suggested using the right – granted to the KVs by the Prussian General Mining Law – to distribute their membership over several smaller sickness funds, perhaps one per each mine. As of 1903, furthermore, an amendment to the Health Insurance Law obligated sickness funds of all sorts to grant sick pay for a minimum of 26 weeks.[92]

Accident insurance (1884) and invalidity and old-age insurance (1889) provided *Reich* pensions, for which miners were eligible regardless of whether they already received a KV pension. To carry accident insurance, so-called employers' liability insurance organizations (*Berufsgenossenschaften*) were formed. The one responsible for the mining sector was the *Knappschafts-Berufsgenossenschaft*, which, despite its name, was strictly separated from the KVs.[93] Covering a particular sector, each organization was to be exclusively financed by employers, which definitely raised their non-wage labor costs above the level of costs borne by workers.[94] Indeed, the Accident Insurance

---

90 See Wilhelm I. (1883), p. 47.

91 See Ministerium für öffentliche Arbeiten (1883), pp. 77–79.

92 See Gottschalk (1912).

93 Boyer (1995) provides a comprehensive history of that mining-specific organization.

94 Accident insurance was a conceptual derivative of the employers' liability law of 1871, the first *Reich* law to somehow affect KVs. Originally, according to paragraph two, a factory or mine owner was liable for an employee if he met with an accident while working. If the employee already was KV member *and* the employer had contributed at least one third, those contributions would be subtracted from the compensation for damages the employer had to pay according the 1871 law; see Kah (1907), p. 64 and pp. 96–97. As Kah (1907), pp. 73–74, points out, an "employee" was understood to be a clerk with oversight function, and not an ordinary worker. According to Karwehl (1907), pp. 26–27, mine owners installed so-called liability funds in many places as a reaction to the law. Those funds were managed by the KVs, financed by employers' payments, and were used to pay accident casualties' compensations beyond those they regularly received from the KVs. Employers were said to have raised those additional funds to prevent legal trials about whether they were liable in a particular case of accident. The logic behind probably was that, at the end of a legal trial, the employer was to pay a compensation to the accident casualty that was likely to be higher than what he paid *voluntarily* into a liability fund. Precisely, an employer that founded such a liability fund seems to have exchanged an uncertain and potentially high loss in the future due to legal trial for a certain and presumably smaller loss at present. However, those reactions did not directly relate to rules the law set out, but to suggestions the Prussian secretary of trade made. According to Bülow (1910), p. 34, the *Märkischer KV*, for example, obliged employers to pay 100 thalers to the liability fund whenever a KV member met an accident through no

Law stated explicitly in paragraph eight that the KVs' obligation to produce insurance coverage, as done before, became all but obsolete:

> "The obligation of registered funds and of other sickness, burial, invalidity and miscellaneous funds to grant those workers and technical civil servants, and their dependants and survivors, benefits in case of job-related accidents, [...] is not affected by this law."[95]

But the accident insurance law contained, as did the invalidity insurance law, rules that allowed KVs to count their own pensions in some way against *Reich* pensions. Article eight of the Accident Insurance Law puts this as follows:

> "Insofar as, because of such obligations, benefits are granted in cases in which this law provides a legal benefit to the supported, the amount of the latter benefit is fully transferred to those funds, parishes or poor relief society that granted those benefits."[96]

Although this paragraph's meaning is not that clear, KVs could obviously cut their payments to the amount of the *Reich* pension, such that they had to pay only the difference – if there remained one. So the KV benefit could, at the maximum, be zero and, thus, the maximum cut was 100 percent. In this context, Franz Milde, in his 1916 insurance-legal study, pointed to a change that came with the amendment to the Accident Insurance Law of June 30, 1900 (the *Gewerbe-Unfallgesetz*). He stated that Reich pensions since were to be counted against KV pensions to a maximum of 50 percent.[97]

Regarding invalidity, the law of 1889 recognized KVs as legal carriers of Bismarckian invalidity and old-age insurance, provided that they met the requirements laid down in article five, concerning financing, and were of a size that was suggested to be kind of minimum efficient;[98] as Lauf (2010) mentions, this size was 100 000 or more insurants. In fact, there were some twenty KVs that took the opportunity to become a "special insurer" (*besondere Kasseneinrichtung* or, respectively, *Sonderanstalt*): The *Allgemeine KV Bochum* and the *Saarbrücker KV* (both Prussia), the *Allgemeine Knappschaftspensionskasse Sachsen*, and a conglomerate of KVs from Prussia – for example, the *Clausthaler KV* – and other German states, such as the *Anhalt KV*. The *Allgemeine KV Bochum* was the result of a merger of three KVs in 1890: the *Märkische KV*, the largest KV in Prussia; the *Essen-Werden KV*, the third-largest; and the *Mülheimer KV*. As Lauf (2010) points out, this merger was directly intended to meet article five and also to allow the newly-merged, larger KV to become the provider of Bismarckian insurance for its own KV members. Besides cost advantages, the main reason for taking over *Reich*

---

fault of his own. This happened 151 times in 1872. Though, if we consider that the *Märkische KV*'s total revenue in that year was around 523 000 thalers, an amount of 15 100 thalers, or 2.88 percent, appears a rather meager relief. There are, however, examples of KVs that did not react at all. Jordan (1905), p. 53, provides examples from the *Mansfeld KV*.

95 Woedtke (1885), p. 92.

96 Woedtke (1885), p. 92.

97 See Milde (1916), p. 2.

98 See Gebhard and Geibel (1889), pp. 134–135.

insurance arguably was to gain power and significance as an insurance institution.[99] Furthermore, the *Saarbrücker KV* was the fourth-largest Prussian KV, and the aforementioned conglomerate of KVs from Prussia and elsewhere traded under the name *Norddeutsche Knappschafts-Pensionskasse*; both came into force in 1890–91, as did the *Allgemeine Knappschaftspensionskasse Sachsen*, too.

While these funds provided both *Reich* insurance and KV insurance regulated at the *state*-level, the various other KVs were recognized as "allowance funds" (*Zuschusskassen*), which provided additional welfare to their members. Both sorts of KVs, the special insurers and the allowance funds, were allowed to count their pension benefits against *Reich* pensions. Precisely how they did this was regulated by article 36, which allowed a KV to cut its benefits to the amount of the *Reich* pension (at the maximum) as long as the KV had reduced contributions proportionally in the past.[100]

As mentioned in the subsequent subsection, access to KV pensions required members only to be incapacitated for miners' work. In contrast, access to *Reich* pensions required being permanently incapacitated for any kind of work (regarding invalidity) or having reached the age of 70 (regarding old-age pensions). This means that KV pensions would, on average, arise earlier in a miner's life than his *Reich* pension would. Thus, it seems reasonable to interpret the law as allowing a KV to count its payments against those from Bismarckian insurance only after *Reich* pensions had arisen, too. If so, this were equal to a pension cut occurring sometime over the course of retirement: Assume therefore that a miner had already received an annual KV pension of, say, 300 marks for five years. From the point of time on when he became eligible for a Bismarckian invalidity pension, his KV pension of 300 marks could have been cut, at the maximum, by the amount of the miner's Bismarckian pension. Assuming an annual amount of that pension of, say, 120 marks, the pension cut would have apparently left only 180 marks of the miner's KV pension such that the miner would have earned, instead of 420 marks on the whole, only 300 marks consisting of two pensions. To qualify the whole issue of "cutting" a bit, a KV could have decided not to cut by the full amount of the Bismarckian pension. If contributions had not been reduced sometime in advance, a cut was – at least formally – not allowed, which does not mean that a KV might not have tried to cut. As the following quotes exemplarily indicate, the practice or, respectively, the opportunity of counting *Reich* pensions against KV pensions raised some resistance on the side of insurants providing an incentive not to use that opportunity.

Milde (1916) provides us with a judgement of how KVs reacted to their new maneuvering room and of how workers reacted to boards' decisions not only to cut pensions, but also to complement cuts with reductions in per capita contributions required:

> "Regarding invalidity pensions, many KVs made use of paragraph 36 [...] and counted the Reich invalidity pension fully or partly against the KV pension. This practice, however, provoked members' opposition because they believed themselves to be entitled to full benefits

[99] See Lauf (2010), p. 19–21.

[100] See Gebhard and Geibel (1889), p. 143.

> due to contributing to two insurance funds [...]. They did not care about the fact that contributions were, therefore, reduced. Members' opposition motivated KVs to gradually abolish deductions. For this, and the fact that some KVs raised their often quite low invalidity and widows' pensions without corresponding increases in contributions, many KVs became financially unsound; they were exposed to the threat of being caught unprepared to settle pension liabilities towards [working] members and pensioners."[101]

Köhne (1915), just one year earlier, had pointed in the same direction, contending that the bottom line among KVs finally became not to count *Reich* pensions and KV pensions against each other.[102]

It follows that observable decreases or increases in the absolute level of payments or, alternatively, in the pension level and the contribution rate, calculable from the KV statistics, could have had two origins: demographic pressure in a PAYG system and reactions to *Reich* regulations. However, I think it is appropriate to assume that KV boards were more prone towards legally cutting benefits they had once promised to pay, the more they felt financial pressure – originating in their very operations – weighing on them. According to Lauf (2008), the practice of counting KV pensions against *Reich* pensions was, finally, fully abolished in 1908.[103]

For the sake of completeness, we should take a look at what the RVO and the White-collar Insurance Law of 1911 did. Basically, the 1912 law was intended to adapt the Prussian KVs' scheme precisely to those regulations. According to Gottschalk (1912), the RVO had no major influence on state-regulated KVs, except for making adjustments to the wording of the Prussian General Mining Law's seventh title necessary.[104] It seems as if the combined effect of both *Reich* regulations had two appreciable effects: First, KVs were recognized as surrogate funds (*Ersatzkasse*) for white-collar insurance up to an income ceiling of 4 000 marks; and, second, arbitration (*Schiedsgerichtsbarkeit*), formally established with the first amendment to the Prussian General Mining Law (1906), was expanded and improved. The latter, in particular, was reflected in the foundation of so-called *Knappschaft* insurance offices (*Knappschafts-Oberversicherungsämter*) fulfilling the purpose of overseeing the adjustment to the RVO and of maintaining arbitration if legal arguments between insurants and the KVs occurred.[105] Indeed, the basic 1854/1865 regulations are judged in the literature to have installed the right to receive a benefit proportional to contributions paid.[106] However, it seems as if

---

101 Milde (1916), p. 2.

102 Köhne (1915), p. 12–13.

103 See Lauf (2008a), p. 17. For the probably most detailed contemporary study on the *Reich* laws' effects on the KVs, see Halbach (1906).

104 See Gottschalk (1912), p. 1 163–1 170.

105 See Gottschalk (1912), pp. 1 202–1 205, and Wiessner (1912). To the best of my knowledge, there is no further contemporary, let alone recent, literature on the working of those *Knappschaft* insurance offices. Though, from what Gottschalk says about their obligations, it seems that they were not commissioned with economic regulation.

106 See, for example, Tenfelde (2004), p. 25.

the devices to properly enforce legal claims when KVs were not willing to pay their obligations were only established some four decades after the basic regulations.[107]

2.3. The *Knappschaften* – an ideal historical laboratory?!

The concise regulatory overview has revealed two key facts: i) certain organizational principles were binding for every Prussian KV; and ii) the KVs retained some considerable room for independent decisions with respect to the qualitative and, even more so, the quantitative configuration of the range of services and the financial underpinnings. As we will see below, this, along with the large size differentials, were the main sources of heterogeneity among KVs.

Those who object to the KVs as a useful example from economic history for economists to evaluate the costs and benefits of size might claim: (i) that size might not have been of relevance for KVs, if they are considered "social insurance carriers;" (ii) that the KVs provided occupational social insurance for only one particular sector of the economy, rather than universal coverage; (iii) that social insurance against invalidity, rather than insurance against old-age, was provided; (iv) that aging was no particular threat to the emerging nineteenth-century welfare state; (v) that, in all, the social, political and economic conditions in the nineteenth and early twentieth centuries were quite different from today's; and (vi) that contemporary observers were not really sure about how to understand the KVs' nature properly. Yet, the bottom line of this study is that those objections are negligible since the core economic problems have not changed; in other words, they are timeless.

Regarding the sixth point, which has not been raised previously, the contemporaries discussed whether KVs really were insurance funds that accepted members to have a legal right to receive benefits commensurate to contributions paid or whether they still were charitable funds that granted benefits simply as long as the cash box allowed without being bound to members' legal right. Did KVs provide proper insurance, as commercial life insurers did, they asked, or did they actually retain their nature as charitable, cooperative organizations that should not be seen in the context of insurance? Based on the KVs' regulatory history, we can maintain the claim that KVs should, indeed, be looked upon as providing insurance and social insurance in particular. It seems as if contemporaries answered these questions according to the "insurance technology" used. A statement by Karwehl (1907) clarifies this point:

107 The initial absence of proper legal channels to enforce one's own claims towards the KV, certainly led to many arguments between insurants and their insurers (which actually were arguments between workers and employers); see, for example, Tenfelde (1977) and Tenfelde and Trischler (1986). Unfortunately, the KV statistics do not offer data based on which it would be possible to assess the amount of benefits (pensions or other) that KVs withheld – temporarily or permanently – from their insurants, and which were the subject of legal arguments.

> "There are two opposing perceptions. […]: the KVs merely were charitable organizations, the vital matters of which are subordinated to employers' wishes, even though they found it appropriate to provide Knappschaft comrades with an equivalent for compulsory membership – as in the securing of benefits. The other perception strictly requires the KVs to put finances on a proper actuarial basis. In between these perceptions is the prevailing one advocated by the government. […]: "The author sees KVs as being mere insurance organizations, granting their members certain annual payments – static pensions. This view is not correct. According to their development and present character, KVs are mere cooperative support funds, thus charitable organizations, which comply with their legal obligations according to their financial capabilities and which, depending on circumstances, raise or lower their benefits. […]". […]. In reality, KVs cannot be looked upon as pure insurance institutions."[108]

The fact is, KVs did not use actuarial technology, as commercial insurers had begun to do in order to specify each insurant's premium according to his individual risk situation. Indeed, that is not precisely what contemporaries understood to be "proper insurance." Rather, they reduced the distinction between insurance and what the KVs did to the difference in the applied financing method – funding versus PAYG. It was precisely the former – funding – that, in their view, reflected the essence of insurance in the more narrow sense of the word. The latter, the PAYG method, was, by contrast, seen as a fairly dubious "technology."

In my opinion, however, the KVs did what insurance does in the more narrow – or technical – sense of the word: namely, to give, in exchange for a premium, a conditional promise to grant benefits in the event that certain pre-defined events occur. Which financing mechanism they applied, and whether contributions were specified to cover the average risk in the pool of insurants (and not the individual insurant's risk), is not the pertinent issue here.

To call KVs "social insurers," from today's perspective, requires identifying certain principles that are commonly associated with social insurance and that clearly isolate KVs from private, market-based solutions such as mutual insurers (*Gegenseitigkeitsversicherer*). The mode of KV regulation did, indeed, lead to a handful of features that, on the one hand, were "engraved in stone" over the whole observation period (and beyond) and, on the other hand, might remind us exactly of those principles. These features were obligatory membership and contributions; self-management by the collective of insurants; joint financing by employees and employers; legal entitlements to benefits "broadly" commensurate with contributions; an open risk community as long as the KV survived; and having been publicly licensed insurance providers officially adopting government tasks on its behalf.[109] However, there are features that might not match with our understanding of social insurance: the fact that KVs could slide into a situation

---

108 Karwehl (1907), pp. 86–88.

109 One additional feature, however, is not that obvious from above, namely the degree of intended income redistribution *within* generations to put some better, to the detriment of others. The only value judgment in this direction appeared when the distinction between established and unestablished miners was addressed.

leading to final closure implying the absence of any government guarantee of survival although government tasks were adopted; the fact that actuarial risk in the applied form played a role for KVs at all; and, as a consequence, the fact that specifying risk loadings on top of the contribution payments was relevant. One feature that is the essence of private mutual insurers – policyholders are equal with the policy seller and, thus, altogether own the firm – is definitely not a feature of KVs. To end this short discussion, KVs may be labeled "hybrid organizations" reflecting many principles that would form the basis of Bismarckian and today's social insurance, but also reflecting other principles that remind us of different insurance organizations. This is no contradiction. Rather, it is a matter of acceptance: KVs simply reflect how social insurance began. This is thought-provoking in the sense that certain principles obviously disappeared over the course of the evolution of social insurance.

3. The welfare state evolves: Exploring quantitative developments

A description of the KVs' quantitative development is a prerequisite for attaining a deeper understanding of their operations in the period under observation. What is more, it is a key desideratum of the research on those pioneers of social insurance. In this respect, this chapter serves a dual purpose. After introducing the data sources, it draws a quantitative picture of Prussian *Knappschaft* insurance. This chapter deals in detail with the size and structure of both the memberships and the expenditures and revenues, thereby extending the database for comparative research on the emergence of nineteenth-century welfare institutions. Although the main focus is on Prussia, I broaden the scope by also using some data on Bavarian and Saxon KVs.

3.1. Data[1]

3.1.1. Sources and nature of the data

This study's research design relies on raw mass-data on the KVs' operations as input. Thanks to the Prussians' efforts to produce statistics on many aspects of economic activity, I am able to draw on the rich *Statistik der Knappschaftsvereine des preussischen Staates*, henceforth referred to as "Prussian KV statistics." Starting in 1854, reporting data on 1852, the statistics were compiled and published over the course of the observation period by four more or less distinct institutions: first, between 1854 and 1858, by Rudolf von Carnall with the permission of the Prussian Department of Mining, Metallurgy and Salines (*Ministerial-Abtheilung für Berg-, Hütten- und Salinenwesen*); sec-

[1] This section is based in part on two of my pre-releases: the article entitled "Die Vorteile und Nachteile großer Invaliditäts- und Krankenversicherungen: Eine quantitative Wirtschaftsgeschichte der deutschen Knappschaftsvereine von 1854 bis zur Gründung der Reichsknappschaft,“ with Timothy W. Guinnane and Jochen Streb; and the article entitled "The hazard of merger by absorption: Why some Knappschaften merged and others did not, 1861–1920,“ published originally in the *Zeitschrift für Unternehmensgeschichte*; see Guinnane et al. (2012a), and Jopp (2011a), pp. 77–78.

ond, between 1859 and 1878, by the Prussian Ministry of Trade, Commerce and Public Works (*Preussisches Ministerium für Handel, Gewerbe und öffentliche Arbeiten*); third, between 1879 and 1889, by the Prussian Ministry of Public Works (*Preussisches Ministerium für öffentliche Arbeiten*); and fourth, after 1890, by the Prussian Ministry of Trade and Commerce (*Preussisches Ministerium für Handel und Gewerbe*).[2] I hereafter refer simply to "the ministry" as editor of the KV statistics.

Because the volumes 1855 to 1860 all lack the statistical information required for a detailed analysis, the whole population of Prussian KVs can be quantitatively assessed only from 1861 onwards, when data started to be published regularly, with each year's publication reporting information on the preceding year. The data situation for the years 1921 and 1922 is insufficient compared to preceding years, so the period under quantitative consideration is limited to 1861 to 1920.[3] What the ministry actually compiled and published were the KVs' official reports, which they were obliged by law to send to the ministry. Data cover the entire population of 103 Prussian KVs operating within the period 1861 to 1920 and provide a broad range of information on various aspects such as memberships, revenues and expenditures.

For the subsequent quantitative analyses, especially that in Chapter 5, it is useful to consider the universe of KVs as consisting of five samples (that are, of course, not randomly chosen): (i) The basic sample consists of the cohort of KVs that are observed between 1861 or later and 1920 *and* that survived the observation period without having been involved in any mergers and acquisitions; (ii) the cohort of KVs that absorbed another KV during the observation period (the "mother KV," so to speak) *and* that survived until 1920; (iii) the cohort of KVs that absorbed another KV *and* ceased operation before 1920 due to having been absorbed or liquidated; (iv) the cohort of KVs that were liquidated sometime before 1920 without having ever absorbed another fund; and (v) those KVs that were absorbed by another KV without ever having absorbed a fund themselves. Information on the question of which KVs actually were either merged or liquidated over the course of the observation period are taken directly from the Prussian KV statistics, which provide, along with the pure data cells, some explanatory notes on each KV.

Table A.1 in the Appendix lists the included Prussian KVs by code, name, first observed year, year of liquidation, year of merger and KV into which the respective KV was merged, size in first observed year, and size in the last observed year. The set of

---

2 See v. Carnall (1854–1858), Ministerium für Handel, Gewerbe und öffentliche Arbeiten (1859–1878), Ministerium für öffentliche Arbeiten (1879–1889), and Ministerium für Handel und Gewerbe (1890–1922). Wirth (1962) provides a study on the genesis of the Prussian Ministry of Trade, Commerce and Public Works, while Borchard (1929) especially addresses the Prussian Ministry of Trade and Commerce since 1879. To be precise, the Ministry of Public Works was a spin-off of the Ministry of Trade, Commerce and Public Works.

3 In 1923, all German KVs merged into the *Reichsknappschaft* that continued to publish data on many aspects of its operations. However, the *Reichsknappschaft* is out of scope here.

Prussian KVs is further distinguished by the mining administration regions in which they were situated.[4] (i) Bonn – mostly congruent with the province Rheinland and encompassing, among other things, the important Saar and Aachen coalfields; (ii) Breslau – where the important coalfields of Upper and Lower Silesia were located; (iii) Clausthal – congruent with the province Hannover and containing the Harz coal and ore fields;[5] (iv) Dortmund – mostly congruent with the province of Westphalia and containing the prominent Ruhr coalfields; and (v) Halle – encompassing the province of Saxony and containing important brown-coal and ore fields. Within the geographical boundaries of these administrative units, the KVs were either operated locally – e.g., for a small part of a coal field (and all the enterprises there) or a particular mine within that field – or regionally – for a complete mining area (and, of course, all the enterprises there). Figure 3.1, below, locates the five Prussian mining administration regions as of 1871.

The dataset on Prussian KVs, the very basis of this analysis, is complemented by data on the Bavarian and Saxon KVs. Therefore, I also draw upon the *Statistik der Knappschaftsvereine im bayerischen Staate*, henceforth referred to as Bavarian KV statistics, and the *Jahrbuch für den Berg- und Hütten-Mann* or, respectively, *Jahrbuch für das Berg- und Hüttenwesen im Königreiche Sachsen*, henceforth referred to as Saxon KV statistics.[6] Table A.2 lists the population of Bavarian KVs observable since 1871. Administratively, Bavaria was divided in four mining administration regions until 1882 and three since 1883, referred to, in German, as *Berginspektionsbezirke*. These were Bayreuth, München, Regensburg and Zweibrücken. According to the Bavarian KV statistics, the mining administration region of Regensburg was dissolved in 1883 and the KVs that had belonged to this region were allocated to either Bayreuth or München; see Figure 3.1 for the location of Bavarian mining administration regions within the German Empire, as of 1883.

---

[4] The German mining sector had its own administrative foundation. The *Gesetz über die Kompetenz der Oberbergämter* of June 10, 1861, which, as we know from the preceding chapter, put foundries under the responsibility of trade control, modified the administrative foundation of Prussian mining by introducing those mining regions; the mining administration headquarters there (the *Oberbergämter*) replaced the royal headquarters (*königliche Bergämter*). What I translate as mining administration regions (*Oberbergamtsbezirke*) should not be confused with the general administrative districts (*Regierungsbezirke*) of Prussia.

[5] For the mining administration region of Clausthal which was founded for the annexed province of Hannover, see, especially, Lahmeyer (1881).

[6] See Königliches Finanzministerium Sachsen (1870–1872) and (1873–1921) and Oberbergamt München (1871–1921). The *Jahrbuch für den Berg- und Hütten-Mann* and the *Jahrbuch für das Berg- und Hüttenwesen im Königreich Sachsen* are electronically available, as of July 7, 2012, 10 a.m.: http://www.tu-freiberg.de/~ub/el-bibl/jb_sachsen/jb_ sachsen.html.

Figure 3.1: Locating the Prussian and Bavarian mining administration regions within the German Reich (around 1871 regarding Prussia and around 1883 regarding Bavaria)

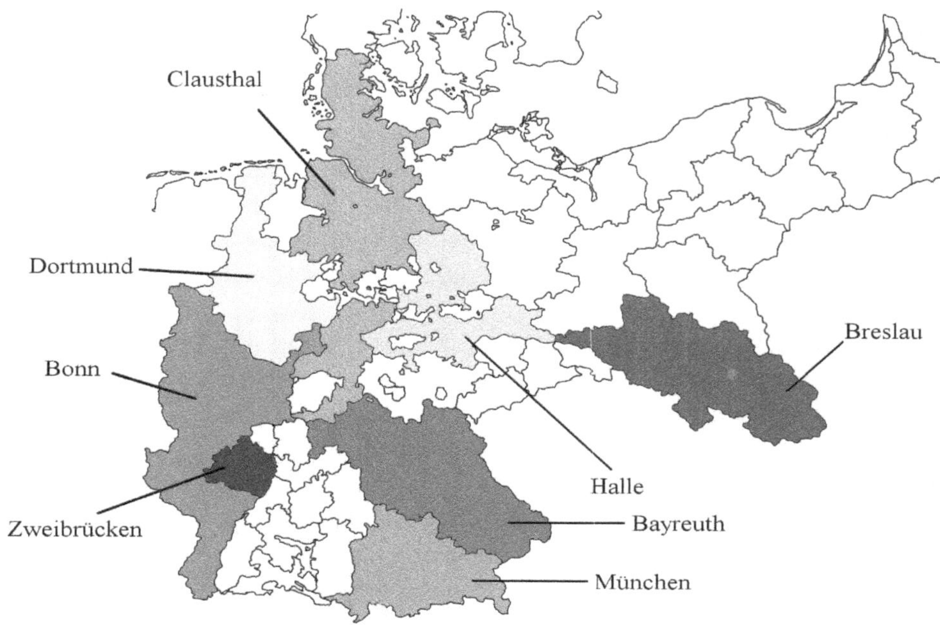

Notes: Demarcation between the Prussian mining administration regions is based on several maps that could be located in the literature (Bonn as of 1871, Clausthal as of 1868, Dortmund as of 1866, and Halle as of 1839). The demarcation of the mining administration region of Breslau follows the demarcation of the Silesian administration districts of Liegnitz, Breslau, and Oppeln. The demarcation between the Bavarian mining administration regions is speculative.

Sources: This depiction is based on Bergamt Halle (1998), p. 29, Fürer (1988), p. 17, Oberbergamt Bonn (1966), p. 22, and Schelter (1992), p.50. The map was kindly constructed by Nolan Ritter (RWI).

Because of the fact that Saxony went a different way in organizing miners' provision for sickness and invalidity, I will not analyze data on Saxon KVs to the extent that I do the Bavarian data; see table A.3 in the Appendix for a list of the Saxon KVs.[7]

---

[7] The Saxon KVs can similarly be linked with several mining administration regions not displayed in Figure 3.1: First, regarding hard- and brown-coal, these were Chemnitz, Dresden, and Zwickau; second, regarding ores, these were Freiberg, Altenberg, Marienberg, and Schwarzenberg. Regarding Figure 3.1, Saxony would be found, more or less, as framed by the mining administration regions of Breslau, Halle, and Bayreuth; see, again, Königliches Finanzministerium Sachsen (1870–1872) and (1873–1921).

As Chapter 2 reported, there were KVs not only in Bavaria, Prussia, and Saxony, but also in the remaining states. Their number, however, was nowhere near as high, and, in addition, their operations seem not to have been captured in as much detail as those of the Prussian, Bavarian and Saxon KVs. Some KV-level and aggregate data can, indeed, be derived from a number of contemporary secondary sources or, respectively, from the source that they presumably drew upon, namely the annual statistical overview in the journal *Der Kompaß: Organ der Knappschafts-Berufsgenossenschaft für das Deutsche Reich* entitled "Die deutschen Knappschaftsvereine im Jahre *t*," covering at the minimum the years 1884 to 1913.[8] In addition to their lower degree of detail,[9] the data presented in these contemporary studies cover only some of the years under observation. Therefore, I will not concentrate on the KVs outside Bavaria, Prussia, and Saxony.

To date, the Prussian, Bavarian and Saxon KV statistics have not been rigorously used to statistically evaluate the KVs' operations.[10] The newly-constructed datasets on the Prussian and Bavarian KVs enable us to conduct an empirical analysis of *Knappschaft* insurance, as outlined in the introduction. The datasets contain quite a large number of statistical variables on many, but not all, aspects of the KVs' business policy. Before turning to the details, four issues must be highlighted for a better understanding of what follows: First, to understand the data, it is important to note that the institutional separation of 1906 is apparent in the statistics beginning with the year 1908; hence, KV operations in 1907 were reported as if bundling (compound insurance) still existed. Thus, data on Prussian (Bavarian) KVs are displayed separately for the sickness and pension insurance sections only from 1908 (1917) onwards;[11] second, data *were not* classified consistently in the period under observation; third, the data displayed are

---

8 See, for example, Simons (1895), who used data not only on Bavaria, Prussia and Saxony, but also on Wurttemberg, Hesse, Brunswick, Anhalt, Saxe-Altenburg, Saxe-Meiningen, Schwarzburg, Waldeck, and Alsace-Lorraine; data cover the years 1885 and 1890 to 1893. Imbusch (1910) and Köhne (1915) similarly used data on all German states – the former for the years 1890, 1895, 1900, and 1905 to 1908 and the latter only for 1913. Bankewitz (1908), who described the development of the Anhaltische KV (Duchy of Anhalt), seems to be an exception insofar as he listed data series covering quite an extensive period, namely 1858 to 1907.

9 We get to know basic information, such as the size of the memberships, membership composition by activity status (number of contributors and number of invalids/widows/orphans), aggregate expenditures and revenues. This is, however, not anywhere near what the Prussian KV statistics offer.

10 Being a seminal contribution to German historical insurance statistics, Borscheid and Drees (1988) do indeed list *aggregate* data series on some states' *Knappschaft* insurance (aggregate membership, revenue, expenditure, and, occasionally, assets), separated by sickness (Kingdom of Saxony, 1885–1937, and Germany as a whole, 1892–1984) and pension insurance (Grand Duchy of Hesse, 1880–1912, Kingdom of Saxony, 1885–1937, and Germany as a whole, only after 1923); see Borscheid and Drees (1988), p. 417, p. 452, p. 490, p. 505 and p. 574. In fact, my study may fill some remaining statistical gaps.

11 Formally, beginning in 1913, the statistics also distinguish explicitly between workers and civil servants. Since this study focuses on worker insurance solely, this distinction does not matter here.

yearly data on the KV-level; and fourth, monetary data were denominated in thaler until 1874 and in marks thereafter, owing to the reform of the German Empire's monetary system immediately after its formation. With reference to Sprenger (1991), I established consistency by converting one thaler into three marks for the years prior to 1875.[12]

According to the way data are presented in the KV statistics, I distinguish three major kinds of problem areas. The first concerns membership size and composition. The second concerns the revenue and expenditure side or, more technically, a KV's "profit and loss statement." The third refers to the asset balance sheet. In the following, I concentrate on the Prussian KV statistics. The Bavarian data are, however, equivalently structured, though not so detailed.[13]

Regarding membership data, there is an impressive mass of primary variables available.[14] It is helpful to bear in mind that data can be distinguished into membership stock and membership flow quantities and by decades prior to the legal reform of 1906 and the remaining years thereafter. Principal categories of membership stock quantities under which various series can be subsumed are "working members" and "pensioners" or, respectively, "sick members." Membership flow quantities may be assigned to categories "in- and outflows of working members" and "in- and outflows of pensioners" or, respectively, "in- and outflows of sick members." To name only few, the statistics provides series on, for example,

- the number of working miners, of sick miners, of invalids, of widows, and of orphans at year-start and year-end;
- the number of working miners by subsector or, respectively, age-group;
- the number of invalids and of widows by age-group;
- inflow of sick miners by reason ("accident and treated at home" or "accident and treated at hospital");
- outflows of active members, invalids, and widows by age-group and reason (e.g., "due to having become invalid" or "due to out-migration" in case of working miners, "due to death" or "due to recovery" in case of sick miners, or "due to death" in case of invalids and widows); and
- the number of sick days (reimbursed with sick pay or treated by physician.

12 See Seeger (1968) pp. 18–22, for an overview of the relevant legal steps – e.g., the Coin Act (*Münzgesetz*) of 1873 and the Banking Law (*Bankgesetz*) of 1875; see Sprenger (1991), p. 187, for the conversion.

13 In this shortened version of my study, I have dropped a number of tables that schematically display the membership balance, the asset balance, and what I call "the profit and loss statement;" the latter term should not be taken too seriously in accouting terms. The reader interested in details is kindly referred to Jopp (2012b, 2013).

14 What I refer to as "primary variable" is a variable not created by transformation of an underlying variable (e.g., the year-to-year growth rates of a series) or by transformation of two or more variables in combination (e.g., mid-year working membership as the ratio of working membership measured at the first of January and that measured at the last of December).

Some peculiarities need to be pointed out: (i) Until 1907, relevant membership information was separately reported for established and unestablished miners; (ii) there are two inconsistencies in reporting age-group sizes of the working membership, insofar as no age-group information about unestablished miners was published, and the age classification regarding established ones changed in 1889 (from six to ten classes); (iii) until 1907, a number of KVs distinguished full invalids from semi-invalids, whereby the latter usually made up only a very small fraction of all invalids;[15] (iv) until 1907, it was reported how many pensioners received a Bismarckian accident pension, but not how many received a Bismarckian invalidity pension; (v) only after 1908, it was explicitly reported how many members of the pension section paid acknowledgement fees – i.e., the payment by which they could maintain entitlements already accumulated, but not collect new entitlements; (vi) some variables capturing the more exact influence of war on memberships are, at times, reported; and, finally, (vii) after 1908, it was also stated how many invalidity pensioners were re-integrated into working membership, with the figures always remaining quite low.

Furthermore, regarding the KVs' revenues and expenditures accounted for in what I call the "profit and loss statement," the statistics is equally rich and, again, we have to distinguish the way the statistics were organized prior to the reform of 1906 from the way afterwards, when both main insurance sections had been institutionally separated in each KV. This separation is definitely reflected in the KVs' accounting framework, although this does not prove they did not cross-subsidize in one direction or the other. The four major revenue categories are the following: income from regular social insurance contributions by workers *and* employers; several case-specific fees; capital income; and miscellaneous sources. There are some inconsistencies in the statements, too. Established miners' contributions, for example, were merged with those of unestablished workers after 1888. Typically, income sources such as "rent," "extraordinary sales" and "donations" were only infrequently reported and, thus, were of minor importance. Before the 1906 reform, expenditures were broken down into pension benefits, health-related benefits, miscellaneous benefits such as funeral pay, and operating costs. After the reform in 1906, the statistics became especially more detailed with regard to operating costs. Although the use of the PAYG mechanism implies that expenditures are perfectly matched by revenues, net surpluses or net deficits different from zero did actually occur among KVs.

Beyond that, there was the KVs' asset balance displaying items such as "cash holdings," "bank deposits" (only after 1907), "interest-bearing assets/papers," "immovable property," "movable property," and "reinsurance deposits" (pensions, after 1907). Note that the statement of liabilities was much less detailed than the left-hand side of the asset balance. In particular, the number of KVs stating "debt on immovable property" as

15 This is the reason why I do not explicitly account for them below.

well as “other debt” was typically rather low; only after 1907 did the proportion of KVs stating one sort of debt or the other increase.

3.1.2. The data’s potential and limitations

Finally, regarding the data’s potential and limitations, I think that the Prussian KV statistics, as compiled by the ministry, are, first, a reliable source for mass data on the KVs and, second, a source that enables us to draw a vivid and appropriate picture of the miners’ funds’ business operations. The data’s potential lies in their adequacy for basic economic and statistical analyses. In particular, this holds since the data display a reasonable extent of heterogeneity among cross-sections, which makes comparisons between small and large or surviving and exiting KVs possible. Formally, however, two points are worth noting: First, I cannot exclude the possibility that a KV deliberately reported doctored data to the administration; and second, I cannot be certain that the editor, while compiling the data, did not make some accidental transcription or typing errors.[16]

The data’s limitations simply stem from the fact that they still lack certain information that would be required if one wished to conduct an *even more* detailed analysis. First, they lack data on the insurant-level; this rules out the use of an individual risk model and makes performing behavioral analyses – of ex-post moral hazard (whether induced by hidden action or hidden information), for example – more difficult, not to mention testing certain assumptions implicit in insurance models (e.g., that individuals forming the collective of insurants are identical and independent). Second, they lack data on the KVs’ pricing behavior that go beyond age-unspecific, per capita figures; this rules out performing much more rigorous analyses of the KVs’ pricing behavior and its distributional stance from *intra-* and *inter*generational perspectives as well as certain actuarial analyses based on a detailed reconstruction of the way revenues (especially contributions) and benefits (especially pensions) were scaled. Thus, the limitations set by the data are apparent if a certain sort of economic-statistical analysis, or a certain degree of detail, cannot be realized. So in case well-trained economists have the impression sometime that a more detailed analysis might be appropriate or preferable, they may consider that we have reached the data’s limits. To the best of my belief, I will give the reader a fair warning if I have the impression that the data do not allow to go one step further.

---

16 A comparison of data from the available original business reports of the *Allgemeine KV Bochum* and the *Saarbrücker KV* with the Prussian KV statistics did not result in findings that would undercut the statistics’ reliability; see Saarbrücker Knappschaftsverein (1879–1920) and Allgemeiner Knappschafts-Verein zu Bochum (1891–1920).

3.2. German mining at the time – an overview[17]

As the subsequent chapter highlighted, German miners of the 1850s could look back at roughly 600 years of experience with mutuality in risk provision. Like few other industries – e.g., the shipping and the railroad sectors – the mining industry with its KVs pioneered the field of organizing social insurance against the contingencies of life, the consequences of which everyone feared, and no one could shoulder on his own. Almost from the start, mine owners were part of the system as co-sponsors.[18]

There is no doubt that mining, especially deep mining for coal, was among the most perilous occupations, connected with a high ex-ante probability of becoming involved in accidents and suffering from severe chronic diseases (e.g., silicosis) or epidemics (e.g., the hookworm) – in part because matters of work safety and hygiene were put on the agenda only little by little.[19] Since KVs were an integral part of the German mining industry, their business policy and the economic and social challenges they faced can be better understood against the background of the industry's secular trends and peculiarities. Thus, some general contextualization is in order before turning to the KVs' quantitative economic history. The following paragraphs set the stage by focusing on issues such as production, productivity and technology.

To begin with stylized growth facts on the industrializing German economy as a whole, aggregate figures draw a picture of steady growth and prosperity. Between 1851 and 1870, net national product (NNP) grew at an annual rate of 1.56 per cent. During the German Reich until 1913, NNP grew even faster, at an annual rate of 2.67 per cent.[20] Yet, due to World War I and post-war inflation, the economy contracted in the following years.[21] Ritschl and Spoerer estimate annual growth of gross national product (GNP) at minus 3.76 percent from 1914 to 1920.[22] The "leading sector complex" of the railroad and heavy industries, replaced after 1890 by the chemical industry and electrical and mechanical engineering, is commonly identified as one of the most important drivers of that secular pre-war growth.[23]

Regarding mining activity, Germany can be associated, essentially, with hard-coal production. Hard coal certainly became the single most important natural resource in the nineteenth century and most miners – hence, most KV members – were engaged in extracting it. However, extractive activities concentrated not only on coal, but also on

17 An earlier version of this subsection has been published as part of the article "The hazard of merger by absorption: Why some Knappschaften merged and others did not, 1861–1920;" see Jopp (2011a), pp. 78–83.
18 See Bartels et al. (2009), pp. 200–208.
19 See Bluma (2009a, 2009b) and Boyer (1995), pp. 114–116.
20 Calculated with data from Burhop and Wolff (2005), p. 651–652.
21 On the Great Inflation, see Holtfrerich (1980).
22 See Ritschl and Spoerer, (1997), p. 51. Data on NNP during 1914 to 1920 are not available.
23 See Holtfrerich (1973), p. 165–168, and Tilly (1990), pp. 50–58.

other resources that were diverse in both their properties and their geological circumstances of production. In the following, I refer to these various resources as "subsectors" of the mining sector. According to German national statistics, we can identify the following subsectors: (1) hard-coal, (2) brown-coal, (3) iron-ore, (4) miscellaneous ores, (5) halite, (6) salts, (7) pyrite, and (8) the rest.[24] Tables 3.1 and 3.2, depict, for selected years, some basic data on the production of the most relevant resources and labor input.

Table 3.1: Resource extraction and utilized labor in German mining, 1850–1930

Year	Output in 1 000 tons			Labor input in 1 000 miners		
	Hard-coal	Brown-coal	ores	Hard-coal	Brown-coal	ores
1850	5 251	1 541	1 209	34.7	6.3	27.5
1861	14 133	4 622	2 399	86.4	18.7	41.0
1871	29 373	8 483	4 872	150.0	23.7	56.4
1881	48 688	12 852	8 949	186.3	25.6	69.7
1891	73 716	20 530	12 197	283.2	35.7	67.7
1901	108 539	44 479	18 178	448.0	58.5	71.7
1913	190 109	87 233	37 340	686.3	77.5	70.0
1920	131 356	111 888	8 710	713.3	136.5	57.5
1930	155 935	144 693	8 510	525.3	63.7	28.5

Notes: Until 1914, the data covered the German customs area. Here, the ores includes iron-, zinc-, lead- and copper-ore.

Sources: My own calculations based on data from Fischer (1989), p. 1, p. 49, p. 121, p. 182, p. 211, and p. 250, and Fischer (1995), p. 1, p. 39, p. 190, p. 267, and p. 284.

Let us turn to Table 3.1 first. It conveys a general impression of how the German mining sector expanded by displaying, on the one hand, tons of hard coal, brown coal, and ores (here: iron, copper, zinc and lead ore, the most relevant ones with respect to KVs) extracted and, on the other hand, the number of mineworkers. German hard coal (brown coal) output rose from merely 5.2 (1.5) million tons in 1850 to a remarkable 190.1 (87.2) million tons in 1913. The data also clarify that aggregate production of the ores most relevant for the KVs declined after 1913. Labor input did so even earlier, beginning at the turn of the century.

Table 3.2, in addition, illuminates the Prussian mining sector's leading position within the German customs territory. As early as 1818, hard-coal production in Prussia exceeded one million tons.[25] In 1861, production had already reached 11.7 million tons,

[24] See Fischer (1989) and Fischer (1995). Miscellaneous ores include zinc-, lead-, copper-, silver-, manganese-, mercury-, cobalt-, nickel-, antimony-, arsenic-, and alum- ores. Salts include potash-salt, kainite, and boracites. The rest includes, for example, tarmac, tungsten, graphite, and tin.

[25] See Kiesewetter (1989), p. 230.

accounting for about 83 percent of German hard-coal output. Both output in tons and the percentage share in the entire economy's production increased further towards 1913. In particular, aggregate Prussian production grew at an average annual rate that exceeded the growth of the economy, as measured by the NNP; it grew by 8.59 percent per year from 1851 to 1870 and 4.87 percent from 1871 to 1913. Moreover, while Prussian brown-coal production was about 3.3 million tons in 1861 (57.3 percent of German production), it reached 70 million tons in 1913 and 92 million tons in 1920. However, the series on aggregate ore production tells a slightly different story in that tons of ore extracted indeed increased toward 1913, but Prussia's share in German ore production decreased.

Table 3.2: Output and labor input in Prussian mining in percent of German totals

	1861		1887		1913		1920	
	Output	Workforce	Output	Workforce	Output	Workforce	Output	Workforce
A) In milllion tons / in 1 000 employed miners								
Hard coal	11.7	68.2	54.6	191.3	180.1	640.2	127.0	708.0
Brown coal	3.3	10.7	12.7	23.3	70.0	59.9	91.9	133.6
Ores	1.5	35.3	5.4	65.8	7.2	57.6	5.5	55.1
B) In percent of German totals								
Hard coal	83.2	78.9	90.5	88.0	94.7	80.3	96.7	99.2
Brown coal	71.8	57.3	79.9	79.2	93.3	77.2	82.1	97.9
Ores	62.5	72.8	49.5	87.6	19.3	68.6	63.2	95.0

Notes: Ores include iron, zinc, lead, and copper ores.
Sources: My own calculations based on data from Fischer (1989), p. 1, p. 49, p. 121, p. 182, p. 211, and p. 250, and Fischer (1995), p. 1, p. 39, p. 190, p. 267, and p. 284.

We must bear in mind that since a KV was tied exclusively to the mining activity of one particular area – often a rather small one – its long-term growth path was directly linked with that area's prosperity, stagnation or decline. A general problem for a KV was, of course, that its area's future economic development was a priori unknown. In retrospect, output data from Fischer's statistics help to identify which subsectors in which region grew steadily or experienced, after some time, decline. So, for example, hard-coal production grew continuously in all mining administration regions, except for Halle where production was generally quite low at the beginning (around 57 000 tons in 1861) and became even lower (around 8 500 tons in 1913). Where brown-coal was relevant (all mining administration regions but Dortmund), production boomed likewise, and especially since the 1890s. Regarding iron-ore, (i) production in the mining administration region of Breslau followed an inverse U-shaped pattern with a peak between 1885 and 1890; (ii) production, more or less, stagnated in Dortmund; and (iii) it grew steadily in the other three mining administration regions. Furthermore, while copper-ore production stagnated in Bonn and declined abruptly in Breslau after 1884, there was a secular modest increase in Clausthal and Halle. Finally, zinc-ore and lead-ore production, at first, grew, but then declined in all relevant regions (zinc-ore: Bonn and Breslau; lead-ore: all but Halle). The only exception was the mining administration region of Clausthal, where production fluctuated around a trend with slightly negative slope; fluctuations, however, became larger in the second half of the observation period. So, the producers of ores especially – and, consequently, the KVs "linked with" ore deposits – seemed to have faced special pressure.

Apart from the path-breaking technological innovation of deep-coal mining by use of vertically sunk shafts (implemented in the late 1830s) and feedback effects from other sectors, it is commonly agreed that the Prussian mining reform (1851–1865) was responsible for the take-off and long-term growth of the German mining industry. The reform fits well into the picture of a general liberal "turn in economic and financial policy" in Germany, drawn, for example, by Tilly or Fischer.[26] Until the reform, the entire mining sector was under direct control of the mining administration according to absolutist-mercantilist policy. But with the reform, mine owners received entrepreneurial freedom very quickly, with, for example, the design of labor contracts, investment decisions or the right to exploit deposits underground even if one did not own the land at the surface (*Bergbaufreiheit*). According to liberal ideals, the regulatory environment within which mining entrepreneurs acted can be seen as very investment-friendly. Considering the fact that sinking shafts and developing resource deposits was a very risky long-term undertaking requiring huge financial input, the reform liberated very productive forces.

The expansion of production capacity, as another measure of the mining sector's growth, can be measured in four ways: by the number of mines developed, by average mine size, by the change in the number of installed steam engines or, respectively, steam engine horsepower (or horsepower of electrical engines), and by increases in the workforce. First, the overall number of mines in Prussia was 1 902 in 1867, peaked in 1873 with 2 675 and finally fell to 1 673 in 1920.[27] These highly aggregate numbers, in isolation, certainly imply a long-term reduction of mine capacity after the 1873 crisis (*Gründerkrise*). On a disaggregated level, by subsector, this trend holds for coal and

---

26 See Fischer (1961a, 1961b, 1961c) and Tilly (1990), p. 48, on this topic.

27 Calculated from Ministerium für Handel, Gewerbe und öffentliche Arbeiten (1859–1878), Ministerium für öffentliche Arbeiten (1879–1889), and Ministerium für Handel und Gewerbe (1890–1922). Concretely, I used information on the number of mines reported in the KV statistics. As "mines," I took all production facilities for hard-coal, brown-coal, ores, halite, and stones.

ores, but was the opposite for halite and stone pits. Second, however, this downward trend in the pure number of mines in the most relevant subsectors was accompanied by an increase, partly a sharp one, in average mine size. For example, the average hard-coal (brown-coal; iron-ore) mine's size, measured by miners employed, increased from about 340 (40; 40) workers in 1867 to more than 2 280 (308; 200) workers in 1920.[28] Third, with regard to hard-coal mining as the single most important subsector, the number of installed steam engine – e.g., for purposes of drainage or raising coal from increasing depths – increased notably. Installed horsepower in Ruhr coal mining, for example, amounted to 69 000 in 1871 and increased to 995 000 by 1909.[29] Fourth, as Table 3.2 shows, production capacity in hard-coal, brown-coal and ore mining in terms of the workforce increased notably, as well. The growth of the workforce is probably the best indicator of production-capacity expansion since mining was a very labor-intensive activity, despite various approaches to mechanizing production and implementing labor-saving techniques.[30] Table 3.2 illustrates that extensive growth of labor input. Among the main hard-coal fields, those in the Ruhr area consumed the most labor, and prosperity induced migration from areas such as the eastern parts of the Reich, which had positive effects on the availability of skilled labor.[31]

In line with the expansion of the mining sector was wage growth. Walther G. Hoffmann estimates average nominal annual wage income across all subsectors and regions at 523 marks in 1861 and 1 496 marks in 1913, which yields a growth trend of 2.04 percent per year. Wages nonetheless varied over the business cycle, as the coefficient of variation of 25.5 percent of Hoffmann's wage series implies.[32] At the coalfield level, in particular, there were clear wage differentials. Take, for example, the case of hard-coal mining in the Ruhr, the Saar and Silesia. While a miner in the Ruhr earned, on average, 772 marks in 1886, miners in the Saar and Upper Silesia earned 809 and 490 marks, respectively. As of 1893, average annual earnings in the Ruhr exceeded those in the Saar, and the differential with Silesia persisted.[33]

Labor productivity increased in the long term in almost all subsectors, even though there were longer stagnant periods, as well as a decline in tons of resources mined per employee. Besides labor productivity, total factor productivity is an alternative measure of productivity. Recent findings of constant total factor productivity in the Ruhr between 1881 and 1913 suggest that worsening geological conditions had offset the technological progress from a number of important innovations related to the production

---

28 See previous footnote.

29 Ministerium für Handel, Gewerbe und öffentliche Arbeiten (1872), *Statistischer Teil*, p. 38, and Ministerium für Handel und Gewerbe (1910), *Statistischer Teil*, p. 83.

30 See Burghardt (1995), pp. 137–151.

31 See Holtfrerich (1990) on capacity in the steel industry.

32 See Hoffman (1965), p. 461, for the time series.

33 See Ministerium für öffentliche Arbeiten (1885–1889) and Ministerium für Handel und Gewerbe (1890–1922).

process. Those innovations affected, in particular, the transport of coal from the coal-face to the surface (e. g., the mechanical conveyor belt, slides or underground locomotives), but did not so much affect the cutting of coal itself (coal cutting supported with mechanical hammers).[34] One should keep in mind that geological condition – measured, for example, by the average depth in which extraction took place – worsened over time, at least in hard-coal mining, and, thus, for about 55 to 75 percent of Prussian miners (see Table 3.2). According to Huske's data on the Ruhr coalfields, I have estimated the average shaft's depth at 134 meters in 1850, 172 meters in 1861, 523 meters in 1913, and 555 meters in 1920.[35]

Finally, mining was among the most hazardous occupations at the time, surpassed only by working as haulers, millers or in quarries, according to Boyer.[36] It was the *Knappschafts-Berufsgenossenschaft* – one of the many employers' liability-insurance associations carrying out Bismarck's Accident Insurance Law (1884) – that compiled substantive statistics on both fatal and non-fatal accidents in German mining. Those statistics show that the number of first-time reimbursed cases of accident per 1 000 insured miners per year numbered 6.6 in 1886 and increased to about 15 in 1913.[37] Across all occupations, however, that number was merely 2.83 in 1886, 6.97 at its peak in 1905, and 4.80 in 1913.[38]

3.3. Number of carriers[39]

To begin the quantitative overview of the KVs, let us focus on the number of independent insurance carriers operating per year in Prussia and elsewhere. It is hard to find data on how many funds actually operated before the mining reform took place. As of 1852,

---

34 See Burghardt (1995), pp. 103–131, and, for the total factor productivity estimates, Burhop and Lübbers (2009), p. 510.

35 See Huske (1998) for the raw data on depth. Clark and Jacks (2007) have recently made the empirical case for depth being very important for total factor productivity in coal mining.

36 See Boyer (1995), p. 39.

37 Boyer (1995), p. 38. Giga (1979), p. 178, presents data on fatal accidents in Ruhr coal mining. These data show an inverse U-shaped pattern of fatal accidents per 1 000 insured from the perspective of the single most important coal field. Trischler (1988), p. 115, reports accidents per 1 000 (fatal and overall) in the Ruhr coal area, too. Overall accidents per 1 000 did increase between 1888 and 1912, from 82 to 162.5. Herewith, Ruhr coal mining lay above the average for the whole mining sector (1888: 74.2 overall accidents per 1 000; 1912: 140.3 overall accidents per 1 000).

38 Calculated with data taken from Khoudour-Castéras (2008), p. 235.

39 An earlier version of this subsection is forthcoming as part of the article „Die Vorteile und Nachteile großer Invaliditäts- und Krankenversicherungen: Eine quantitative Wirtschaftsgeschichte der deutschen Knappschaftsvereine von 1854 bis zur Gründung der Reichsknappschaft"; see Guinnane et al. (2012a).

at the beginning of the regulatory regime's reform, the Prussian KV statistics report 53 active funds under the mining administration's direct control.[40]

Table 3.3 depicts the number of German KVs for selected years within the observation period. As of 1861, after the *Knappschaft* Law had been enacted, there were 71 Prussian funds. This number grew due to many entries and peaked at 91 in 1871, the year the German Empire was founded. After that, the number diminished steadily due to many mergers and numerous terminal liquidations, which clearly outnumbered the five entries that occurred since 1861.

Table 3.3: Number of German *Knappschaften*

Year	Bavarian KVs	Prussian KVs	Saxon KVs		Other KVs
			Pension funds	Sickness funds	
1861	–	71	–	–	*(1)*
1866	–	77	–	–	*(1)*
1871	37	91	53	–	*(2)*
1876	41	87	53	–	*(6)*
1881	40	83	41	–	*(7)*
1884	38	83	40	–	18
1886	39	75	29	84	18
1891	40	74	3	74	18
1896	41	73	3	67	12
1901	41	73	3	61	17
1906	28	72	3	58	20
1911	28	65	3	51	21
1913	26	62	3	51	21
1916	26	56	3	46	n.a.
1920	12	44	2	40	n.a.

Notes: The Bavarian KV statistics are available from 1871 onwards, and they distinguish between active and non-active KVs; the latter did not report contributors, but pensioners to be financed. "Other KVs" is the likely sum of those KVs situated in the Kingdom of Wurttemberg, the Grand Duchy of Hesse, the Grand Duchy of Bruns wick, the Duchy of Anhalt, the Duchy of Saxe-Altenburg, the Duchy of Saxe-Meiningen, the Principality of Waldeck, the Principality of Schwarzburg, the Principality of Reuß and, finally, Alsace-Lorraine. "n.a" means "not available." The figures in brackets in the last column are all but certain.

Sources: My own depiction based on Bankewitz (1908), p. 1, Imbusch (1910), pp. 127, Karwehl (1907), pp. 142, Köhne (1911), pp. 92–96, Verband der Bergarbeiter Deutschlands (1909), pp. 48–53, Königliches Finanzministerium Sachsen (1870–1872) and (1873–1921), Martell (1916), Ministerium für Handel, Gewerbe und öffentliche Arbeiten (1861–1878), Ministerium für öffentliche Arbeiten (1879–1889), Ministerium für Handel und Gewerbe (1890–1922), Oberbergamt München (1873–1921), Pietsch (1911), pp. 1–27, Rabenau (1909), Simons (1895), pp. 14–21, and Tecklenburg (1877).

[40] Carnall (1855), pp. 86–91.

As Table 3.4 reports in this regard, the KV statistics reveal 20 mergers and 22 liquidations. Here, it is worthwhile to distinguish between two kinds of mergers: a situation in which at least one KV was absorbed by a pre-existing "mother" fund that continued to exist afterwards (commonly referred to as "acquisition" in economics); or one in which two or more KVs were merged to create an entirely new fund. All but four of the mergers mentioned above were of the former sort. Shortly before the *Reichsknappschaft* was founded in 1923, 44 KVs still operated in Prussia. As pointed out in Subsection 1.2, contemporary observers, as well as the mining administration, were not merely open-minded about mergers, but demanded them again and again.

Table 3.4: Mergers and liquidations among Prussian and Bavarian KVs by decade

Decade	Prussian KVs			Bavarian KVs		
	Mergers	Involved KVs	Liquidations	Mergers	Involved KVs	Liquidations
1861–1869	1	2	–	–	–	–
1870–1879	7	13	4	5	11	1
1880–1889	3	10	2	1	2	–
1890–1899	1	3	2	–	–	–
1900–1909	4	8	2	3	15	4
1910–1919	4	15	12	2	14	–
1920	–	–	–	1	6	1

Sources: My own calculations based on Ministerium für Handel, Gewerbe und öffentliche Arbeiten (1861–1878), Ministerium für öffentliche Arbeiten (1879–1889), Ministerium für Handel und Gewerbe (1890–1922), and Oberbergamt München (1873–1921).

The number of KVs in Bavaria was always lower than in Prussia. In 1871, the first year for which data are available, 37 Bavarian funds operated. Although the number fluctuated in the following years between 36 and 43, absolute concentration[41] among Bavarian KVs occurred as well, beginning at the turn of the century. While twelve mergers occurred involving 48 KVs, six KVs withdrew from business by liquidating themselves. It is noteworthy that a lot of KVs in both Prussia and Bavaria were absorbed around the *Gründerkrise* of 1873. Beyond that, the majority of liquidations in Prussia, as well as a quarter of mergers in Bavaria, took place during the First World War or immediately afterwards. This phenomenon might be explained by the fact that many KVs temporarily suffered losses of contributors due to enlistments, while the number of sick days and pensioners to be paid further increased. This, in turn, reflects, to a certain extent, the higher frequency of accidents due to employment of many inexperienced prisoners of war. Liquidation seems to have been the natural consequence of a more and more pre-

41 One has to distinguish between concentration as state-measured at a particular point in time and concentration as the process leading to changes in the states.

carious financial situation.[42] Moreover, with regard to the Saxon KVs, we have to distinguish between miners' pension funds and miners' sickness funds; the latter were reported as such as of 1886.[43] Table 3.3 illustrates their development separately. As with Bavaria and Prussia, the Saxon KVs underwent a concentration process, too. To take a single example, 27 pension funds were merged into the *Allgemeine Pensionskasse Sachsen* in 1890, which aimed to retain not only its position as the miners' carrier of *Knappschaft* insurance, but also to become a legitimate special carrier (*besondere Kassenreinrichtung*) of Bismarckian invalidity and old-age insurance. The same happened in Prussia with the merger of the *Märkische*, *Essen-Werden'sche* and *Mülheimer KVs* into the entirely new *Allgemeiner KV Bochum*.[44] Lastly, determining how many KVs operated in the remaining German states is not an easy task, at least regarding the period prior to 1884 which is not well-documented in this respect. The figures in the sixth column of Table 3.3 should be taken as the best available estimates.[45]

Finally, to broaden the scope, I also take into account what we know about the number of other sorts of mutual social insurance funds. This comparison should simply give an impression of whether the number of KVs was rather high or rather low in the national context. Besides the miners' funds that focused equally on provision of sickness and pension benefits, what other sorts of funds existed? To answer this question, it is useful to look at three distinct periods. The first is the period before the Prussian Industrial Code (*Allgemeine Gewerbeordnung*) was enacted on January 17, 1845. In this period, private associations of, for example, artisans and journeymen were quite common – and had been since mediaeval times. Referred to as (Prussian) provident or guild funds, these associations were once based on compulsory membership, but they eventually were dismantled by the state authority of their ability to force people into member-

42 See Boyer (1995), pp. 50–51, and Lauf (1994), p. 54. According to the Prussian KV statistics, 38.4 (40.6, 35.1, 26.3) percent of the regular overall working membership of 1915 (1916, 1917, 1918) were actually classified as war participants.

43 Before 1886, sickness funds were not reported as *Knappschaft* insurance-related, but generally as "support funds."

44 See Lauf (2009a), p. 19. In addition, the *Saarbrücker* KV and the *Norddeutsche Knappschaftspensionskasse* were approved as special insurers, as well; please consult Chapter 2 once more for this.

45 By drawing on Simons (1895), pp. 14–21, and Köhne (1915), pp. 92–95, we can determine the number of German KVs exemplarily for 1885 and 1913. Accordingly, Prussian KVs accounted for about 47 (54) percent of all German KVs in 1885 (1913), while having been responsible for 88 (90) percent of all KV members. The corresponding figures for Bavarian KVs were 24 percent with regard to KVs and 1.5 percent with regard to insured miners. The figures were the same for both 1885 and 1913. Saxon pension funds accounted for roughly 17 (3) percent of funds in 1885 (1913) and, respectively, 6.8 (3.3) percent of insurants. As of 1885, the fourth-important state in terms of the number of KVs and their share in miners was Hesse: There were eight KVs operating covering 4.6 percent of all miners; in 1913, still five KVs covering 4.3 percent of miners existed there, but the position of the fourth-important state was lost to Alsace-Lorraine (eight KVs covering 6.9 percent of German miners).

ship.[46] The second period began with the Industrial Code of 1845, when the Prussian state enabled local authorities to impose compulsory membership in industrial provident funds (*gewerbliche Unterstützungskassen*) first on artisans and journeymen, and soon on all tradesmen. Alongside these compulsory funds, voluntary factory and support funds were established, and the Law on Industrial Provident Funds (*Gesetz betreffend die gewerblichen Unterstützungskassen*) of April 3, 1854 authorized district governments, in addition to local authorities, to install compulsory funds.[47] Finally, the third period began in 1883, with the implementation of Bismarckian health insurance, which defined local funds (*Ortskrankenkassen*), factory funds *(Betriebskassen*), building-trade funds (*Baukassen*), guild funds (*Innungskassen*), registered voluntary funds (*eingeschriebene Hilfskassen*), parish funds (*Gemeindekassen*) and the miners' funds as the legal carriers of sickness insurance. Hence, all but the miners' funds did arguably evolve as a means to protect members against sickness, although support of survivors and funeral pay were also addressed.[48]

As of 1860, and according to Frerich and Frey (1993), 2 998 sickness funds were recorded for Prussia. Their number further increased to 4 901 in the 1880.[49] Extending the view, Wagner-Braun (2002) reports how sickness funds that operated throughout Germany around 1871, when the Empire was founded, were distributed over states. Of some 8 354 funds, 56 percent were situated in Prussia.[50] According to Hennock (2007), the number of sickness funds acknowledged by Bismarckian health insurance amounted to 18 776 (miners' funds excluded) in 1885, rose to 23 109 in 1911, just before the RVO, and collapsed, due to the RVO's requirements, to 8 681 in 1920. The local, factory and parish funds together (each accounting for several thousand funds) made up 86.2 percent in 1885 and 90.3 percent in 1911; local and factory funds, which still were recognized by the RVO, accounted for 84 percent in 1920.[51] Against this background, the pure number of KVs may not be that impressive at all.

---

46 See Hennock (2007), p. 151, and Frerich and Frey (1993), p. 9.

47 See Hennock (2007), p. 154, Reininghaus (1983), and Frerich and Frey (1993), pp. 56–57.

48 See Hennock (2007), p. 158, for the English terms, Tennstedt (1983), pp. 311–314, and Tennstedt (2011). In addition to these mutual sickness funds, the Accident Insurance Law of 1884 established various employers' liability insurance associations (*Berufsgenossenschaften*) organized by sector, and the Invalidity and Old-age Insurance Law established 31 territorial insurance institutions (*Provinzial- oder Landesversicherungsanstalten*) and ten special insurance institutions (*besondere Kasseneinrichtungen*); see Frerich and Frey (1993), pp. 97–101.

49 See Frerich and Frey (1993), for the data.

50 See Wagner-Braun (2002), p. 58. The percentage figure should be considered with some caution because the data on the various German states do not all refer to the same year, but to 1868, 1872, 1873, *or* 1874.

51 Compare Hennock (2007), p. 164.

3.4. Aggregate membership[52]

To substantiate the KVs' economic and social significance within the whole German economy, I begin by presenting time series evidence on aggregate membership and the relative scope of the miners' insurance. Therefore, Table 3.5 illustrates the long-term development of the number of working miners and of inactive pensioners by state, and for selected years.

Table 3.5: Number of *Knappschaft* members by sort and state

	1861	1871	1881	1891	1901	1913	1920
A) Prussia							
Contributors	118 925	226 788	289 388	429 099	636 722	750 334	1 013 927
Pensioners	21 058	48 355	86 805	132 090	156 321	214 415	327 640
Index overall membership (1913=100)	14	29	39	58	84	100	139
B) Bavaria							
Contributors	–	4 977	5 585	7 456	10 198	14 242	20 557
Pensioners	–	1 415	1 809	1 972	2 604	3 907	4 380
Index overall membership (1913=100)	–	35	41	52	71	100	137
C) Saxony							
Contributors	–	24 784	24 231	29 002	30 310	32 748	51 405
Pensioners	–	9 845	10 925	12 080	13 720	15 287	17 984
Index overall membership (1913=100)	–	72	73	86	92	100	144

Notes: Figures relate entirely to the KVs' worker insurance. The membership in Saxon sickness funds corresponds, by and large, to membership in pension funds. Annual membership is given in 1 000s of miners.

Sources: My own calculations based on Ministerium für Handel, Gewerbe und öffentliche Arbeiten (1861–1878), Ministerium für öffentliche Arbeiten (1879–1889), Ministerium für Handel und Gewerbe (1890–1922), Oberbergamt München (1873–1921), and Königliches Finanzministerium Sachsen (1870–1872) and (1873–1921).

[52] An earlier version of this subsection is part of the article „Die Vorteile und Nachteile großer Invaliditäts- und Krankenversicherungen: Eine quantitative Wirtschaftsgeschichte der deutschen Knappschaftsvereine von 1854 bis zur Gründung der Reichsknappschaft;" see Guinnane et al. (2012a).

Regarding overall membership, *Knappschaft* insurance evidently thrived alongside mining output. [53] The overall number of members in Prussia increased enormously, from about 140 000 to more than 1.3 million over the course of the observation period, which corresponds to an average geometric growth rate of 3.9 percent per year.[54] Regarding Bavaria, aggregate membership grew at 2.8 percent, from about 6 400 to more than 24 000 members. Finally, the collective of Saxon miners increased from about 34 500 in 1871 to more than 73 000 members. This evidence once more reflects the rapid expansion of the German mining sector as a whole.

Let us turn to the systems' relative scope. I measure it in two ways. On the one hand, I compare KV membership with the German working population. To me, it is unclear whether the KVs' widows were actually accounted for in the working population. Thus, I look only at the ratio between the KVs' male membership over the age of 16 and the working population, which may slightly underestimate coverage. On the other hand, I compare the KVs' coverage to the entire German population, which eliminates the issue of widows. In terms of the working population, the coverage of Prussian KVs climbed from about one to four percent towards 1920. Alternatively, considering the entire membership, including widows and orphans below age 15, and comparing it with the entire German population, the Prussian KVs' coverage was about 0.67 percent in 1871, then slightly more than doubled by 1913 (1.44 per cent), and finally reached 2.17 percent in 1920.[55] The degree of coverage of Bavarian and Saxon KVs was much smaller, as Table 3.6 shows.

Considering the coverage of Bismarckian social insurance after its stepwise introduction beginning in 1883, the miners' own social insurance program was, of course, small. However, far from being "universal" at the beginning, Bismarckian health insurance covered less than ten percent of the population (or around 21 percent of the working population), and coverage rose to about 21 percent (or 44 percent of the working population) before the First World War due to expanding statutory protection to further worker categories not covered initially; by 1920, coverage still did not exceed 28 per-

---

53 As mentioned in the introduction, I focus exclusively on the KVs' worker insurance; the distinction between worker and white-collar insurance becomes important only as of 1911.

54 Note that after 1907, I relate to the aggregate membership of the Prussian KVs‘ pension insurance section. Aggregate membership in their sickness insurance section was usually higher, implying that not all former unestablished miners received access to both insurance sections. Looking at the median (arithmetic mean) ratio of sickness section membership to pension insurance membership reveals that membership in a KV's sickness insurance section was between 17 and 23.6 percent (23 and 28 percent) higher during 1908 to 1913 and likewise after the war; during the years 1914 to 1917, however, membership decreased to a level below that of the pension insurance section (around 95 percent or, respectively, 96.5 percent, depending on whether the median or arithmetic mean is addressed).

55 Data on the working population are taken from Khoudour-Castéras (2008), pp. 234–237, who refers to the working population as the "economically active population," or EAP. Data on the German population are taken from Rothenbacher (2002), pp. 288–290.

cent of the population (or 64 percent of the working population). Coverage of both Bismarckian accident insurance and invalidity and old-age insurance was initially eight percent in the former program (of the working population: 17.8 percent) and 22 percent, as of 1897, in the latter program (of the working population: about 49 percent). While the scope remained, all in all, the same for invalidity and old-age insurance with the German population as a reference point (28 percent in 1920), more and more working people gained from being protected (65.3 percent of EAP in 1920).[56] But what about the period before Bismarckian insurance was implemented? Here, we can compare the relative size of *Knappschaft* insurance with the coverage of the mutual sickness funds introduced in the previous subsection. Prussian industrial provident funds, as of 1864, covered some 457 635 people, or approximately 1.17 percent of the German population. As with KVs, aggregate membership was notably extended and made up 1.86 percent in 1880, at the end of the pre-Bismarckian insurance era.[57]

Table 3.6: Coverage of *Knappschaft* insurance compared to Bismarckian insurance

Year	Coverage – overall membership as percentage of German population					
	Bavarian KVs	Prussian KVs	Saxon KVs	Bismarckian health insurance	Bismarckian accident insurance	Bismarckian invalidity and old-age insurance
1871	0.016 %	0.67 %	0.08 %	–	–	–
1876	0.018 %	0.76 %	0.08 %	–	–	–
1881	0.016 %	0.83 %	0.08 %	–	–	–
1886	0.016 %	0.93 %	0.08 %	9.7 %	7.9 %	–
1891	0.019 %	1.13 %	0.09 %	13.8 %	36.2 %	n.a.
1896	0.020 %	1.18 %	0.08 %	15.1 %	33.4 %	22.0 % (1897)
1901	0.022 %	1.39 %	0.08 %	17.0 %	33.2 %	23.2 % (1900)
1906	0.024 %	1.49 %	0.08 %	19.1 %	38.9 %	23.1 %
1911	0.026 %	1.36 %	0.08 %	20.8 %	43.5 %	24.3 %
1913	0.027 %	1.44 %	0.08 %	20.2 %	43.3 %	24.4 %
1916	0.030 %	1.57 %	0.06 %	n.a.	n.a.	24.8 % (1915)
1920	0.040 %	2.17 %	0.12 %	27.6 %	43.5 %	28.3 %

Notes: "n.a." abbreviates "not available."
Sources: My own calculations; see Figure 3.2, Khoudour-Castéras (2008), pp. 234–237, Frerich and Frey (1993), p. 106 and Rothenbacher (2002), pp. 288–290, for the data.

56 EAP data are taken for 1886 to 1913 from Khoudour-Castéras (2008), pp. 234–237, and for 1920 from Sommariva and Tullio (1987), p. 235.

57 Extracting the Prussian census population for 1871 and 1910 from Rothenbacher (2002), Appendix Table D.2, and comparing the number of KV members with the figures, we find that Prussian KVs covered roughly 1.04 and 2.1 percent, respectively, of the state's population. Regarding Bavaria, the percentages for 1884 and 1910 are 0.11 and 0.2 percent, respectively.

Turning to the KVs' membership composition, several questions arise: (i) Exactly how many miners were employed in the subsectors covered? (ii) Given the fact that many, but not all, KVs discriminated between established and unestablished miners until the reform of 1906 eliminated this choice, how large was the actual share? (iii) What was the age composition of KVs? (iv) What was the composition of the KVs' collectives by activity status – i.e., active contributor or inactive pension recipient? The first two questions will be answered immediately, while the latter two questions deserve special attention. I will return to them in the subsection on demographics.

Some of the contemporaries – Karwehl (1907), for example – expressed the opinion that the different subsectors underlying miners' insurance would reflect different degrees of hazard to the life of the miner, and that it would be preferable to have mixed KVs (see the quotation in Subsection 1.2).[58] His thought seems to be grounded in the argument that differences in the geology surrounding the extraction of particular resources translate into different relative accident, invalidity or mortality frequencies, all of which have direct implications for the risk that a KV's collective of insurants represents and that must be insured and financed. I will comment further on this idea later, particularly in Chapter 4. For the moment, I only convey an impression of how many miners worked in the different subsectors and whether there were obvious differentials in mortality or, concretely, in relative frequencies of death.[59]

Table 3.7 reports overall membership by subsector. As expected, the majority of Prussian miners – strictly speaking: of contributors – were engaged in extracting hard coal, and hard-coal's share successively increased towards 1920. While brown-coal's share was quite constant at six to eight percent in all but a few years towards the end of the observation period, employment in ore, stone and salt production lost significance. The share of workers attributable to the "related industries," however, declined from 11.4 percent in 1867 to 5.8 percent in 1920; this decline was a direct consequence of the ores' declining importance.[60]

58 See Karwehl (1907), pp. 74–75.

59 One might think of taking the relative frequency of becoming invalid into account as well. But this measure would involve measuring indirectly two further issues that may complicate interpretation: moral hazard and differentials in the strictness of eligibility rules. In contrast, focusing exclusively on mortality as the measure of comparative subsector risk seems to involve measuring indirectly work-safety differentials. This is something we must live with. Nonetheless, mortality should arguably be de-linked from the moral hazard and eligibility-rule issues.

60 Among all the 103 Prussian KVs under observation, there are not many cases in which a fund's membership arose from only one subsector; this was the case for salines and most of the steel-works-related KVs. In fact, 18 KVs, among them the presumably best-known and largest ones – the *Märkische* and *Essen-Werden'sche* KV (since 1890: *Allgemeiner* KV *Bochum*) and the *Oberschlesische*, *Niederschlesische*, and *Saarbrücker* KV – provided insurance coverage predominantly for hard-coal miners. Nine (11, 14, five and two) KVs were focused mainly on brown coal (iron ore, miscellaneous ores, Stones, and halite); and 15 funds can only be identified as truly mixed ones, where no subsector accounted for the vast majority of their members.

Table 3.7: The percentage shares of the KVs' contributors by subsectors

Year	Hard-coal	Brown-coal	Iron-ore	Misc. ores	Stones	Halite	Salines	Related industries
A) Prussia								
1867	55.4 %	8.6 %	8.9 %	13.4 %	1.1 %	0.4 %	0.8 %	11.4 %
1871	55.8 %	7.1 %	9.0 %	15.2 %	0.8 %	0.3 %	0.6 %	11.2 %
1881	57.2 %	7.0 %	8.6 %	15.0 %	0.8 %	0.9 %	0.4 %	10.1 %
1891	62.5 %	7.4 %	5.7 %	11.4 %	1.0 %	1.4 %	0.3 %	10.3 %
1901	68.4 %	8.1 %	3.7 %	7.7 %	0.9 %	2.1 %	0.2 %	8.9 %
1911	73.4 %	6.8 %	2.5 %	4.9 %	0.8 %	4.2 %	0.6 %	6.8 %
1916	74.7 %	6.1 %	3.0 %	5.0 %	0.5 %	2.9 %	0.1 %	7.7 %
1920	73.4 %	10.4 %	2.3 %	3.2 %	0.2 %	4.6 %	0.1 %	5.8 %
B) Bavaria								
1871	56.8 %	1.9 %	10.8 %	0.9 %	1.9 %	2.7 %	9.7 %	15.2 %
1881	59.3 %	1.6 %	9.8 %	5.2 %	1.9 %	2.2 %	8.6 %	11.3 %
1891	66.7 %	2.2 %	9.2 %	1.1 %	1.4 %	1.5 %	11.5 %	6.3 %
1901	75.7 %	1.3 %	7.1 %	0.8 %	1.0 %	1.6 %	8.0 %	4.4 %
1912	32.1 %	38.9 %	11.6 %	0.7 %	1.2 %	0.8 %	2.5 %	12.1 %
1916	29.4 %	39.2 %	8.8 %	0.2 %	1.0 %	0.6 %	3.0 %	17.7 %

Sources: See Table 3.5.

Table 3.8: Death of active Prussian miners per 1 000 per subsector compared to the national level

Year	Crude death rate of active miners by subsector					Crude death rate of the Population
	Hard-coal	Brown-coal	Ores	Salt	Steelworks	
1867	11.6	11.7	12.2	9.7	11.3	2.6
1871	13.7	10.4	12.1	10.1	11.9	2.7
1881	10.1	8.1	12.6	7.0	7.6	2.5
1891	7.9	6.2	9.7	6.7	9.6	2.3
1901	7.3	8.2	8.2	7.4	6.2	2.1
1911	5.4	7.5	7.3	4.1	4.6	1.7
1916	27.2	18.2	15.0	18.6	21.3	1.9
1920	8.9	5.9	5.9	8.1	5.0	1.5

Notes: The crude death rate per KV per year has been calculated as [(deaths of active miners/sum of active miners)*1 000].

Sources: My own calculations; see Figure 3.2 and Rothenbacher (2002), pp. 289–291.

The picture for Bavaria seems to have been different in that hard-coal first gained, then heavily lost, share to brown-coal. Actually, a look into the statistics reveals that two KVs – the *Hohenpeissenberg* KV and *Miesbach* KV – switched the subsector to which they were tied; until 1908, both funds were reported as insuring hard-coal miners and, as of 1909, as insuring brown-coal miners.[61] Beyond that, and in contrast to what happened in Prussia, steelworkers' share increased after a phase of stagnation to the level of the 1870s.

Linking with the evidence on accidents discussed in Subsection 3.2, Table 3.8 reports, on the one hand, the number of deaths among 1 000 contributors to KVs per subsector and, on the other hand, the deaths per 1 000 in the German population. Crude death rates suggest once more that mining per se was extremely hazardous. The comparison with death rates on the national level is, of course, imperfect since national figures do include more than just the economically-active male population. Nonetheless, the rates give an indication of the comparatively high probability of death while working as a miner, as compared to the economy-wide average. Figures also indicate differences across subsectors, diminishing rates toward World War I and a heavy increase in rates during war.[62] Diminishing rates may have been a result of improved medical treatment or improved work safety, although Boyer concludes that, since 1871, safety regulations had not improved much beyond pre-*Reich* standards.[63]

Observable are differentials in average mortality over subsectors ranging between zero and four deaths per 1 000. Omitting the war period, it is noteworthy that the subsector "ores" seems to have been the relatively most hazardous subsector. That those crude death rates were even four or five times higher in war than before it might especially be explained by the fact that more experienced professional mineworkers were partly substituted for relatively less or non-experienced workers (e.g., prisoners of war) due to enrollment for military service. Because it seems reasonable to think of a miner's "work experience" as reflecting *ceteris paribus* a lower probability of him causing or, at least, being involved in severe accidents, higher rates follow from the increased average probability in the memberships of being involved in a deadly accident.

---

61 Unfortunately, I cannot say whether this is a typo in the statistics or a real switch in product.

62 Note that deaths were either due to accident or due to disease. From the Prussian KV statistics, it is indeed possible to calculate both incidences separately. For 1871 (1881, 1891, 1901, and 1913), for example, the conditional frequency of deaths by accident was 1.8 (1.6, 1.3, 1.8, and 2.0) given that at least one death occurred. Compared to those rates, average mortality due to "other reason," disease first and foremost, was much higher for 1871 (1881, 1891, 1901, and 1913) with 11.3 (9.6, 6.1, 4.5 and 8.1) deaths per 1 000 working members. For comparison, consider, for example, Mills (2010), p. 164, who provides us with rates of accident mortality regarding British coal and metalliferous mining industries for 1875–1914. The "coal rates" decreased from about 2.3 deaths by accident per 1 000 to 1.0. The "metalliferous rate" decreased from 1.6 to about 1.2.

63 See Boyer (1995), p. 116. Up to here, the current paragraph is part of the article "The hazard of merger by absorption: Why some Knappschaften merged and others did not, 1861–1920;" see Jopp (2011a), p. 83.

In ending this subsection, let us have a look at the proportion of unestablished members. Pointed out in Chapter 2, the act of discriminating between members with more rights and those with fewer rights had been pursued long before the mining reform of 1851–1865 took place. It was only the reform of 1906, in fact, that eliminated this distinction, which had raised concerns among both the disadvantaged membership and contemporary observers. In fact, among Prussian KVs, the unweighted mean share of unestablished miners decreased from nearly 45 percent in 1861 to 36 percent in 1906. However, the standard deviation (1861: 23.6 percent; 1906: 25.0 percent) indicates that there were marked differences in the unestablished ratio across KVs. Weighting each observation with the KVs' share in overall membership is a simple way to detect differences between small and large funds. Since the weighted mean (48.6 percent in 1861 and 37.5 percent in 1906) is higher than the unweighted statistic in each selected year – significantly in some years and rather modestly in others – we can conclude that the share tended to be higher for larger KVs. Regarding Bavaria, the unweighted share decreased from about 45 percent in 1871 to 29.9 percent in 1906; the weighted share is 46.3 percent and 37.9 percent.

3.5. *Knappschaft* size, growth and concentration[64]

As we saw in Chapter 1, the KVs' operations were a highly debated topic in the nineteenth and early twentieth centuries – as were international institutions such as the friendly societies, the fraternal lodges and the industrial sickness funds, to name only the most prominent examples from the literature. In contrast to those schemes, the size of a KV's pool of insurants was at the center of that historical discussion. While the "absolute" dimension of concentration was illuminated above, I now explore the *relative* concentration – i.e., the shift in the share of insurants that the KVs held. My aim is to supply hard facts substantiating a major design flaw that contemporaries were certain to have detected: fragmentation of membership and an extremely skewed size distribution of KVs, which allowed few funds to thrive and made many even worse off.

What might be a reasonable measure of KV size? The literature on firm size in general, and on insurance in particular, offers various sets of measures that might be applied. In general, firm size at a certain point in time may be measured by some physical

[64] An earlier version of this subsection is part of the article „Die Vorteile und Nachteile großer Invaliditäts- und Krankenversicherungen: Eine quantitative Wirtschaftsgeschichte der deutschen Knappschaftsvereine von 1854 bis zur Gründung der Reichsknappschaft;" see Guinnane et al. (2012a). Furthermore, it contains parts of the published article "The hazard of merger by absorption: Why some Knappschaften merged and others did not, 1861–1920" and also parts published in the article "Insurance, Size and Exposure to Actuarial Risk: Empirical Evidence from Nineteenth- and Early Twentieth-Century German *Knappschaften*;" see Jopp (2011a), pp. 85–88, and Jopp (2011c).

quantity, such as the number of employees or the output in pieces or, alternatively, by some financial quantity, such as capital shares, sales, profit or value-added. While each measure has some desirable properties, each also has shortcomings. In the end, the choice of measure seems to be a matter of data availability.[65] The insurance literature, in particular, proposes to measure the size of an insurer by premium income, expenditures on claims, the number of sold insurance policies or the number of insured individuals.[66]

Here, as in the econometric analysis below, the main measure of KV size used is the number of contributing miners; pensioners are left out. I argue that this measure is appropriate to highlight the true financing power upon which a KV relied. If, for any reason, the contributor base *ceteris paribus* eroded, we can immediately conclude that the financial pressure on a KV increased. Therefore, this choice of a measure seems to be appropriate for the particular research design of this study.[67]

Following Eisen (1991), who addresses the question of how to measure an insurer's minimum efficient size (MES) in any insurance market, I first analyze the two statistical quantities he proposes as workable proxies: the arithmetic mean ("MES 1") and the median ("MES 2") of the empirical KV size distribution; in shorthand, I refer to the arithmetic mean simply as "average KV size." The larger (smaller) the average KV size was, compared to the median KV size, the more skewed the distribution was to the left (right);[68] the larger the difference, the more unequal was the size distribution over KVs. As a matter of the thriving mining sector, on the one hand, and the absolute concentration process taking place among KVs due to mergers and liquidations, on the other hand, the average KV must have grown larger and larger. We can observe precisely that trend among Prussian KVs in Figure 3.2. While Figure 1.1 already addressed the very unequal distribution of memberships on the fund-level at the *Knappschaft*'s re-start, Figure 3.2 is indicative of how the story continued. The data also reveal that the median KV size was considerably and persistently smaller than the average KV size, and that this differential even expanded over time, as the following shows: The average KV was

---

65 See, for example, Penrose (1995), pp. 199–200, and Degner (2011), p. 132. Physical output, for example, might not be properly comparable across industries, and output measured in monetary units is subject to price movements that may inflate output beyond reason.

66 See Doherty (1981), Ackermann (1983), Kürble and Schwake (1984), p. 115, Mordi (1985), Eisen (1991), pp. 268–271. The question of how to measure an insurer's size is, thus, interwoven with the question of how to measure an insurer's outputs and inputs. Dohery (1981), for example, pp. 391–394, goes into more detail about measurement difficulties in this respect. I will come back to this issue in the subsequent parts of this chapter, as well as in Chapter 4.

67 Actually, many KVs had no real staff in the sense of employed labor. Medium-sized or larger KVs might have employed an insurance mathematician or somebody who was familiar with economics, but also hospital staff or a pharmacist; board members and Elders were certainly "employed," and technical civil cervants are likely to have been paid by the state. If the focus here were different – if, say, we were interested in the KVs as reflecting networks of social relations – it might be useful to measure KV size by including pensioners and staff.

68 See Eisen (1991), pp. 268–271.

already four times as large as the median KV in 1861 and about ten times as large at the peak in 1913. However, because some smaller KVs exited during war, the ratio developed back to its starting value, which was still impressive. The average Prussian KV began with 1 675 contributing members and climbed to 4 442 in 1886, 10 271 in 1906 and, finally, 23 044 in 1920. In contrast, the median KV was initially responsible for no more than 449 working members, and the median contributor base still did not exceed 1 747 by 1916.[69]

Figure 3.2: Average and median size among Prussian KVs

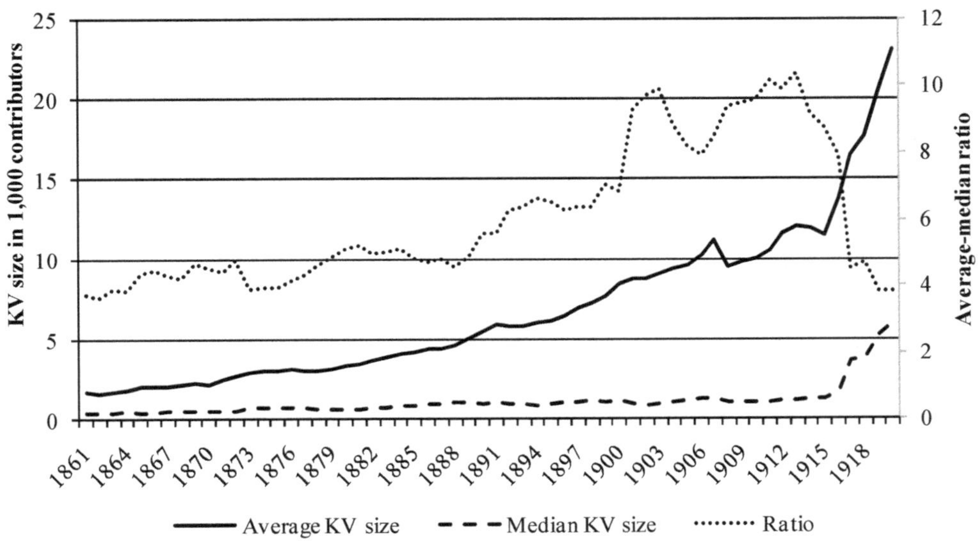

Notes: Size is measured in terms of the number of contributing members.
Sources: My own calculations based on Ministerium für Handel, Gewerbe und öffentliche Arbeiten (1861–1878), Ministerium für öffentliche Arbeiten (1879–1889), Ministerium für Handel und Gewerbe (1890–1922).

Regarding Bavarian KVs, the picture was, all in all, similar, although the figures were not that impressive: The average (median) Bavarian KV relied upon 135 (57) contributors in 1871, upon 160 (83) in 1884, and more than 6 800 (8 100) in 1920.[70] Thus, Bavarian KVs were significantly smaller. Looking at the Saxon pension funds, it can be observed that the average (median) KV size was 459 (84) in 1871, 10 423 (1 644) right

[69] For illustrative purposes, look at Tables A.1 in the Appendix. It lists all observed Prussian KVs also by their initial size when entering the data set and by their final size, just before leaving the data set.

[70] The Bavarian KV statistics did report some KVs as inoperative, meaning that they had no contributing members – but actually had pensioners to be taken care of. Therefore, average and median KV size solely relate to those groups of KVs that were reported as being in operation, thus as having at least one contributor.

after the merger of 1890 (see above), and, ultimately, 27 199 (27 199) in 1920. Hence, the development in Saxony resembles Prussia's more than Bavaria's.

So far, we know that the size distribution among KVs evidently was quite skewed to the left, towards smaller operating sizes. However, I want to delve even deeper into the Prussian KVs' annual size distributions. Therefore, Figure 3.3 reports deciles of KV size. In fact, ten percent of Prussian funds in the respective years persistently operated with 51 or fewer contributors; the picture would remain the same if we were to depict the whole time series. 30 percent of KVs per year always fell short of about 500 contributors. To push this further, let us consider the minimum and maximum KV sizes to be observable: The minimum KV size always lay in the range of one to 16 contributors; the maximum, in contrast, climbed steadily from 18 895 (1861) to more than 136 314 (1890) to, finally, to 411 585 contributors (1920) and was always determined by the *Märkische KV* (1861–1889) or the *Allgemeine* KV *Bochum* (1890–1920).

Figure 3.3: Deciles of KV size (Prussia)

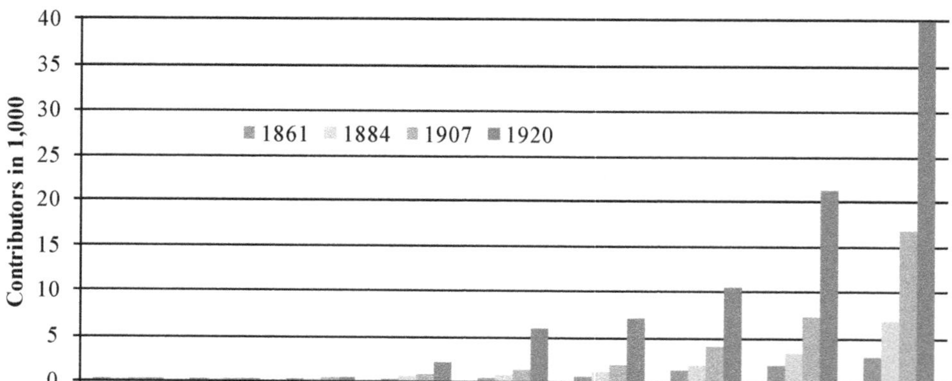

	1	2	3	4	5	6	7	8	9
1861	51	89	220	323	458	668	1295	1928	2997
1884	51	193	285	565	798	1219	1980	3268	6858
1907	31	293	504	782	1333	2015	4130	7385	16847
1920	21	243	414	2229	5999	7205	10642	21189	40808

Deciles

Notes: *Knappschaft* size is measured in terms of contributing miners. The deciles are displayed in the data cells below the graph. The first decile on the left, for example, which is the ten-percent decile, implies that ten percent of all KVs observed in 1861 had 51 contributors or less, while 90 percent of KVs had more contributors.

Sources: See Figure 3.2.

Finally, the high dynamics in the KVs' size distribution can also be underlined by turning to the number of KVs per size class per year. In the following, I assign each observed *Knappschaft* year – of which we have 4 450 over the period 1861 to 1920 for

Prussia (60 years times 103 cross-sections minus the sum of unobserved years per KV) and 1 664 over 1871 to 1920 for Bavaria – to one of six size classes. Measured in terms of the number of contributors, these are 1–199, 200–999, 1 000–4 999, 5 000–9 999, 10 000–49 999, and 50 000+. Figure 3.4 shows the development of the relative number of KVs by size class for Prussia.

The percentage of KVs operating in the smallest size class relatively quickly declined from its peak of 32 percent in 1868 to 20 percent at the end of the 1870s, but then more or less stagnated at between 15 and 21 percent, and increased again during wartime; the percentage of contributors and pensioners falling into this size class was, of course, marginal. The relative number of KVs in sizes classes 200–999 and 1 000–4 999 declined from about 43 percent to 15 percent or, respectively, about 28 percent to 11 percent. The latter observation is consistent with the fact that many smaller and, especially, firm-related KVs successfully prevented the loss of autonomy by being merged into larger KVs. Moreover, while the third-largest size class (5 000–9 999) first gained, then lost share in members, the second-largest faced a steady decline. However, those classes' share in the number of KVs increased in the long term. Finally, the largest size class gained share, too. Not visible from the figure is that among Bavarian KVs the smallest size class clearly lost share (around 80 percent in 1871 and, finally, about 33 percent), while the number of KVs operating in size class 1 000–4 999 soon dominated (zero percent in 1871 and 45 percent in 1919). The two largest size classes became relevant only during wartime and towards the end of the observation period.

Figure 3.4: Relative number of Prussian KVs by size class

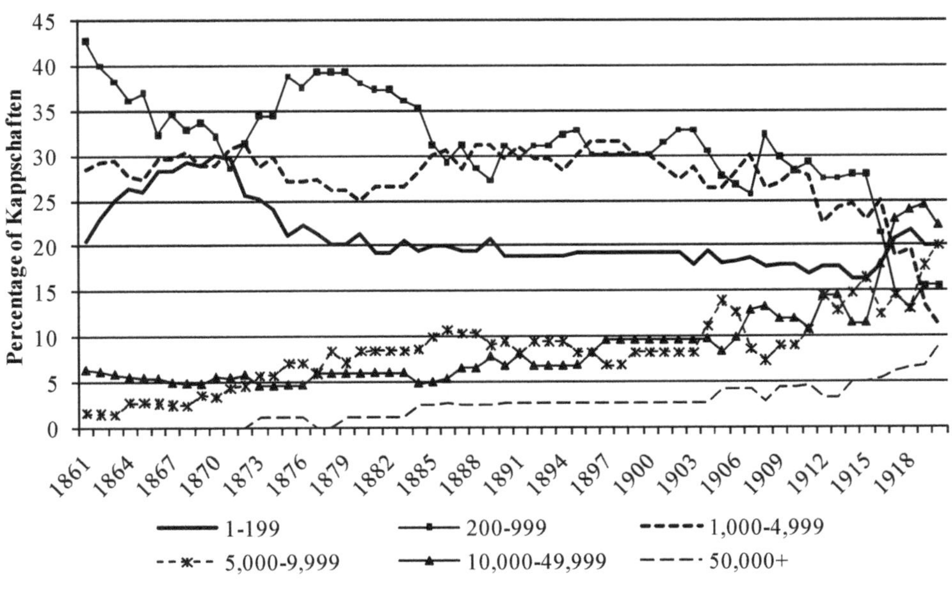

Sources: See Figure 3.2.

Having unraveled the extreme size distribution of the Prussian KVs, we can now ask: What about their growth performance? Over the long term, it was quite uneven. Some KVs were subject to stagnation and even shrinkage, while other funds thrived. As discussed above, this was very probably so because each KV was tied either to one mining area, which could be subject to exhaustion and structural decline over time, or – what might have been even more restrictive – to the fate of one particular mining firm. Table 3.9 shows that of 103 observed Prussian KVs, 40 exhibited a negative average annual growth rate during their operation. 63 KVs, in contrast, were quite dynamic in the long run and grew by at least 1.9 percent per year. Note that size is still measured here in terms of contributors; thus, "growth" refers directly to an increasing number of contributors. In addition to showing long-term growth performance in terms of the categories "stagnant" and "dynamic," the table also shows that growth performance was obviously different - on average – for different KV sizes.[71]

Table 3.9: Summary of long-term growth patterns among Prussian KVs

Size class	Long-term growth pattern of KVs			
	Stagnant		Dynamic	
	Number	Mean growth rate	Number	Mean growth rate
Small (≈ < 500 contributors)	24	-5.52 % p.a.	16	1.88 % p.a.
Medium (≈ 500–12 000 contributors)	15	-3.40 % p.a.	37	3.30 % p.a.
Large (≈ > 12 000 contributors)	1	-2.97 % p.a.	10	3.64 % p.a.
Sum	*40*		*63*	

Notes: KV size is measured in terms of contributors.
Sources: See Figure 3.2.

Although the baseline measure of KV size is the number of contributors, there is an alternative measure that will turn out to be useful in analyzing actuarial risk in Chapter 4. That is the mean number of working members per production unit (mine, saline or steelworks, respectively). Thanks to the Prussian KV statistics, data on the number of such production units per KV are available. The average production unit in hard-coal mining (60 contributors per mine in 1867 and 661 in 1920) and steel production (83 contributors per steelworks in 1867 and 883 in 1920) was significantly higher than for the remaining subsectors (brown-coal: seven and 103, respectively; iron-ore: 15 and 57, respectively). What is more, not only was the average production unit larger, but it also

71 I chose to define only these three size-class intervals here because nearly all KVs easily fit into one of these three classes. However, for some exercises to be done below, this classification is too crude and, thus, will be modified.

grew faster. In all, compared to the production processes in brown-coal mining, iron-ore mining and so forth, available figures underline that hard-coal mining and steel processing were soon based on large-scale operations, while the average non-hard-coal mines and non-steelworks plants had only between ten and 70 working members, depending on product and time.

One question naturally arises after having described KV size: How small or large were the KVs, in general, compared to similar insurance institutions at the time like industrial provident funds, provincial funds, local funds or the like? Referring again to Frerich and Frey (1993), the average industrial provident fund in Prussia had 110 members in 1860 and 171 in 1880, which was much smaller than the average KV size. According to Hennock (2007), the average *Reich* law sickness fund (exclusive of KVs) took care of slightly more insurants, 229 in 1885 and 598 in 1911. However, the data that Hennock and Frerich and Frey present also indicate that the average local fund was the largest fund among the ones introduced above, with about 416 members in 1885 and some 1 490 in 1911. While I find that all funds grew in membership over time, the miners' funds did, on average, outperform them in terms of size.[72] Note, however, that all these average fund size data do not enable us to infer the size distribution among those funds, which would be preferable indeed; therefore, we would at least need to know the median fund size.

It should be clear now that if we want to assess the empirical relationship between size and insurance fund performance by using data on nineteenth-century German funds, we would do well to choose Prussian KVs since we know that they show the variation required in the dataset to infer information about the threshold separating actuarial disaster from actuarial stability. The data on Prussian KVs enable us to assess the question of whether collectives of insurants in the range between ten and, say, 1 000 members were appropriate or whether far larger collectives of insurants of maybe 50 000 or of more than 100 000 were a prerequisite to initializing positive scale effects.

3.6. Demographic issues

The Prussian KVs aged, as a rule. Recall from the institutional overview in Chapter 2 that the Prussian KVs used the PAYG method, which implies that, within a given period, claims costs were directly financed by contributions of the working membership. That is, the financing burden put on contributors would reflect, to a certain extent, the proportion of working to non-working membership and its changes over time. The fact that invalids and survivors exclusively consumed insurance benefits made them eco-

[72] See Frerich and Frey (1993), p. 57, and Hennock (2007), p. 167, for the data. The average size of a territorial insurance institution carrying invalidity and old-age insurance was much higher than that of the average KV.

nomically dependent on the KV – and, after all, on the solidarity promises implicit in the KVs' PAYG schemes (the "intergenerational contract between the miners"). This dependency, which broadly increased over time, implied a sort of "social" aging rather than aging in the "biological" sense. My main measures of that process are, thus, the invalids-to-contributors ratio (ICR), the widows-to-contributors ratio (WCR), and the orphans-to-contributors ratio (OCR). The latter two form the survivors-to-contributors ratio and, together with the ICR, the overall pensioners-to-contributors ratio (PCR).[73]

Recall, also, that in this context, economists consider the contribution rate charged on an insurant as exactly a function of the number of pensioners per contributor and the ratio between the average pension benefit and the average economy-wide wage, or "pension level."[74] Commonly, the PCR is approximated by the old-age dependency ratio, although the two are not necessarily equal. This, however, would be no serious problem to accuracy as long as coverage is universal.[75] The dependency ratio is the measure of choice to pinpoint how a population has aged over the past and, thereby, to substantiate the economic consequences of that aging process for the pension system. Aging *ceteris paribus* drives the pensioners-to-contributors ratio up. In this case, the contribution rate must *ceteris paribus* increase, too. Apart from raising the contribution rate, however, budgetary balance might be maintained by lowering the pension level, or by sharpening eligibility criteria instead – e.g., raising the legal retirement age, which has a delayed effect on the pensioners-to-contributors ratio.[76]

3.6.1. Economic dependency

Figure 1.1 in the introductory chapter has already pinpointed the rule that we can observe among Prussian KVs – namely, that the number of invalids and survivors per 100 contributors rose in the long term. If, at the very start of the KVs' insurance system, four invalids had had to be financed by 100 working members – quite a low number – the same number of contributors would have financed, on average, 14 invalidity pensioners in 1913; the number of dependent survivors more than doubled, as well, by the First World War (1861: ≈ 14; 1913: ≈ 30). With this in mind, it is more than just a guess to assume that KVs would have certainly been exposed to financial-adjustment pressure. Extending the perspective on Bavarian KVs, Figure 3.5 shows how economic

---

73 In equations they are measured as follows: (i) $ICR_t=[(invalids_t/contributors_t)*100]$, (ii) $WCR_t=[(widows_t/ contributors_t)*100]$, and (iii) $OCR_t=[(orphans_t/contributors_t)*100]$; it holds that (iv) $PCR_t = ICR_t + WCR_t + OCR_t$.

74 See, for example, Clark (2004). Here, subsidization by the state is neglected for the sake of simplicity.

75 See, for example, Verdugo (2006). The old-age dependency ratio equals the ratio of the working-age population, usually aged between 16 and 65, divided by the non-working population beyond age 65.

76 See Schmähl (2001) for analysis of possible adjustment paths in a PAYG pension system.

dependency among those funds developed. In fact, the picture is notably different in that the burden with pensioners first increased, then decreased, on average, or fluctuated around a more or less stationary trend after 1902.

Figure 3.5: Average number of invalids and survivors per 100 contributors among Bavarian KVs

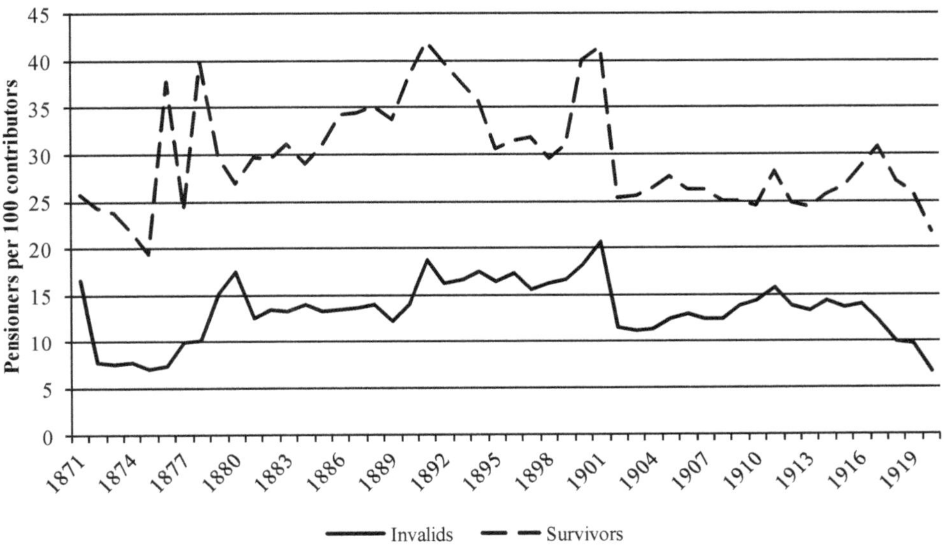

Notes: Survivors include widows and orphans.
Sources: My own calculations based on Oberbergamt München (1873–1921).

Table 3.10 illustrates the aging problem among the Prussian KVs in more descriptive detail, focusing on five-year periods and the ICR and SCR separately. Reported are minimum and maximum values, as well as the lower quartile, upper quartile and median values. As we can see from the table, all quartiles constantly increased. It should be highlighted that there almost always were KVs operating with no invalids and only a few survivors to be financed. However, these KVs were exclusively among the smallest. The range between the upper quartile and the maximum indicate that 25 percent of KVs always faced a *significant* burden with pensioners.

A closer look at the KVs ranging in that upper 25 percent reveals that they were mostly smaller ones. It seems helpful at this stage to once again draw a clear line between the KVs' experience and Bismarckian invalidity and old-age insurance's experience after 1889, for which data covering the period 1902 to 1913 point to a PCR of only between six and seven pensioners per 100 working members.[77]

[77] I used data from Khoudour-Castéras to calculate the PCR of the Bismarckian system; see Khoudour-Castéras (2008), pp. 236–239.

Table 3.10: Descriptive statistics on the components of the pensioners-to-contributors ratio among Prussian KVs

Year	Minimum	Lower quartile	Median	Mean	Upper Quartile	Maximum
A) Invalids per 100 contributors						
1861–65	0.00	0.66	2.32	4.03	5.87	24.91
1866–70	0.00	1.32	3.78	5.91	7.21	55.08
1871–75	0.00	1.68	4.21	7.22	7.30	111.40
1876–80	0.00	3.09	5.50	9.71	9.46	153.90
1881–85	0.27	3.35	5.35	8.81	9.63	101.15
1886–90	0.00	3.80	6.32	9.24	10.46	78.02
1891–95	0.00	4.09	7.50	9.87	11.22	67.68
1896–00	0.99	5.28	8.44	11.96	15.96	62.35
1901-05	0.00	5.61	9.41	11.90	15.88	60.54
1906–10	0.00	6.63	10.54	13.48	15.33	95.29
1911–15	0.00	8.58	11.95	14.49	20.66	48.70
1916–20	0.00	6.39	10.50	13.11	16.62	53.33
B) Survivors per 100 contributors						
1861–65	0.00	3.98	11.61	14.08	18.78	74.48
1866–70	0.00	7.89	15.10	49.24	20.71	2 913.39
1871–75	0.91	11.51	16.88	21.73	23.24	157.24
1876–80	0.00	14.62	20.39	27.74	29.64	236.01
1881–85	1.90	14.31	21.65	27.41	32.01	156.67
1886–90	0.00	12.57	21.67	24.83	32.26	98.91
1891–95	5.24	12.96	22.21	26.13	30.27	117.55
1896–00	3.45	12.18	20.96	26.08	32.01	125.10
1901–05	3.41	12.06	19.36	26.63	30.07	202.78
1906–10	3.64	12.58	21.75	29.64	31.45	406.41
1911–15	4.20	13.34	24.24	27.39	32.83	120.45
1916–20	0.00	19.66	27.50	29.40	35.57	95.04

Sources: See Figure 3.2.

Although the KVs' experience is by far not representative of the entire German economy at the time, there definitely was a parallel to the challenge of the mature late-twentieth-century welfare state.[78]

---

78 The estimates on the ICR and SCR presented so far have to be considered as lower- bound estimates of economic dependency. In other words, one could make the case for an even worse historical scenario among KVs than is done above. Why is that? The answer is quite simple: Recall the distinction between established and unestablished miners. Contemporaries were convinced that the latter were usually treated relatively unfairly in an actuarial sense. Data from the KV statistics for the period 1861–1888 show that unestablished miners usually contributed less marks per capita to

Furthermore, in order to get an impression of whether smaller Prussian KVs suffered comparatively more from the burden of pensioners, Table 3.11 provides a comparison of weighted and unweighted cross-section averages of the PCR and ICR. While the average financing burden on contributors increased over time, the annual standard deviations highlight that KVs were definitely hit differently by this process of increasing dependency; note that the percentage of KVs facing more than 40 pensioners per 100 contributors increased from about ten percent in 1861 to 30 percent and more after 1894. Weighting ratios with KV size and comparing them to their unweighted version brings attention to the fact that, indeed, the smaller KVs struggled comparatively more.

Table 3.11: Weighted and unweighted economic dependency among Prussian KVs

Cross-section averages	1861	1871	1881	1891	1901	1913	1920
A.) Unweighted							
Pensioners per 100 contributors	17.4 *(20.1)*	26.4 *(24.2)*	34.9 *(37.9)*	32.8 *(30.5)*	34.9 *(23.2)*	38.2 *(26.7)*	35.8 *(17.7)*
thereof invalids per 100 contributors	3.8 *(5.9)*	6.3 *(9.9)*	8.8 *(14.7)*	8.9 *10.9)*	11.4 *(9.3)*	13.6 *(9.5)*	11.1 *(8.6)*
B.) Size-weighted							
Pensioners per 100 contributors	18.6	22.2	32.3	30.8	24.5	28.6	32.2
thereof invalids per 100 contributors	4.4	4.5	7.5	8.8	9.6	11.2	9.0
Number of cross-sections	71	91	83	74	73	62	44

Notes: Annual standard deviations are in brackets. The KVs' size is measured in terms of the number of contributors.
Sources: See Figure 3.2.

A closer look at the data reveals that a large difference between the smallest KVs (< 200 contributors) and the rest existed; this holds for both Prussian and Bavarian funds. While both small and large Prussian KVs started at about the same level of relative burden in 1861 (≈ three to seven regarding invalids, and 11 to 18 regarding survivors), the small KVs' average ICR increased much faster, to 25 at the end of the "founders'

the KV than did established ones. In this sense, the ratio between established and unestablished miners should matter for the financing power of the collective of insurants. Taking only established members into account, over which the financial burden of pensioners was distributed, we have a simple upper bound estimate of the burden at hand. As of 1861, the upper bound ICR (SCR) was about nine invalids (26 survivors) per 100 established contributors, and it increased steadily to 23 (50) before the unestablished status was abolished. That abolishment meant, in a sense, that members who had formerly contributed rather weakly to the KV now contributed equally. So, the drop in the level of the ICR and SCR after 1907 is indicative of this equalization.

crisis" (*Gründerkrise*) in 1879, and to between 15 and 25 thereafter, until 1920. In contrast, the difference between all size classes except the smallest were not so considerable, at least not with regard to the ICR; regarding the ICR, they arrived at between five and 15 in 1920, and, regarding the SCR, at between nine and 31.

For reasons of comparison, I have introduced the experience of the Bismarckian system until 1913. But what about today's schemes? Table 3.12 tells in a very stylized way of the aging problems of a set of countries, with all but the U.S. adhering to the tradition of the Bismarckian welfare state with its emphasis on insurance.[79] The table depicts dependency ratios for selected years, as well as average annual growth rates, taking into account all years between 1950 and 2008.

Table 3.12: Economic dependency among modern economies

Country	1950	1970	1990	2008	2030	Average annual growth from 1950 to 2008
Germany	16.2	24.3	23.6	32.8	51.2	1.22 %
France	19.5	23.9	24.4	28.2	45.3	0.64 %
Italy	14.3	19.6	25.0	33.0	47.5	1.45 %
Spain	13.0	18.0	23.4	26.7	39.3	1.25 %
Austria	17.3	25.6	24.5	27.5	43.8	0.81 %
Belgium	18.1	24.1	24.8	28.8	44.7	0.80 %
Sweden	17.0	23.3	30.8	30.2	41.4	1.00 %
United States of America	14.3	18.7	20.9	21.1	35.3	0.67 %

Notes: Displayed are old-age dependency ratios (population aged 65 and older per 100 individuals aged between 20 and 64). Growth rates are geometrically averaged.

Sources: United Nations (2008), cited according to OECD Database, http://www.oecd.org/ document/0,3746,en_2649_201185_46462759_1_1_1_1,00.html, EXCEL-file on keyword "demography" and indicator "old age support ratio."

From the 1950s on, old-age dependency increased in all countries by 0.64 percent to more than 1.4 percent per year. In comparison, the median average annual growth rate of the PCR among the population of Prussian KVs was 2.8 percent. The PCR in German old-age insurance amounted to about 39 in 1993 and 46 in 2006. Compared to these figures, the average PCR among KVs was equally high and certainly not negligible. This is due to the fact that membership expansion in among KVs was bounded from above by focusing on exactly one sector of the economy (the "principle of occupa-

[79] Countries are selected according to Disney (2004), p. 287.

tional insurance"). This created the very conditions that make the KVs a natural case study regarding the challenges of the modern, mature welfare state.[80]

There is an alternative measure of economic dependency in a pension insurance system that we also can assess – namely, the average pension duration – that is, the average number of years a pensioner received pension payments from entry into retirement until death. I start with some general considerations about this measure before presenting evidence on the KVs. First, let us consider an old-age scheme in which a threshold age (of, say, 60 or 65) separates "being old and economically dependent on the working population" from "not being old and still capable of earning one's living with one's own manpower." This definition may be the result of societal or political consensus. If we observe an increase over time in the average number of years for which an old-age pension would be paid, we would do well to attribute the additional years to rising life expectancy – i.e., aging – provided that the average effective age at retirement remained constant. However, this straightforward interpretation of rising average pension duration would not hold if the effective retirement age simultaneously decreased. In that case, the increase in the average pension duration might be attributed, at least to a certain extent, to the incentive structure of the system's institutional design. This, then, would lead us away from the idea of aging in the biological or social sense and towards the question of whether incentives to retire earlier than legally scheduled might be at work. In the case of KVs, in particular, a decreasing effective retirement age might, instead, indicate that the geological conditions surrounding the production process became increasingly worse and, thus, the miner's ability to work was exploited earlier than it once had been, resulting in earlier inability to work. In any case, a decreasing effective retirement age would definitely leave us with an identification problem.[81]

Fortunately, we have for KVs annual data on the average pension duration of those invalids and widows who died in the respective year (not to be confused with the average pension duration of those invalids and widows who formed the surviving stock). Unfortunately, data are reported only for the period 1900 to 1920, but his should still enable us to assess the extent to which miners faced rising longevity. Figure 3.6 dis-

---

80 Figures on German old-age insurance calculated from Deutsche Rentenversicherung Bund (2008), p. 14 and p. 169. I divided the number of *pensioners* by the number of active insurants and multiplied by 100. An alternative measure would be to divide the number of *pensions* by active insurants, which actually yields a higher burden (1993: 59; 2006: 68). The difference is simply due to the fact that one pensioner could have received more than one pension.

81 For post-war Germany, for example, the average pension duration in statutory old-age insurance amounted to 9.9 years in 1960 and increased to 17.3 in 2007; these figures relate to the entirety of insurants (males: 9.9 years in 1960 and 15.3 in 2007; females: 10.6 years in 1960 and 19.4 in 2007). At the same time, the effective average retirement age with regard to old-age pensions followed a U-shaped pattern, with 64.7 in 1960, 62.2 in 1998 and 63.1 in 2007 (invalidity pensions: 56.0 in 1960, 51.5 in 1998, and 50.0 in 2007). So, the lengthening of the average pension duration is certainly due to a combination of a rise in longevity and increased early retirement; see, again, Deutsche Rentenversicherung Bund (2008), p. 105 and p. 123.

plays descriptive statistics on that measure regarding invalids. Note that the statistics are calculated after having eliminated all zero values, which had almost always occurred among the small KVs, as they sometimes did not see a single invalid die over the course of a year. Thus, the description is conditional on the fact that at least one invalid had died, for whom, then, the average pension duration was reported. The median average pension duration increased slightly, from about seven years in 1900 to ten in 1920. The minimum and maximum values tend to reflect the experience of smaller KVs, for which the variation in the series was generally much higher. However, the median scenario, as a rule, is representative of the larger KVs and, thus, of the majority of insured miners. The evidence, as it stands, supports the interpretation that miners were, indeed, subject to rising longevity, as, by the way, was the German population as a whole. Note, however, that it is not possible to assess average pension durations by age group, which, I concede, would certainly draw a more differentiated picture.

Figure 3.6: Average pension duration among Prussian KVs

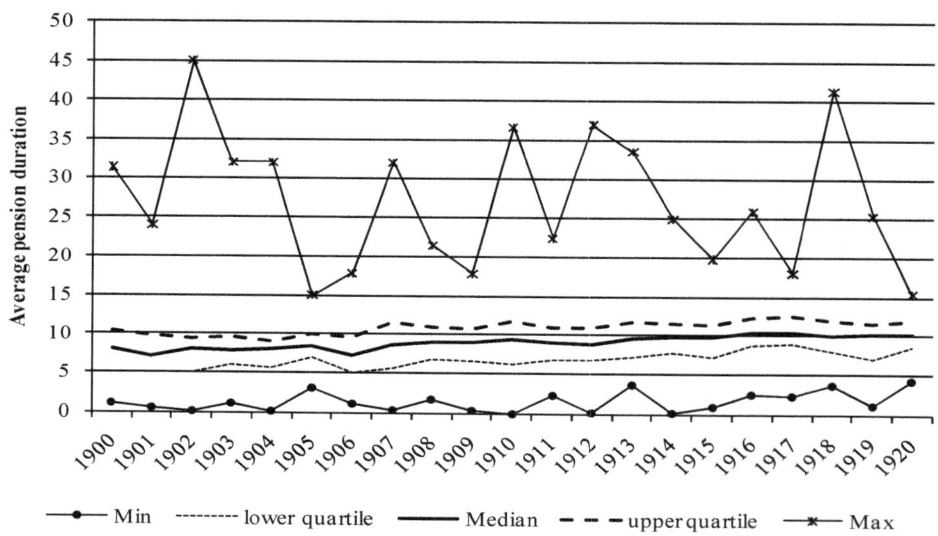

Notes: Figures relate to those miners who died over the course of the year and not to those who formed the remaining stock.
Sources: See Figure 3.2.

One might object that, without further evidence, a drop in the effective retirement age would be as likely a cause as increasing longevity for the lengthening of the average pension duration. Yet, Figure 3.7, presenting descriptive evidence on the average effective age at retirement, supports the view that the development of the average pension duration indicates increased longevity. The median effective retirement age stayed rather constant or decreased only slightly over time. The variation implied by the series of minima and maxima again reflects what occurred among small KVs, as opposed to what happened to the majority of insured miners. The median age at retirement – that is, when one became eligible for an invalidity pension – was 51 in 1868, 54 in 1913, and then dropped again towards age 51.

Figure 3.7: Descriptive statistics on the average effective age at retirement among Prussian KVs

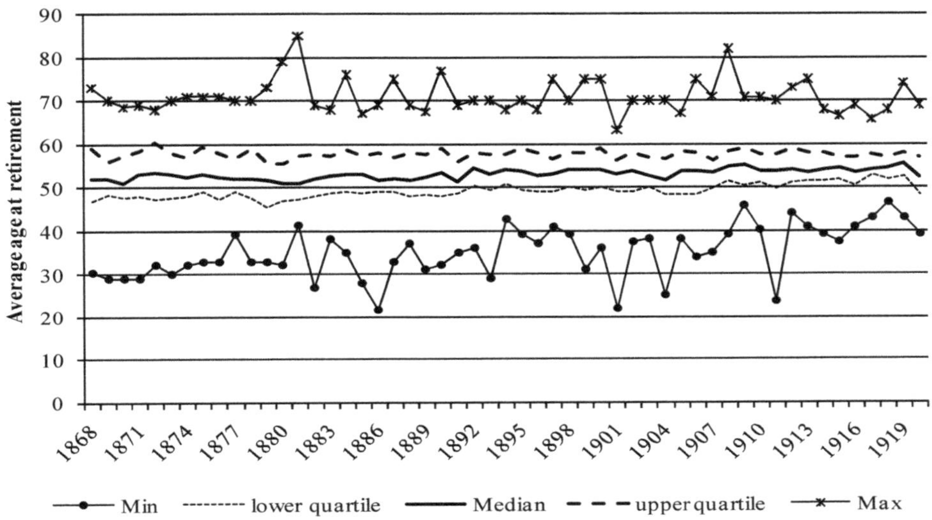

Sources: See Figure 3.2.

3.6.2. Determinants of economic dependency

Understanding the determinants of economic dependency – especially the rising burden from pensioners per contributing member – brings us to the determinants of the size of a KV's collective of insurants; "determinants" should be understood here in a more narrow and, thus, technical sense of the word. Analogous to a demographic equation that explains the development of a whole population according to its technical determinants, the size of the contributor base, or the stock of invalids, at the end of a period *t* can principally be written as

(3.1) $Contributors_t = contributors_{t-1} + inflows_t - deaths_t - out\text{-}migration_t - invalidity_t$

and, respectively,

(3.2) $Invalids_t = invalids_{t-1} + invalidity_t - deaths_t - reactivation_t$.

While the left-hand-side variables represent membership-stock quantities measured at the end of period *t*, the right-hand-side variables reflect both membership-flow and membership-stock quantities: (i) the number of working members having become eligible for an invalidity pension over the course of the period (*invalidity*); (ii) the number of working members having been reactivated as a working member (*reactivation*); (iii) the number of working members or invalids having left the KV due to death (*deaths*); (iv) the number of working members having left the KV for other reason (*out-migration*); and (v) the number of working members and invalids at the end of the preceding year. In the following, I will not consider the number of reactivated invalids since, if reactivations occurred, the number of them was extremely low, after all; as mentioned in Chapter 2, the possibility of being reactivated due to improved ability to work did certainly exist. Hence, "determinants" in a broader sense can be understood as the factors that might drive, in turn, inflows, mortality, and invalidity. For example, determinants of the frequency with which miners became invalid in a broader sense certainly were working conditions (or, respectively, work safety) and eligibility criteria (or, respectively changes in those).

Technically speaking, to establish a rising ICR, a net increase in the number of invalids over contributors is to occur during the course of a year. I do not intend to present a formal, sophisticated model to explain the ICR and SCR. Rather, my intention simply is to take a look at the time series on determinants of both the number of contributors and the number of pensioners, which can be calculate from the statistics. Therefore, some "demographic rates" have to be established, stating the relative frequency of certain events of interest (e.g., deaths or having become invalid). While the definition of the events of interest – their numbers establish the nominator – is no major problem, determining how many person-years principally were at risk – this is the denominator – is. The relative frequencies I want to observe are (i) the probability of becoming invalid over the course of a year (*IP*), (ii) the probability of dying over the course of a year (*DP*), and (iii) the probability of out-migrating over the course of a year (*OMP*), or in equations:

(3.3) IP^E = *Inflow of established invalids* / [*established miners at the end of the year + inflow of invalids + deaths of established miners + miscellaneous outflow ofestablished miners*] ,

(3.4) IP^U = *Inflow of unestablished invalids* / [*unestablished miners at the end of the year + inflow of unestblished invalids + deaths of unestablished miners*],

(3.5) DP^I = *Deaths of invalids* / [*invalids at the end of the year + deaths of invalids*],

(3.6) DP^W = *Deaths of widows / [widows at the end of the year + deaths of widows + re-marriages + miscellaneous outflows of widows]*,

(3.7) DP^E = *Death of established miners / [established miners at the end of the year + inflow of established invalids + deaths of established miners + miscellaneous outflow of established miners]*,

(3.8) DP^U = *Death of unestablished miners / [unestablished miners at the end of the year + inflow of unestablished invalids + deaths of unestablished invalids]*,

(3.9) OMP^E = *Outflow of established miners / [established miners at the end of the year + inflow of established invalids + deaths of established miners + outflow of established miners]*.

Note that the period index *t* has been omitted here. Superscripts *E*, *U*, and *W* denote established and unestablished miners or, respectively, the invalids that came from either of those two groups, and widows. I define "person-years at risk" as all persons-years in which workers were exposed to the possibility of death, becoming invalid or out-migrating.[82] In what follows, I first consider the "fertility side" of the KVs' membership by focusing on the mean inflow rate of established and unestablished miners and the invalids that came from the working membership. I calculated the inflow rate as the number of incoming miners over the course of a year divided by the number of miners existent at the beginning of that year times 1 000. This measure should not be confused with the ones above since it is no true probability of occurrence of the event "new miner enters the KV."[83]

Column two in Table 3.13 tells us that the (unweighted) mean inflow of established miners initially fluctuated around 127 per 1 000 existing miners averaged over the years 1867 to 1870. The sharp increase for the period 1908–1913 is, of course, due to the reform of 1906/1907, when the unestablished status was eventually abolished. That the inflow rate remained rather high after 1913, and especially for 1919 and 1920, might

---

82 In this framework, a miner who already was a KV member at the beginning of year *t* and who survived that year, represented one full person-year. In contrast, if that miner died at the end of June, he would have been exposed to the risks of death, becoming invalid and out-migrating only for half a year; he represented 0.5 person-years. For the reason that the KV statistics does report the annual number of the events "death," "becoming invalid," and "out-migration," but not the timing – i.e., how many events did occur on which day or in which month – I count every miner as exactly one person-year. This procedure has the consequence that the denominator I estimate is, to some extent, biased in that it is larger than it would be if we knew of every KV member precisely when one of the aforementioned events occurred to him over the course of a particular year. Thus, I call the calculated probabilities "lower-bound estimates" since they underestimate the true risk.

83 That is due to the fact that the denominator does not consist of the person-years exposed to the "risk" of joining a KV – e.g, the whole male population between the age of 16 to 45.

reflect the recently improved conditions regarding membership turnover (see Chapter 2). The mean inflow rate regarding unestablished miners depicted in column three fluctuated more strongly, showing some kind of cyclical pattern that matches the consideration that unestablished workers – as long as they were day-laborers or had just entered the mining sector and, so, were young and had to undergo a waiting period – were used to satisfying labor-demand peaks, but also to reduce capacities immediately if necessary in a downswing.[84] Clearly, these series do not bring out all the differences between small and large, or stagnating and thriving, KVs. Table 3.14 just gives some impression of how differently KVs performed in attracting new members by showing the inflow rate by growth pattern.

With regard to the number of invalids produced over the course of a year, column four in Table 3.13 displays the development of the mean ratio in the population of KVs between inflowing invalids, on the one hand, and exiting invalids, on the other hand. Numbers clearly indicate that – on average and except for the war years – there were more miners who became eligible for an invalidity pension than invalids who died. However, while both series seem to exhibit an only slightly declining trend until 1906/07 or, respectively, 1908/13, they show a clearly declining trend thereafter. So, there apparently was lot of pressure on the ICR's nominator – a pressure that, in part, disappeared towards 1920.

Columns five and six in Table 3.13, in addition, establish how the probability of becoming invalid developed over time.[85] The (unweighted) invalidity rate regarding established miners – that is, IP^E multiplied by 1 000 – initially amounted to approximately 16 per 1 000 working miners at risk and then ranged between 17 and 18 until 1906/07. With the reform of 1906, the invalidity rate significantly dropped to somewhere between seven and 11 invalidity cases per 1 000 (or five and 13 if we address the annual cross-sectional means one by one). This could reflect advances in medical treatment, as well as uniformly tightened eligibility rules; the latter seems more probable since the drop came immediately.[86]

---

84 Displayed figures hide the fact that some values for the cross-sectional mean inflow rate are negative; this, in turn, is a consequence of the fact that the KV statistics do not explicitly report how many unestablished miners out-migrated. The inflow calculated is, thus, net of out-migration.

85 Much effort was spent in the contemporary literature to calculate the probability of becoming invalid and to explore whether there was an underlying law governing the process; see Behm (1863), Küttner (1881, 1888), Caron (1882), and Morgenbesser (1882).

86 It seems likely that KVs used the reform as an opportunity to do so. The reform law itself still had not regulated the eligibility issue. Unfortunately, due to the data situation, we are missing a measure that would tell us how tight the KVs' eligibility rules generally were. Such a measure is, indeed, available for Bismarckian insurance, for which we have the proportion of workers that qualified for certified invalidity among all workers who applied. Hennock (2008), pp. 202–204, shows that Bismarckian insurance seems to have been quite generous in granting invalidity in its first decade since the percentage of applications refused was rather modest. But even this measure is only second-best since it does not tell how many workers were discouraged by the rules from applying at all.

Table 3.13: Demographic fundamentals at a glance (Prussian KVs)

Period	Mean inflow rate of contributors		Mean ratio between inflow of invalids and deaths of invalids	Mean invalidity rate	
	E	U		E	U
1867–1870	127	9	2.2	16	1
1871–1875	243	78	2.3	18	2
1876–1880	114	50	2.1	18	5
1881–1885	119	101	1.7	17	3
1886–1890	247	82	2.0	18	1
1891–1895	96	26	2.0	19	3
1896–1900	131	139	1.9	19	3
1901–1907	131	91	1.8	19	1
1908–1913	735	–	1.6	11	–
1914–1918	223	–	1.2	9	–
1919–1920	354	–	1.0	7	–

Period	Mean out-migration rate	Mean crude death rate of contributors		Mean crude death rate of invalids	Mean crude death rate of widows
		E	U		
1867–1870	131	13	10	96	32
1871–1875	219	14	8	109	38
1876–1880	87	11	8	98	37
1881–1885	173	11	9	110	40
1886–1890	82	10	8	105	43
1891–1895	99	10	9	107	38
1896–1900	76	7	8	88	44
1901–1907	97	7	6	94	37
1908–1913	243	5	–	76	50
1914–1918	159	17	–	76	44
1919–1920	197	8	–	83	31

Notes: Rates display unweighted cross-sectional averages per period of incidences per 1 000 person-years. "E" and "U" denote "established" and "unestablished" miners. The mean inflow rate of contributors is calculated as the number of incoming miners over the course of a year divided by the number of miners existent at the beginning of that year times 1 000; the mean ratio between inflow of invalids and deaths of invalids as (sum of incoming invalids/sum of deaths of invalids); the mean invalidity rate according to equation (3.3) and (3.4); the mean out-migration rate according to equation (3.9); the mean crude death rate of contributors according to equation (3.7) and (3.8); the mean crude death rate of invalids and the mean crude death rate of widows according to equation (3.5) and (3.6).

Sources: See Figure 3.2.

Table 3.14: Demographic fundamentals by growth pattern (Prussian KVs)

Growth pattern	1867–1880	1881–1890	1891–1900	1901–1913	1914–1920
A) Inflow rates, established plus unestablished miners					
Small stagnant	86 (49)	100 (146)	93 (151)	167 (26)	128
Medium stagnant	792 (115)	139 (51)	168 (26)	1 283 (213)	267
Small dynamic	41 (3)	74 (74)	109 (123)	199 (32)	118
Medium dynamic	1 852 (80)	138 (93)	135 (66)	296 (104)	318
Large dynamic	2 199 (11 162)	700 (56)	98 (62)	357 (36)	301
B) Invalidity rates, established miners only					
Small stagnant	23	19	22	22	11
Medium stagnant	13	21	26	19	11
Small dynamic	10	9	14	12	10
Medium dynamic	12	13	16	13	7
Large dynamic	16	17	15	13	7
C) Rates of out-migration, established miners only					
Small stagnant	47	50	44	69	61
Medium stagnant	98	105	101	178	156
Small dynamic	56	39	35	64	86
Medium dynamic	64	56	61	123	155
Large dynamic	33	34	34	117	153

Notes: Rates are rounded. Rates display incidences per 1 000 person-years. Inflow rates regarding unestablished miners are in brackets; for the period 1901–1913, rates on unestablished miners refer to 1901 to 1907 only.

Sources: See Figure 3.2.

Regarding unestablished miners, the mean invalidity rates were generally lower, implying, on the one hand, that access to pensions was certainly far more restricted for them, which is consistent with what was said in Chapter 2; on the other hand, the lower rates also imply that unestablished miners were utilized differently in the production process, such that their risks for disease and severe accidents were lower than for established miners. In addition, Table 3.14, once again, reports additional period means by growth patterns.

It seems as if the relative incidence of "becoming invalid if being an established miner" was, by tendency, higher in stagnating KVs, with between 13 and 26 incidences per 1 000, compared to nine to 17 if focusing on dynamic KVs and the pre-war years. However, there seems to be no systematic difference across size classes if focusing on all funds that, on average, thrived.[87]

[87] There is another question regarding invalidity and the factors that drive application rates: Were invalidity rates at least partly driven by the business cycle? The basic idea behind this question is

Furthermore, the issue of migration, or labor turnover, can only be assessed in terms of out-migration probabilities (or, if multiplied by 1 000, in terms of out-migration rates) according to equation 3.9. Unfortunately, useful data to reconstruct migration flows from KV *i* to KV *j*, and vice versa, are not available, so this source of growth in size and the potential refreshment of the stock of insurants cannot be reasonably assessed. Starting in the 1870s, miners' migration from one *Knappschaft* area into another was certainly a relevant demographic phenomenon (for example, the movement of miners from the eastern parts of the *Reich* towards the Ruhr area), probably due to structural changes within the mining sector (exhaustion of local deposits, for example, or wage differentials).[88] Column eight in Table 3.13 displays that mean out-migration rate among Prussian KVs by period. In fact, the rate fluctuated considerably. However, much of the fluctuation cancels out if we construct a KV size-weighted series of the out-migration rate. So fluctuations are due mainly to variation across smaller funds.[89] Re-organized by growth pattern (see Table 3.14), we find that the rate of out-migration broadly increased in all cases, but stagnant KVs could not, ultimately, compensate for losses by way of turnover.[90]

Now that I have focused on inflows and out-migrations, I turn to mortality trends. First, columns eleven and twelve in Table 3.13 show mean periodical death rates for invalids and widows (the probability of dying over the course of a year according to equations (3.5) and (3.6) times 1 000). Note that these rates are "crude" – that is, age-unspecific. Until the middle of the 1890s, the (unweighted) death rate if you were an invalid notably fluctuated around 100 deaths per 1 000 at risk and then declined to between 76 and 94. However, the fluctuations in the series are once more due to the experience of smaller KVs. A weighted series would show that the crude death rate remarkably declined over almost the entire observation period, from 90 deaths per 1 000 to 50 deaths per 1 000 in 1913 (60 deaths per 1 000 is the weighted average for the period 1914–1920). Since medical-technical progress is generally said to be one of the major drivers of a secular decline in mortality, we may do well to consider the decline as re-

---

that workers might use invalidity insurance as unemployment insurance in economic downswings. I tried to answer this question in this study's long version. Evidence suggests that the invalidity rate increased in an upswing and decreased in a downswing. This is consistent with the consideration that labor utilization should rise in an upswing; that miners used invalidity insurance as surrogate for unemployment insurance seems thus rather unlikely.

88 Regarding labor turnover among German miners, see, for example, Brüggemeier (1983), pp. 28–30 and pp. 60–62, and Tenfelde (1977), pp. 231–236.

89 Since neither the number of out-migrated miners nor the number of inflows was reported by age in the statistics, it is not possible to reasonably calculate the number of out-migrations by age group. However, since miners usually had to be between 20 and 45, or 50 at the maximum, to be able to join the established membership, it is likely that those who moved to another KV were in their late twenties or thirties, rather than in their forties.

90 Note that, in Table 3.14, categories *B* and *C* only account for established miners. So if a comparison with category *A* led to a net increase in KV size for the stagnant KVs, this is due to neglecting unestablished miners.

flecting advances in medical treatment of the typical mining-specific diseases and injuries that would, years before, have led to an earlier death. The crude mean death rate among widows followed, in contrast, a stagnating trend of about 33 per 1 000 if the weighted series is addressed.

Beyond that, columns nine and ten in Table 3.13 address the development of the crude mean death rate among the working membership. Both series exhibit a clear declining trend before the First World War, implying either that working as a miner had, in general, become less perilous or that medical treatment had advanced such that fewer miners died. However, the former certainly appears not to have been the case if we consider that (i) work safety regulations were put through only little by little; and (ii) geological conditions – at least in hard-coal mining – got worse. The deep-cutting effect of war on mortality among the KVs membership is likewise obvious.

Finally, a more accurate picture of mortality is drawn by Table 3.15, which presents age group-specific probabilities of death for invalids and established miners. Therefore, equation (3.7), referring to the established working membership, is slightly adjusted, in that the number of out-migrations was proportionally distributed across age groups 26–35, 36–45, and 46–55.

Table 3.15: Weighted mean age-specific death probabilities of invalids and contributors among Prussian KVs

	Invalids					Contributors				
	Age group					Age group				
	16–25	26–35	36–45	46–55	56+	16–25	26–35	36–45	46–55	56+
1867–70	8.0%	7.6%	7.8%	8.0%	11.9%	0.6%	0.8%	1.2%	1.6%	3.5%
1871–75	6.4%	6.3%	7.5%	8.3%	13.6%	0.6%	0.8%	1.2%	2.1%	3.8%
1876–80	6.3%	6.9%	6.9%	7.5%	13.9%	0.5%	0.7%	1.1%	1.7%	2.9%
1881–85	6.2%	6.2%	6.7%	8.0%	12.2%	0.5%	0.7%	1.1%	1.6%	3.1%
1886–90	6.3%	6.5%	5.9%	6.3%	11.1%	0.4%	0.5%	1.0%	1.4%	1.9%
1891–95	7.5%	5.6%	5.3%	6.2%	10.3%	0.4%	0.5%	1.0%	1.4%	2.2%
1896–00	6.0%	4.7%	4.8%	5.4%	10.4%	0.4%	0.6%	1.3%	1.4%	2.2%
1901–05	5.7%	4.6%	4.4%	5.2%	11.3%	0.3%	0.4%	0.8%	1.1%	1.9%
1906–07	4.9%	3.9%	4.0%	4.5%	9.1%	0.4%	0.4%	0.6%	0.9%	1.9%

Notes: Data on contributors refer to established membership.
Sources: See Figure 3.2.

With regard to invalids, all age groups reflect a declining trend, even though the group of oldest invalids, 66 years and beyond, shows the smallest negative slope. From age 36 to 66+, moreover, the probability of dying increased with age. Invalids aged 26 to 35 were an exception since their mortality was equal or even higher. A reasonable explanation for this negative mortality dividend appears to be the following: If a miner had become permanently incapacitated for work in his relatively early years, this probably occurred due to severe injury and not to a job-related disease, which took time to develop. So, the argument runs, severe injuries significantly reduced longevity. In addition, all age groups of established miners show a decline in mortality, as well, especially among the oldest group, aged 56 and beyond.

3.6.3. Measuring the age structure of the *Knappschaften*

In the context of KVs, the PCR and its components addressed above are not directly indicative of the memberships' age structure measured at different points in time.[91] Therefore, I want to finish this subsection by presenting additional evidence on this issue. Thus, to begin with, let us have a look at Figure 3.8, which depicts the age structure of established membership aggregated over all Prussian KVs; unfortunately, the KV statistics does not report the age structure of unestablished membership.

Figure 3.8: Age structure of aggregate established membership (Prussian KVs)

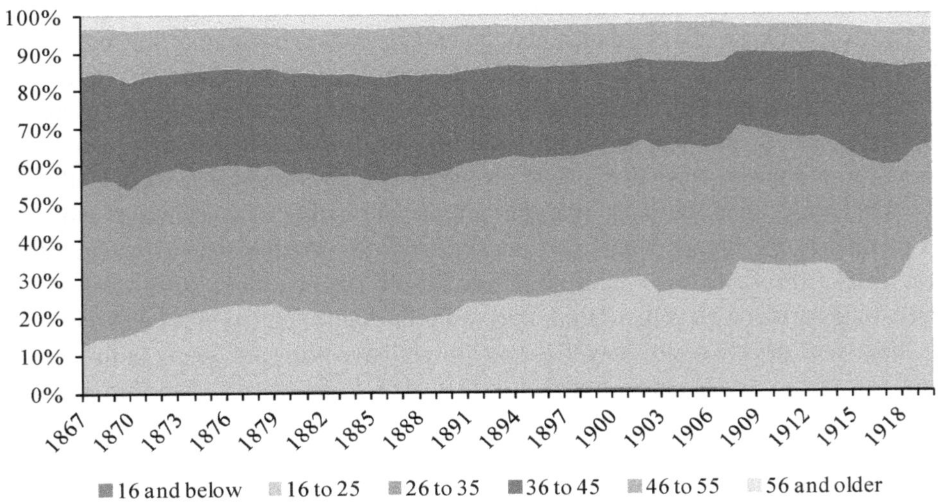

Sources: See Figure 3.2.

On the whole, the percentage share of the youngest age groups together – that is, miners aged below 16 and miners aged between 16 and 25 – increased from unspectacular 12 percent in 1867 to nearly forty percent in 1920, Although the age group "26 to 35" lost

91 This is simply due to the fact that invalidity and survivorship did not only occur at higher ages, but also at lower ages, whereas in a typical old-age scheme, economic dependency is defined as meeting a threshold age of 60 or 65. In the latter case, the old-age dependency ratio is definitely indicative of a system's age structure.

share (≈ 43 percent in 1867 and still 22 percent in 1920), the increasing significance of the group of younger miners is apparent. This effect in the aggregate is, not the least, due to the secular prosperity in the large mining areas (Ruhr, Silesia, Saar), where the large KVs were situated. Evaluated at the KV-level, however, we would find notable differences in the age structure measured by age shares.

Two straightforward measures can be conveniently calculated and, thus, used to pinpoint age-structure differences on the KV level: (i) the young contributors-to-old contributors ratio (or, in short-hand, the youth ratio); and (ii) the average age of the membership (in contrast, for technical reasons, to median age usually applied when population ageing is addressed).[92] The young contributors-to-old contributors ratio exclusively relates to the fraction of established members and is equal in my framework to the ratio between the number of miners aged 16 to 35 and the number of those aged 36 and older, times 100, or

$$(3.10) YOR^{E}_{t} = \left[\frac{agegroup\ 1625_{t} + agegroup\ 2635}{agegroup\ 3645_{t} + agegroup\ 4655_{t} + agegroup\ 56older_{t}} \right] * 100 \cdot$$

A (continuously) decreasing youth ratio is especially indicative of aging “from the bottom” because the base of young members potentially contributing to a KV for some 30 to 40 years would shrink relative to (i) the number of older members to retire earlier than the young miners and (ii) the pensioners who had to be financed. To ensure sustainable finances in the long term, KVs arguably had to have a broad base of young miners at any point in time that would pay their contributions for a sufficiently long time.[93] Otherwise, they faced the risk that a reduced number of contributors would remain to finance the higher number of pensioners. Two additional facts made running miners’ social insurance even more problematic: On the one hand, KVs were locally-focused funds and, on the other hand, they were disconnected from population growth since they were strictly sector-specific. They might have attracted young individuals by holding out favorable insurance conditions. However, in the end, KVs had to rely upon people’s decision for or against entering the mining sector (or, later, the KV area).[94]

Table 3.16 depicts some information on this measure. What we find is that, before 1900, the median ratio amounted to 100 young workers per 100 old workers, or below, and between 126 and 146 per 100 thereafter. In all, the mean does not deviate too much.

---

92 I also incorporated age pyramids by year and by growth pattern into this study’s long version; see Jopp (2012b).

93 In the very long-term, this argument is ambivalent insofar as every recruited young miner (unless he died early) became part of the elderly fraction one day. The more young men were recruited, the more severe would probably be the age shock decades later.

94 That certainly depended on various factors, not just the attractiveness of the KVs’ social insurance. These factors included family tradition, expected economic growth of the mining sector, wage differentials, working opportunities with respect to other sectors, severity of work, or their own health status.

The sharp increase from 1891/1900 to 1901/13 is, of course, due to the abolishment of unestablished memberships; apparently, unestablished memberships were generally younger than established memberships. In addition, the mean youth ratio is shown by growth pattern, too. In fact, there is a clear difference in observable levels: small, stagnant KVs usually saw, before the reform, not 50 young miners per 100 old and large, dynamic funds saw about 140 per 100 old. The reform seems to have reduced the gap, but thriving medium and large KVs still were naturally advantaged.[95]

Finally, the average age of a KV's established membership has been calculated according to equation (3.11):

(3.11) AGE^E = [*age group (16–25)**20.5 + *age group (26–35)**30.5 + *age group (36–45)**40.5 + *age group (46–55)**50.5 + *age group (56+)**60.5] / *Established contributors*.

Table 3.16: Measures of the age structure (Prussian KVs)

	1867–1880	1881–1890	1891–1900	1901–1913	1914–1920
A) Young contributors-to-old contributors ratio					
a. overall					
Median ratio	95	96	100	126	146
Mean ratio	103 *(71.3)*	99 *(61.2)*	106 *(60.5)*	136 *(100.6)*	153 *(115.6)*
b. by growth pattern					
Small stagnant	48	58	56	85	127
Medium stagnant	99	103	113	132	188
Small dynamic	92	91	97	116	112
Medium dynamic	128	111	119	157	168
Large dynamic	137	133	146	176	165
B) Average age by fraction of membership					
a. overall					
Contributors	38 *(5)*	38 *(5)*	37 *(4)*	36 *(5)*	36 *(5)*
Invalids	56 *(7)*	58 *(6)*	59 (5)	59 *(5)*	60 *(4)*
Widows	50 *(6)*	52 *(5)*	53 *(4)*	55 *(3)*	53 *(4)*
Overall	41 *(6)*	42 *(6)*	42 *(5)*	41 *(6)*	40 *(6)*

Notes: Standard deviation in brackets. Orphans are excluded from "overall average age."
Sources: See Figure 3.2.

95 Note that, before 1908, the youth ratio only refers to established memberships. Insofar, regarding the information sorted by growth pattern (which, in turn, refers to overall working membership), it was well possible that KVs classified as "stagnant" could have experienced a rising youth ratio.

Table 3.17: Average ages by growth pattern (Prussian KVs)

	1867–1880	1881–1890	1891–1900	1901–1913	1914–1920
A) Small stagnant					
Contributors	44 *(6)*	44 *(6)*	42 *(4)*	42 *(5)*	42 *(6)*
Invalids	61 *(7)*	62 *(5)*	63 *(5)*	64 *(4)*	65 *(4)*
Widows	53 *(6)*	55 *(5)*	57 *(3)*	58 *(2)*	58 *(3)*
Overall	47 *(6)*	48 *(6)*	48 *(5)*	48 *(5)*	48 *(6)*
B) Medium stagnant					
Contributors	37 *(3)*	37 *(3)*	36 *(2)*	35 *(3)*	32 *(2)*
Invalids	55 *(6)*	56 *(5)*	57 *(5)*	58 *(4)*	60 *(2)*
Widows	48 *(5)*	51 *(4)*	54 *(3)*	55 *(2)*	54 *(3)*
Overall	40 *(4)*	41 *(5)*	43 *(4)*	43 *(4)*	38 *(4)*
C) Small dynamic					
Contributors	38 *(4)*	38 *(3)*	38 *(4)*	36 *(4)*	36 *(3)*
Invalids	56 *(9)*	60 *(5)*	60 *(6)*	60 *(5)*	58 *(5)*
Widows	48 *(7)*	51 *(4)*	54 *(3)*	55 *(2)*	54 *(3)*
Overall	41 *(5)*	43 *(5)*	42 *(5)*	40 *(4)*	40 *(4)*
D) Medium dynamic					
Contributors	36 *(4)*	37 *(4)*	36 *(4)*	35 *(3)*	34 *(3)*
Invalids	56 *(5)*	56 *(6)*	57 *(5)*	58 *(4)*	59 *(4)*
Widows	49 *(5)*	51 *(4)*	52 *(4)*	54 *(3)*	51 *(3)*
Overall	39 *(4)*	40 *(4)*	40 *(5)*	39 *(4)*	38 *(5)*
E) Large dynamic					
Contributors	35 *(2)*	35 *(2)*	35 *(2)*	34 *(3)*	33 *(2)*
Invalids	54 *(4)*	55 *(3)*	57 *(2)*	57 *(2)*	58 *(2)*
Widows	50 *(3)*	51 *(2)*	52 *(3)*	53 *(2)*	51 *(2)*
Overall	38 *(2)*	39 *(1)*	39 *(2)*	38 *(3)*	37 *(2)*

Notes: Standard deviation in brackets. Orphans are excluded from "overall average age."
Sources: See Figure 3.2.

As the equation stands, it refers to the period 1867–1888, for which we have data on only *six* age groups (those few miners aged below 16 are incorporated in the age group 16–25). I weight age groups with the age at the class midpoint. Regarding the period 1889–1920, I used the *ten* age groups since reported. I did the same regarding invalids and widows.

Results on the average age among KVs are also depicted in Table 3.16. While the average age among established miners modestly decreased over the displayed periods, from about 38 to 36 years, the average age of invalids and widows increased notably, arguably reflecting the "biological side" of the aging process. The drop in the average age of widows after 1914 seems to reflect the fact that a lot of younger married miners lost their life during war, leaving behind relatively younger wives. Overall average age was apparently driven by the age trend among contributors and widows.[96]

Reviewed by growth pattern, the picture becomes even more detailed. Table 3.17 displays the corresponding data. Two observations must be highlighted: First, regarding cross-sectional differences, small KVs had, on average, the oldest overall memberships as well as sub-memberships, and the small stagnating KVs' memberships were even older than memberships of the dynamic ones. Largest KVs, as a consequence of prosperity and, presumably, actuarial advantages, had the youngest memberships. Second, over time and as we would expect regardless of growth patterns, the average age among invalids, but also among widows, shows a secular increase providing evidence for "aging from above" in the KVs.

3.7. The expenditure side

In this subsection and the subsequent one, I want to point out secular financial trends among KVs, or breaks in those. I begin with investigating the expenditure side. KVs arguably ran expenditure-oriented schemes, meaning that they first determined the monetary amounts of benefits to be paid per insurance case, and then looked at how many cases would probably occur and the amount of revenues required to keep their books balanced. I think it is appropriate to start with the aggregate social budget of KVs and the functional distribution of costs before turning to the issue of per capita generosity.

3.7.1. Aggregate social budget

To visualize the significance, or simply the scale, of an economy's social security spending, especially as compared to other economies, it is common practice to form the ratio between expenditures and a measure of the economy's overall economic performance – usually the gross domestic product (GDP).[97] According to this procedure, we can express overall insurance benefits paid by Prussian KVs as a percentage of German NNP (GDP data are not available). Not surprisingly, the percentage is extremely low when levels are considered, but the change over time compared to NNP is quite impressive; the percentage steadily increased until 1903/04, from 0.05 percent to 0.17 percent, implying that the KVs' insurance output, if approximated by factual payments, grew for decades faster than the economy. This is certainly consistent with the mining sector's

96 Reviewed by size class, these trends still hold, even though differences in the levels are obvious.

97 See, for example, Lindert (1994), p. 10, Lindert (1996), p. 2, and Cutler and Johnson (2004), p. 90.

status at the time as one of the major driving forces of Germany's economic growth.[98] What about a comparison with Bismarckian social insurance, in particular? Since its implementation, expenditures increased from roughly 0.3 percent, measured against NNP, to about 1.5 percent in 1913.[99]

Furthermore, a direct comparison between Prussian KVs and Bismarckian insurance might help to get a better idea of the volume of the KVs' benefits. So let us consider the proportion of KV expenditures on health and, respectively, pensions to the corresponding benefits that the Bismarckian system provided. In fact, as of 1885, KV pension expenditures were six times higher than the initial Bismarckian expenditures on accident pensions; surely, the public accident scheme was just in the warm-up phase. As of 1889, when invalidity and old-age insurance were implemented on the national level, Bis-marckian pension expenditures had, however, caught up in that they were about as high as KV pension expenditures. Between 1890 and 1900, the proportion of KV expenditures declined from parity to about 12 percent. Since the turn of the twentieth century and towards 1913, the ratio then remained quite constant at between ten and 12 percent. With regard to health expenditures, KVs ranged, from the start, from seven to nine percent of Bismarckian expenditures on health, neglecting the downward spike in the immediate post-KV-reform years (i.e., 1907–1910); with 1911, the proportion returned to that range.[100]

These figures convey an impression of the relative scale of the KVs' schemes, but would be even more meaningful if we compared, at the same time, relative memberships. So, while Prussian KVs ultimately spent about 11 percent of the volume of Bismarckian pensions, they relied on the financing power of only between 3.6 percent (in 1892) and 5.2 percent (in 1907) of the Bismarckian contributor base. That clearly implies that the financing burden put on the average contributing miner was definitely higher.[101] We can do the same comparison for health insurance. Here, we find that the relation was slightly more proportional, in that KVs spent an amount on health that was between the aforementioned seven to nine percent, while they relied on a contributor

---

98 As an alternative to German NNP, we can also measure KVs' total benefit expenditures against Prussian national income as calculated by Hoffmann and Müller (1959), pp. 86–87. As of 1861 (1871; 1890; and 1913), the ratio was 0.053 (0.077; 0.171; and 0.262) percent.

99 Taking an international perspective, Germany set a new standard with the introduction of social insurance on the national level, regarding social policies for the masses and, thus, how people's welfare was broadly raised. Peter Lindert, though, argues that – if focusing on redistributive social spending activities (not such social spending that was based on the principle of "insurance" – other countries, such as Denmark and Norway, have to be considered forerunners of the welfare state. It is not my intention to delve into this discussion; see Lindert (1994), pp. 9–10.

100 Calculated with data from Khoudour-Castéras (2008), pp. 234–241, and the Prussian KV statistics.

101 Surely, the average wage in mining exceeded the average wage in the economy as a whole. As the discussion of per capita generosity below will indicate, this differential cannot be not simply explained by these wage differentials.

base that was between 7.8 percent (1885) and 5.0 percent (in 1911) of Bismarckian health insurance's membership.[102] I will come back to this point when per capita generosity, especially in relation to income, is addressed.

I proceed with introducing the series on the main expenditure items explicitly: invalidity pensions, survivorship pensions, daily sick pay, and medical treatment, including drugs and physicians' salaries. Table 3.18 provides a comparative overview of aggregate real expenditures of Prussian and, for comparative reasons, of Bavarian and Saxon KVs for those categories.

While all Prussian KVs together spent about 1.8 million marks on pensions in 1861, spending increased significantly to 39 million marks in 1913; this equals growth of a little more than 2 000 percent (or more than 3 000 percent regarding invalidity pensions and more than 1 200 percent regarding survivorship pensions). The amount spent in 1919, about 10.6 million marks, already reflects the transition from suppressed inflation to undisguised inflation, depreciating the real value of insurance benefits.

Note, in particular, that expenditures on survivorship pensions exceeded those on invalidity over the first decade of the century. Thus, the pension structure, as assessed by the aggregate, definitely changed over the period. Initially, only about 30 percent of pensions were associated with invalidity, but this relation exactly reversed as time went by. Besides, real pension expenditures of Bavarian KVs ranged far below, and Saxon KVs spent between one and, at the peak in 1913, about five million marks on pensions. In 1861, real health expenditures of Prussian KVs were roughly equally high as those on pensions, and still totaled 36 million in 1913. Bavarian KVs' health expenditures ranked below the Prussian KVs' on pensions, and what is reported on Saxony refers to the sickness funds, which have to be treated separately from Saxon pension funds.

Now that we have an idea of the aggregate scale of KV expenditures, and of pension and health expenditures separately, let us turn to the functional division of insurance benefits. Major cost categories to be distinguished are: i) pensions; ii) all kinds of sickness-related benefits; iii) miscellaneous benefits including, among others, funeral pay and extraordinary (i.e., charitable) payments; and iv) all costs related to operating the KVs, including administrative costs in the more narrow sense (wages, rent) and miscellaneous operating costs (such as maintenance of buildings).

Tables 3.19 to 3.21 describe the functional division of total costs in detail; results are separately displayed by the compound insurance period (until 1907) and the pension and sickness insurance sections that emerged thereafter (1908-1920). In addition to the percentage figures, I also report the nominal and real aggregate budget over all KVs.

[102] See preceding footnote.

Table 3.18: Aggregate real pension and health expenditures compared (in 1 000 marks and 1913 prices)

Year	Bavarian KVs		Prussian KVs		Saxon KVs		Bismarckian invalidity and old-age insurance
	I	S	I	S	I	S	
A) Pension expenditures							
1861	–	–	855	933	–	–	–
1866	–	–	1 348	1 430	–	–	–
1871	19	35	1 875	2 107	1 049		–
1876	58	74	3 781	3 564	1 412		–
1881	92	94	5 426	4 698	1 520		–
1884	110	108	6 602	5 696	1 624		–
1886	122	118	7 927	6 729	2 775		–
1891	121	122	9 692	7 105	3 974		1 563
1896	210	154	13 305	8 724	3 637		35 324
1901	280	175	16 937	10 496	3 883		106 406
1906	426	201	17 839	11 948	3 859		150 703
1911	543	233	26 357	11 752	4 450		169 356
1913	578	238	26 909	12 263	5 141		181 000
1916	340	172	16 252	10 105	2 393		–
1919	142	89	6 118	4 496	2 660		–
	Sick pay	Health care	Sick pay	Health care	Sick pay	Health care	Bismarckian health insurance
B) Health expenditures							
1861	–	–	786	932	–	–	–
1866	–	–	1 074	1 277	–	–	–
1871	11	26	1 308	1 707	–	–	–
1876	27	66	2 007	2 178	–	–	–
1881	29	72	1 855	2 515	–	–	–
1884	36	82	2 220	3 165	–	–	–
1886	58	85	3 461	3 315	987		77 914
1891	71	107	4 759	4 310	1 095		118 535
1896	101	130	6 366	5 022	1 112		152 579
1901	129	164	12 662	7 457	1 261		217 288
1906	182	180	12 528	12 165	1 326		288 840
1911	245	234	17 342	17 597	1 804		407 944
1913	243	256	16 722	19 935	1 976		459 889
1916	143	128	8 109	10 418	1 017		–
1919	165	124	9 540	10 632	1 252		–

Notes: Thalers were converted into three marks (1861–1874). The price series used here is, again, that of Hoffmann (1965). The series, which runs until 1913, has been extended to 1920 using indices displayed in Sommariva and Tullio (1987). "I" and "S" abbreviate "invalids" and "survivors"

Sources: My own calculations; see Table 3.5 and Khoudour-Castéras (2008), pp. 236–237.

Table 3.19: Percentage distribution of aggregate expenditures among Prussian KVs by type (compound insurance, 1861–1907)

Year	Invalidity pensions (full)	Invalidity pensions (semi)	Widows' pensions	Orphans' pensions	Physician costs	Medical treatment	
1861	19.8 %	0.7 %	16.9 %	5.5 %	7.6 %	14.9 %	
1867	19.1 %	0.6 %	15.5 %	6.3 %	6.4 %	13.1 %	
1871	22.2 %	0.1 %	17.0 %	8.1 %	7.2 %	13.1 %	
1881	33.0 %	0.7 %	19.3 %	9.9 %	5.7 %	9.9 %	
1891	33.3 %	0.8 %	16.9 %	8.1 %	4.8 %	10.3 %	
1901	32.9 %	0.3 %	14.2 %	6.4 %	5.0 %	9.6 %	
1907	30.3 %	0.2 %	13.1 %	5.3 %	6.2 %	13.5 %	
Year	**Sick pay**	**Funeral benefits**	**Extraordinary Support**	**Education support**	**Administration**	**Miscellaneous Expenditure**	**Aggregate real (nominal) budget in marks**
1861	18.9 %	1.2 %	1.9 %	4.5 %	6.0 %	2.1 %	4 196 275 *(3 021 318)*
1867	15.0 %	1.6 %	1.6 %	13.5 %	4.6 %	12.4 %	6 970 025 *(5 220 549)*
1871	15.6 %	1.7 %	3.5 %	3.6 %	4.6 %	3.1 %	8 386 211 *(6 692 196)*
1881	11.5 %	1.3 %	0.8 %	2.8 %	3.4 %	1.4 %	16 059 812 *(13 152 986)*
1891	16.8 %	1.1 %	0.7 %	0.4 %	2.8 %	3.8 %	28 371 806 *(23 605 343)*
1901	24.9 %	1.2 %	0.8 %	0.1 %	2.7 %	1.8 %	50 904 550 *(42 912 536)*
1907	21.2 %	1.1 %	0.7 %	0.1 %	3.0 %	5.0v	67 392 424 *(62 001 030)*

Sources: See Figure 3.2.

Table 3.20: Percentage distribution of aggregate expenditures among Prussian KVs by type (pension insurance section, 1908–1920)

Year	Invalidity pensions	Widows' pensions	Orphans' pensions	Widows' compensation due to re-marriage	Miscellaneous pension Benefits
1908	61.4 %	24.3 %	3.8 %	0.2 %	0.8 %
1913	62.1 %	23.3 %	5.0 %	0.1 %	0.8 %
1916	54.3 %	25.3 %	8.5 %	0.1 %	0.7 %
1920	27.2 %	15.0 %	6.2 %	0.4 %	10.9 %

Year	Pension refunds to other KVs	Medical treatment	Funeral benefits due to miner's death	Funeral benefits due to dependant's death	
1908	0.0 %	2.7 %	0.6 %	0.1 %	
1913	0.1 %	2.3 %	0.6 %	0.1 %	
1916	0.1 %	1.8 %	0.8 %	0.1 %	
1920	0.0 %	4.6 %	0.5 %	0.2 %	

Year	Refunds to foreigners	Miscellaneous support	Maintenance of buildings	Administration	Arbitration tribunal costs
1908	0.0 %	0.3 %	1.1 %	3.7 %	
1913	0.0 %	0.3 %	0.3 %	4.3 %	
1916	0.3 %	0.3 %	0.4 %	3.8 %	
1920	0.0 %	0.2 %	2.4 %	15.4 %	

Year	Interest on debt	Taxes	Health examination of joiners	Miscellaneous expenditure	Aggregate real (nominal) budget in marks
1908	0.1 %	0.0 %	0.0 %	1.0 %	40 112 174 *(36 702 639)*
1913	0.0 %	0.0 %	0.2 %	0.4 %	43 354 481 *(43 354 481)*
1916	1.6 %	0.0 %	0.1 %	1.8 %	29 927 819 *(53 870 075)*
1920	2.5 %	0.1 %	0.2 %	13.8 %	11 419 390 *(119 218 430)*

Notes: Percentages on the sickness insurance section are in brackets. A value of 0.0 usually implies a percentage of between 0.0 and 0.1 percent.

Sources: See Figure 3.2.

Table 3.21: Percentage distribution of aggregate expenditures among Prussian KVs by type (sickness insurance section, 1908–1920)

Year	Sick pay if stayed at home	Sick pay if stayed at hospital	Women in childbed	Hospital treatment	Physician costs
1908	15.4 %	4.8 %	0.0 %	31.4 %	18.5 %
1913	34.4 %	7.5 %	0.0 %	26.3 %	12.9 %
1916	32.0 %	7.4 %	0.1 %	27.8 %	13.1 %
1920	29.3 %	5.4 %	0.5 %	25.2 %	9.0 %
Year	Miscellaneous medical treatment	Funeral benefits due to miner's death	Funeral benefits due to dependant's death	Extraordinary support	Maintenance of buildings
1908	15.9 %	2.7 %	1.3 %	0.7 %	2.4 %
1913	10.7 %	1.7 %	0.6 %	0.5 %	0.4 %
1916	9.5 %	2.3 %	0.5 %	0.5 %	0.3 %
1920	9.0 %	1.5 %	0.2 %	0.3 %	0.3 %
Year	Administration	Interest on debt	Taxes	Miscellaneous expenditure	Aggregate real (nominal) budget in marks
1908	5.9 %	0.0 %	0.0 %	0.7 %	23 606 565 *(21 600 007)*
1913	4.5 %	0.0 %	0.0 %	0.3 %	39 856 939 *(39 856 939)*
1916	5.5 %	0.6 %	0.0 %	0.4 %	20 610 783 *(37 099 409)*
1920	5.8 %	0.0 %	0.0 %	13.3 %	33 903 449 *(353 952 008)*

Notes: Percentages on the sickness insurance section are in brackets. A value of 0.0 usually implies a percentage of between 0.0 and 0.1 percent.
Sources: See Figure 3.2.

Major baseline trends are as follows. First, the share of invalidity pension expenditures significantly increased from 1861 to the early 1880s, from about 20 to 33 percent, and thereafter remained rather constant. Keeping in mind the break-down of the distribution, the obvious decline in pension costs as a percentage of overall costs was due to survivorship benefits rather than invalidity benefits. The share of widows' pensions declined only after the introduction of Bismarckian invalidity and old-age insurance. The calculations prove that invalidity pensions accounted for the single most important benefit category. Furthermore, the share of physician costs varied between five and eight percent, while the share of costs for medical treatment followed a somewhat U-shaped pattern. Daily sick pay in the aggregate benefit mix, at first, declined from 19 percent to only 11 in 1881, but, due to the regulations concerning Bismarckian health insurance, increased again towards 1907 and then accounted for the second-most important benefit item. In fact, the share of categories such as invalidity pensions for the semi-invalid, funeral pay, extraordinary support, and education support were low. Presumably reflecting economies of scale, the share of aggregate administration costs declined from six percent to around three.

Turning to the pension-insurance section of the post-reform period and neglecting the inflationary year 1920, the shares of the most important benefits – namely invalidity and widows' pensions – were rather stable, while orphans' pensions increased due to war. The picture definitely changes regarding the last observed year, in which the share of invalidity, but also widows' pensions, dropped considerably. Administration costs in the narrow sense were actually higher than before 1907. This holds especially if we consider operating costs in broader terms – that is, taking into account maintenance of buildings, taxes, interest, arbitration tribunal costs, and costs of initial health examinations. Beyond that, regarding the aggregated sickness insurance section, we find that the share of sick pay expenditures increased significantly, but that of healthcare costs (hospital treatment, physician costs, and miscellaneous medical treatment) declined.

After having established the functional division of costs from the perspective of the aggregate, I now assess differences across KVs and deviations from the reference. Therefore, Table 3.22 reports mean shares of the main benefit categories and of operating costs at the KV level. To clarify the variation across KVs, annual standard deviations are provided in brackets. Moreover, in order to make a comparison across the reform possible, I have calculated the share for the years 1908–1920 as though a KV's insurance sections were not unbundled, but were still run as a compound scheme. The annual, cross-sectional mean shares of invalidity pension expenditure were, in fact, a bit lower than the reference shares highlighted for the aggregate. However, this does not hold for survivors' pensions before 1908. In particular, the mean share of operating costs is higher when evaluated across KV-level data than when evaluated in terms of the aggregate.

Table 3.22: Descriptive statistics on the functional division of expenditures on the KV-level (Prussia)

Year	Invalidity pensions	Widows' pensions	Orphans' pensions	Sick pay	Health care	Operating costs
A) Compound insurance period						
1861	17.3 %	15.7 %	4.8 %	17.4 %	30.5 %	10.0 %
	(12.1)	(9.4)	(2.4)	(9.9)	(15.3)	(8.0)
1867	21.8 %	16.9 %	7.0 %	16.5 %	21.5 %	7.8 %
	(23.2)	(11.7)	(4.6)	(10.4)	(12.1)	(5.5)
1871	22.0 %	17.7 %	7.0 %	17.8 %	26.9 %	6.3 %
	(21.0)	(11.3)	(4.3)	(9.8)	(13.3)	(4.4)
1881	26.0 %	20.8 %	7.9 %	13.8 %	22.2 %	5.5 %
	(15.7)	(9.2)	(4.6)	(7.5)	(10.5)	(3.7)
1891	25.8 %	21.3 %	5.3 %	17.0 %	20.9 %	4.9 %
	(14.0)	(12.8)	(3.4)	(10.3)	(8.7)	(4.2)
1901	27.3 %	19.6 %	3.8 %	19.0 %	20.6 %	4.2 %
	(15.2)	(8.8)	(2.3)	(10.0)	(8.9)	(3.1)
1907	26.2 %	18.8 %	3.3 %	19.2 %	23.1 %	4.3 %
	(12.2)	(10.1)	(1.9)	(9.7)	(8.5)	(3.0)
B) Period of separated pension and sickness insurance sections						
1908	25.6 %	18.9 %	2.9 %	24.9 %	21.7 %	5.8 %
	(12.1)	(11.1)	(1.8)	(39.9)	(10.1)	(4.9)
1913	27.7 %	16.2 %	2.7 %	20.1 %	28.9 %	5.1 %
	(13.6)	(9.0)	(1.5)	(9.7)	(32.9)	(2.8)
1916	28.3 %	16.6 %	4.6 %	21.0 %	20.7 %	5.0 %
	(13.6)	(7.7)	(2.4)	(18.0)	(9.1)	(3.3)
1920	10.1 %	6.2 %	1.7 %	25.3 %	32.7 %	7.4 %
	(9.4)	(3.9)	(0.9)	(11.0)	(14.4)	(7.0)

Notes: Annual standard deviations in brackets.
Sources: See Figure 3.2.

3.7.2. Generosity

A crucial issue for any form of charitable or insurance-based welfare institution is per capita generosity, which can be assessed in absolute or relative terms. I will do both in the following – focusing on the major benefits' average value in marks, as well as on their income-replacement properties as measured against earned wages. When I say that I will focus on the "major benefits," I mean the average invalidity and survivorship pensions, as well as daily sick pay per capita. I have decided not to go into depth on less-important benefit items here, such as average funeral pay. As the percentage distribution of expenditures presented above shows, the major benefit categories usually covered 90 percent or more of insurance-benefit expenditures.[103]

I begin by looking at absolute per capita generosity in terms of the KVs' invalidity benefit. Therefore, Figure 3.9 shows the weighted and unweighted mean invalidity pension among Prussian KVs in 1913 prices.

Figure 3.9: Weighted and unweighted real average invalidity pension (in marks) among Prussian KVs

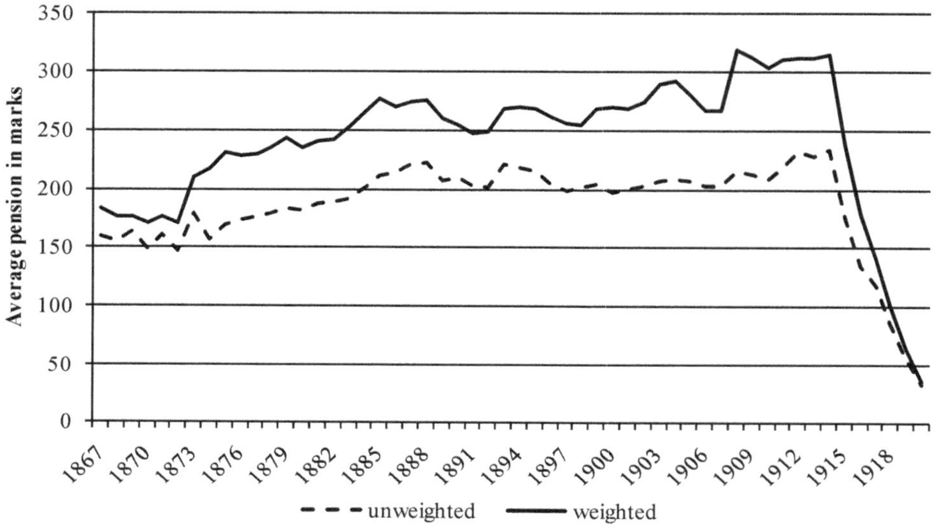

Notes: A KV's share in aggregate contributors is used to weight an observation.
Sources: See Figure 3.2.

The real unweighted pension amounted to a little more than 150 marks in 1867 and climbed to 225 marks at the onset of *Reich* invalidity insurance. Between 1890 and the middle of the First World War, however, generosity stagnated at that level or slightly below it. Pretty clear is that the impact of inflation had set in. In contrast to the unweighted baseline, the KV size-weighted series implies a much higher per capita generosity, between 1874 and 1915, of not less than 50 marks. From the difference in both series, we can conclude that the largest KVs provided the highest absolute pensions. So, in this respect, members in the large and thriving KVs were much better off than mem-

103 In the following, I am not able to distinguish per capita payouts due to whether a miner was an established or an unestablished one. This is because the statistics do not provide such data. I also do not address the differentiation into full and semi-invalidity. The percentage distribution referring to the pre-reform period shows that payouts to semi-invalids were, in fact, negligible from a quantitative viewpoint.

bers of smaller funds – a fact that might signal either differences in preferences or some financing advantage of larger KVs that made a higher per capita generosity possible.

In addition, the widows' and orphans' pensions should be taken into account; in the following, I express the average real widow's pension and the average real orphan's pension as a percentage of the invalidity benefit. Figure 3.10 provides the corresponding mean percentages among KVs over the observation period.

Figure 3.10: Mean survivorship pensions among Prussian KVs as a proportion of the invalidity benefit

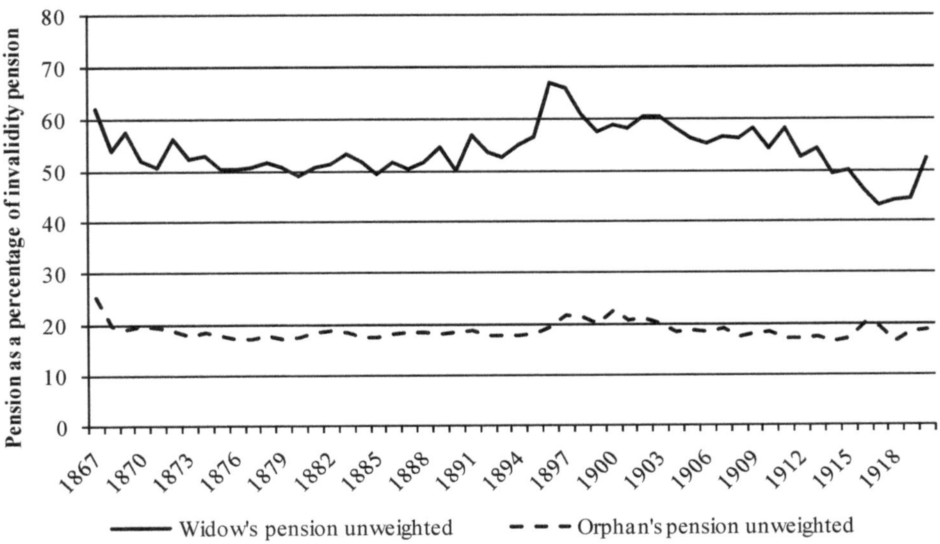

Sources: See Figure 3.2.

While the widow's pension initially amounted to 60 percent or slightly below, it ended up at around 46 to 50 percent. At any rate, besides the war and inflationary years, it generally holds that widows' pensions provided a replacement of between 50 and 60 percent of the breadwinners' invalidity pension. If there were children, the surviving household's pension income would have been raised by 15 to 20 percent per child.

Table 3.23, moreover, rearranges KV-level data on the average invalidity pension by size classes. Comparing evidence by cross-section leads to the conclusion that the average invalidity pension increased with the KVs' size, up to the fifth size class. The average real Bismarckian invalidity pension, in comparison, amounting to between 164 marks, averaged over 1892–1896, and 193 marks, averaged over 1908–1913.[104] So, in

104 See Khoudour-Casteras (2008), p. 237. However, we also have to consider that every Bismarckian invalidity pension contained a *Reich* allowance (*Reichszuschuss*) of 50 marks. What I do not display here in detail is evidence on per capita pensions in Bavaria and Saxony. Besides,

absolute terms, this is broadly equivalent to the median invalidity pension that Prussian KVs granted, meaning that half of all KVs were less generous, while half were more generous.

So far, I have concentrated on pensions in absolute terms. However, as the comparison with the average Bismarckian pension, in particular, suggests, the differences between KV and state pensions might have purely been a matter of differences in wages. Since the mining sector as a whole grew at an above-average rate in terms of production and employment, we would expect to find wages in mining above the national average. Even if the Bismarckian pension were absolutely lower than about half of the KVs' pensions, income replacement rates might not have differed. Table 3.23 also clarifies comparative income-replacement standards within the population of KVs.

Concerning the calculation of income-replacement rates, we have a choice between two approaches: taking either mining-sector wages as a whole or subsector-specific wages as the reference income. While Hoffmann (1965) provides series of the former sort on miners and steelworkers, the *Statistische Mitteilungen über die beim Bergbau Preußens gezahlten Arbeitslöhne und erzielten Arbeitsleistungen* is a statistics framework reporting annual net wage by product and mining area (e.g., hard-coal in the Ruhr, brown-coal in the Harz region, and iron-ore in the Mansfeld mining area).[105] Hoffmann's series has the disadvantage that it ends in 1913 (but cover years back to the 1850s). The *Statistische Mitteilungen*, published in the same journal as the Prussian KV statistics, has the disadvantage, however, that it runs to 1920 in some cases, but does not start before the middle of the 1880s. Weighing both disadvantages, I chose to present baseline calculations of the replacement rate using Hoffmann's series. These are included in Table 3.23 in addition to the absolute average pension in marks.

What we find is that, once again, replacement rates increased over size classes, while the largest KVs do not generally show the comparatively largest rates. Regarding the average level of income replacement, estimates range between a mere 11 percent and close to one-third. The Bismarckian replacement standard, in comparison, ranged between 18 and 20 percent until 1913, which exceeded the rate of many KVs but fell short of what those KVs granted that together represented the majority of miners. So, with respect to the question of whether differences in absolute pensions might simply have reflected wage-level differences, we can say that they did not in this particular context. Rather, differences reflected, in part, a higher "intensity of care" per insurant.[106]

---

the corresponding KV statistics, though, imply an average real invalidity pension among Bavarian KVs of between 141 (1884) and 215 marks (1913), while Saxon KVs granted, for example in 1891, 323 marks and, as of 1913, 301 marks.

105 See Ministerium für öffentliche Arbeiten (1885–1889), Ministerium für Handel und Gewerbe (1890–1922), and Hoffmann (1965).

106 See Holzmann (1988), p. 415–422, for this term.

Table 3.23: The average real invalidity pension, the average real daily sick pay and replacement rates among Prussian KVs by size class (in marks and prices of 1913)

Period	1–199		200–999		1 000–4 999	
	Pension	RR	Pension	RR	Pension	RR
1867–1871	170	17.8%	154	16.4%	154	16.3%
1872–1876	143	13.5%	153	14.8%	160	15.3%
1877–1881	154	17.5%	167	18.9%	199	22.3%
1882–1886	161	16.7%	180	18.2%	226	22.9%
1887–1891	167	15.5%	198	18.4%	222	20.4%
1892–1896	183	15.6%	196	16.6%	215	18.3%
1897–1901	159	12.0%	181	13.8%	207	15.8%
1902–1906	170	13.0%	169	12.7%	221	16.4%
1908–1913	173	12.3%	161	11.3%	246	17.8%
1914–1918	88	–	101	–	156	–
Period	5 000–9 999		10 000–49 999		50 000+	
	Pension	RR	Pension	RR	Pension	RR
1867–1871	162	16.4%	206	21.1%	–	–
1872–1876	180	17.2%	232	21.6%	–	–
1877–1881	209	23.8%	267	30.4%	273	31.5%
1882–1886	226	23.3%	289	29.8%	301	29.0%
1887–1891	202	19.1%	327	30.3%	270	24.0%
1892–1896	207	17.9%	368	31.5%	254	20.8%
1897–1901	155	11.6%	343	25.9%	251	18.8%
1902–1906	192	14.7%	303	23.1%	288	22.9%
1908–1913	231	16.6%	285	20.6%	330	24.3%
1914–1918	159	–	148	–	182	–
Period	1–199		200–999		1 000–4 999	
	Sick pay	RR	Sick pay	RR	Sick pay	RR
1867–1871	0.69	–	0.73	–	0.74	–
1872–1876	0.71	–	0.71	–	0.79	–
1877–1881	0.69	–	0.76	–	0.89	–
1882–1886	0.76	–	0.87	–	1.03	–
1887–1891	0.95	31.6%	1.18	38.9%	1.39	45.1%
1892–1896	1.05	37.0%	1.28	42.0%	1.51	48.6%
1897–1901	1.06	34.7%	1.34	36.0%	1.53	40.0%
1902–1906	0.77	29.3%	1.12	33.5%	1.29	35.2%
1908–1913	1.28	43.3%	1.53	39.4%	1.76	42.7%
1914–1918	0.87	–	0.98	–	1.13	–

Table 3.23 continued

Period	5 000–9 999		10 000–49 999		50 000+	
	Sick pay	RR	Sick pay	RR	Sick pay	RR
1867–1871	0.78	–	0.95	–	–	–
1872–1876	0.86	–	1.02	–	1.18	–
1877–1881	1.02	–	1.12	–	1.19	–
1882–1886	1.07	–	1.15	–	1.17	–
1887–1891	1.22	40.0%	1.49	46.7%	1.34	38.5%
1892–1896	1.33	42.2%	1.60	46.6%	1.69	44.6%
1897–1901	1.46	37.5%	1.59	44.8%	1.92	43.8%
1902–1906	1.55	39.5%	1.53	41.9%	1.32	32.0%
1908–1913	2.08	48.6%	1.92	51.3%	1.93	44.0%
1914–1918	1.40	–	1.35	–	1.26	–

Notes: "RR" abbreviates "replacement rate." Hoffmann's wages on the mining sector as a whole are used regarding pensions. Wages on the economy as a whole are taken from Khoudour-Castéras. Subsector-specific shift wages are used are used regarding sick pay.
Sources: My own calculations; for the data, see Figure 3.2 and Hoffmann (1965), pp. 599–601, Sommariva and Tullio (1987), p. 231–232, and Khoudour-Castéras (2008), pp. 236–238.

The income-replacement rates displayed are definitely not comparable to rates that many present-day schemes, especially in developed countries, provide. Depending on the relative income position, the German system granted, around 1992, a net replacement rate of between 53 and 77 per cent.[107] That observation can be explained by the fact that the concept of an individual's lifetime as split into three stages – youth, employment, and retirement – had not really broken through in the nineteenth century, not even with Bismarckian old-age insurance. So, nineteenth- and early twentieth-century contemporaries may not have identified a pension as self-standing retirement income (because gainful employment during invalidity and family insurance had still played a major role) and, hence, had not advocated high replacement rates.[108] Invalidity pensions were arguably perceived, instead, as "allowance to subsistence" (*Zuschuss zum Lebensunterhalt*). Yet, for the time, receiving a retirement income equaling a replacement rate of 25 percent or more seems not to have been a bad deal.

After having briefly reviewed the pension benefits, I now turn to daily sick pay per capita as one of the two important sickness-related benefit categories. Before assessing the KVs' generosity in this respect, let us have a look at the number of sick days that

107 See Börsch-Supan (1992), p. 539.

108 See Conrad (1991), p. 171, Hardach (2003), pp. 5–7, and Kaschke (2000), pp. 349–352. Kaschke, in particular, argues against a one-dimensional approach towards measuring the welfare-enhancing property of the Bismarckian invalidity pension by just looking at economic replacement rates. He argues that they do not adequately address the complexity of the issue. I certainly do not deny that replacement rates are a straightforward tool. They are a tool, however, that is economically justified, especially in the context of the PAYG method, where the replacement rate (commonly referred to as pension level in this context) is one major determinant of the contribution rate.

had to be financed.[109] Figure 3.11, therefore, introduces mean and median sick days per working member. I will refer to this variable subsequently as the sick-days-to-contributors ratio (SDCR), which is equal to the number of paid sick days per contributor (and an analogue to the PCR). The mean ratio fluctuated between six and eight sick days per contributor until about 1903, but increased thereafter, to about eight days between 1904 and the beginning of the First World War, and up to 13 days at the end of war.

Figure 3.11: Sick days per contributor among Prussian KVs

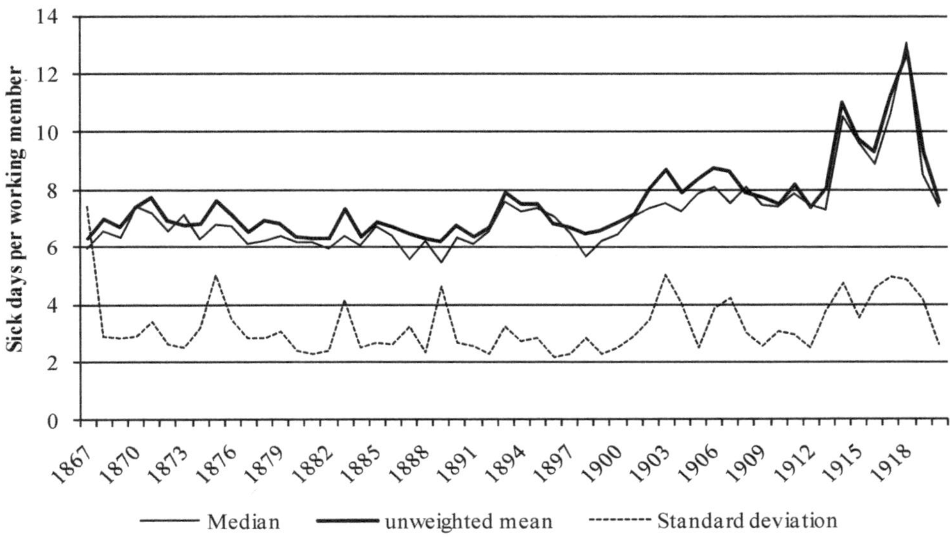

Sources: See Figure 3.2.

From the Prussian KV statistics we can gather that median sick pay in the population of funds was initially about 0.8 marks, but increased to 1.3 marks around 1888 and 2.0 marks in 1913. In comparison, and according to Frerich and Frey (1993), average sick pay in Bismarckian health insurance amounted to not less than one mark nominal in 1885, or 1.29 marks in 1913 prices. Compared to what KVs granted at the median – i.e., one mark – the generosity standard on the national level was indeed higher (about one quarter). For 1910, Frerich and Frey (1993) report an average real amount of sick pay associated with German worker insurance of 1.86 marks per day. KVs actually appear to have, at the median, caught up to this benefit level by that year.[110] In all, available

[109] With regard to Guinnane and Streb (2011) and Murray (2003), sick claims such reflect a mixture of claims related to having been ill and of claims made to take some paid leisure time without being ill at all. The latter might be called malingering or simulating, and definitely refers to incentives for moral hazard.

[110] See Frerich and Frey (1993), p. 102.

data confirm that KVs visibly raised their sick pay benefit as a reaction to the requirement of the 1883 *Reich* Health Insurance Law.

As with pensions, in the following analysis, I assess real sick pay per day on leave by size classes, as well. Therefore, let us turn once again to Table 3.23 that also shows cross-sectional averages of real sick pay for several five-year periods and replacement rates, which reflect the percentage of daily shift wages replaced by the sick-pay benefit. Daily shift wages were taken from the *Statistische Mitteilungen über die beim Bergbau Preußens gezahlten Arbeitslöhne und erzielten Arbeitsleistungen*, introduced above. Reported series are organized analogously to annual wages – i.e., by certain subsectors and, within them, by mining areas. The series have the same deficiency as the annual series: They are reported only as early as the late 1880s, which is why replacement data are lacking for the first four sub-periods displayed.

We can actually observe developments similar to those regarding pension generosity: Sick pay increased from small to large KVs, taking a cross-sectional view. While the average small KV (no more than 199 contributors) initially paid 0.69 marks per sick day, the average large KV (more than 10 000 contributors) paid 0.95 marks. That difference of slightly more than 25 percent even increased over time possibly reflecting the positive effect of secular economic growth on generosity in certain mining areas. Turning to the income-replacement standard of daily sick pay, KVs did not – on average, let it be understood – grant the 50 percent that contemporary observers envisaged.[111]

Finally, let us briefly turn to healthcare expenditures which arose for such services as doctors' visits, drugs, and stays at a health spa or hospital (if a KV owned one). Here, median expenditures draw a quite appropriate picture of per capita generosity. So, evaluated at the median, KVs began in 1867 by spending around seven marks per worker for those services mentioned; this amount increased to 20 marks in 1913, but fell to around 10 marks towards 1920.

3.8. The revenue side

3.8.1. Revenue structure

As in the previous subchapter, I assess the percentage distribution of aggregate revenues first. For the years 1908 to 1920, "aggregate" again refers to the sum of the pension and sickness insurance sections. The overall scale of revenues clearly followed aggregate expenditures, although budgets were not necessarily perfectly balanced, as we will see below. Revenue quantities fall into the four major categories "social insurance contribu-

111 Note that I implicitly assumed that a miner would have worked exactly one shift per day. This was usually between seven and ten hours long, and its length depended on the subsector and mining area.

tions," "case-specific fees" (due to joining the KV or as a disciplinary means), "capital income," and "miscellaneous revenues." In fact, contributions paid by the working membership and employers accounted for the overwhelming majority of revenues, usually between 80 and 90 percent. The importance of fees as a financing tool was, from the start, more than limited since they never accounted for more than two percent of aggregate revenues. Moreover, the share of capital income, which was mostly equal to interest on assets, was quite constant until 1906/07, at six to eight percent. With the reform of 1906, however, the relative importance of interest income regarding the aggregate increased; this is clearly due to the fact that KVs were to raise their reserves to meet the new liability-coverage requirement. Income that falls into the category "miscellaneous" reflected rather irregular revenue due to, for example, donations or extraordinary sales of assets; the contribution of miscellaneous revenues steadily declined towards 1920.

Tables 3.24 and 3.25 display a detailed percentage distribution of revenue sources. With regard to the compound insurance period, the miners' contributions' share was pretty stable, at between 45 and 50 percent. In contrast, employers' contributions ranged below these percentages indicating that both parties did not contribute equally, looking at the aggregate. Furthermore, with regard to the post-reform period, contributions continuously made up the single most important fraction of revenue. Apparently, this holds even more for the aggregate sickness-insurance section.

Finally, information depicted in Table 3.26 helps us to assess the relative importance of revenues sources at the KV-level. As with the functional division of expenditures, displayed mean percentages do not necessarily match the reference percentages obtained by looking at the aggregate. Regarding the standard deviation for interest income, it becomes clear that this source of income may have played quite an important role for the one KV or another. In fact, interest income as a share of total revenues tended to be higher for smaller KVs. Besides, the mean percentage share of employers' contributions measured at the KV-level is lower than the share measured at the aggregate; this points to some outliers in the data.

Table 3.24: Percentage distribution of revenues among Prussian KVs by type (compound insurance, 1861–1907)

Year	Established miners' contributions	Unestablished miners' contributions	Contributions of the suspended	Contributions of the sick	Employers' contributions
1861	49.3 %		0.0 %	0.0 %	38.9 %
1867	25.4 %	17.5 %	0.3 %	0.4 %	34.2 %
1871	28.3 %	17.7 %	0.6 %	0.1 %	35.3 %
1881	30.5 %	13.9 %	0.3 %	0.1 %	41.1 %
1891	48.6 %		0.0 %	0.0 %	42.1 %
1901	49.9 %		0.0 %	0.0 %	41.4 %
1907	47.7 %		0.0 %	0.0 %	40.5 %
Year	**Initiation fees**	**Deductions due to wage increases**	**Punishment fees**	**Interest**	**Rent**
1861	0.3 %	0.0 %	1.0 %	9.2 %	0.0 %
1867	0.7 %	0.0 %	1.0 %	5.9 %	0.3 %
1871	0.7 %	0.1 %	0.6 %	7.0 %	0.4 %
1881	0.4 %	0.0 %	0.6 %	5.0 %	0.3 %
1891	0.7 %	0.0 %	0.4 %	4.2 %	0.2 %
1901	0.2 %	0.0 %	0.3 %	5.5 %	0.1 %
1907	0.3 %	0.0 %	0.1 %	6.3 %	0.3 %
Year	**Miscellaneous revenues**	**Extraordinary sales**	**Donations**	**Aggregate real (and nominal) revenues in marks**	
1861	1.3 %	0.0 %	0.0 %	4 461 078 *(3 243 204)*	
1867	1.7 %	12.4 %	0.1 %	7 580 656 *(5 677 911)*	
1871	2.9 %	6.1 %	0.1 %	9 075 237 *(7 242 039)*	
1881	1.5 %	6.1 %	0.2 %	18 301 283 *(14 988 751)*	
1891	1.3 %	2.3 %	0.0 %	34 823 663 *(28 973 288)*	
1901	0.4 %	2.2 %	0.0 %	64 419 318 *(54 305 485)*	
1907	0.6 %	4.1 %	0.0 %	86 794 638 *(79 851 067)*	

Notes: A value of 0.0 usually implies a percentage of below 0.1 percent.
Sources: See Figure 3.2.

Table 3.25: Percentage distribution of revenues among Prussian KVs by type and insurance section, 1908–1920

Year	Contributions	Initiation fees	Acknowledge-ment fees	Interest on fixed interest-bearing papers	Interest on bank deposits
1908	90.0 %	0.2 %	0.1 %	7.7 %	0.3 %
1913	82.4 %	0.1 %	0.1 %	12.4 %	2.6 %
1916	70.9 %	0.1 %	0.1 %	25.2 %	0.2 %
1920	83.5 %	0.1 %	0.1 %	9.2 %	0.4 %

Year	Rent	Net revenues from own works	Miscellaneous revenues	Aggregate real (and nominal) revenuesin marks
1908	0.5 %	0.0 %	1.1 %	74 279 791 *(67 966 009)*
1913	0.4 %	0.3 %	1.6 %	80 758 796 *(80 758 796)*
1916	0.5 %	0.2 %	2.7 %	34 375 262 *(61 875 472)*
1920	0.1 %	0.1 %	6.5 %	24 906 209 *(260 020 820)*

Year	Contributions	Initiation fees	Acknowledge-ment fees	Interest on fixed interest-bearing papers	Interest on bank deposits
1908	96.5 %	0.0 %	0.3 %	0.1 %	0.0 %
1913	94.8 %	0.0 %	1.7 %	0.1 %	0.3 %
1916	92.3 %	0.0 %	3.8 %	0.1 %	0.4 %
1920	98.1 %	0.0 %	0.6 %	0.1 %	0.0 %

Year	Rent	Net revenues from own works	Miscellaneous revenues	Aggregate real (and nominal) revenues in marks
1908	0.0 %	2.4 %	0.6 %	40 231 965 *(36 812 248)*
1913	0.0 %	2.1 %	0.9 %	4 7439 988 *(47 439 988)*
1916	0.0 %	2.4 %	1.0 %	22 502 310 *(40 504 159)*
1920	0.0 %	0.8 %	0.2 %	48 283 146 *(504 076 042)*

Notes: A value of 0.0 usually implies a percentage that is actually between 0.0 and 0.1 percent.
Sources: See Figure 3.2.

Table 3.26: Descriptive statistics on the major sources of revenues on the KV-level (Prussia)

Year	Miners' contributions	Employers' contributions	Interest income
A) Compound insurance period			
1861	59.2 % (9.4)	35.3 % (7.6)	12.7 % (29.1)
1867	47.3 % (13.6)	31.0 % (10.3)	10.9 % (9.5)
1871	44.4 % (14.5)	31.8 % (12.0)	12.9 % (13.2)
1881	44.3 % (12.9)	33.8 % (12.4)	11.8 % (9.8)
1891	45.6 % (10.8)	35.9 % (13.2)	11.4 % (9.5)
1901	46.5 % (9.9)	36.4 % (9.0)	11.5 % (11.0)
1907	45.1 % (10.0)	37.9 % (9.0)	12.6 % (13.1)
B) Period of separated pension and sickness insurance sections			
1908	42.8 % (8.7)	42.8 % (8.7)	11.2 % (10.8)
1913	42.0 % (5.8)	42.0 % (5.8)	14.2 % (10.6)
1916	37.5 % (6.6)	37.5 % (6.6)	23.7 % (28.3)
1920	43.0 % (7.3)	43.0 % (7.3)	5.2 % (3.4)

Notes: Annual standard deviations in brackets.
Sources: See Figure 3.2.

3.8.2. Social insurance contributions and assets

The single most important source of revenues, by far, was social insurance contributions paid by workers and mine owners. Before 1908, however, they were not necessarily paid on equal terms. In the following, I always focus – if not explicitly stated otherwise – on the *total* social insurance contribution per capita, which consists of the payments of both miners and employers. Table 3.27 gives some indication of the real amount of that average total contribution among Prussian KVs. Displayed by periods are the (unweighted) mean, the corresponding standard deviation, and the median. Data show that the average contribution initially amounted to some 36 marks per year per miner and then rose steadily to nearly 70 marks in 1902/1907.[112] After 1907 the average contribution is reported separately for the pension- and sickness-insurance sections. In the latter section, a lower per capita contribution was charged equal to the initial level of the compound contribution payment of 35 marks. However, in real terms, per capita payments dropped visibly due to price inflation.

[112] While the unweighted and KV size-weighted mean average contributions were not much different over the first two decades (not more than seven to eight marks), the weighted average contribution increased since (1888: ≈ 77 marks) and notably exceeded the unweighted estimate as of 1907 (with about 103 marks). This differential has to be considered as showing that the large KVs raised their contributions per capita significantly faster in relative to smaller funds.

Table 3.27: Real annual per capita contributions among Prussian KVs

Period	Total contribution per capita in marks			Ratio of unestablished miners' to established miners' contributions			Employers' share in total contributions (in percent)		
	Mean	Med.	SD	Mean	Med.	SD	Mean	Med.	SD
1861–66	35.9	24.7	29.5	0.57	0.55	0.24	40.7	40.4	11.4
1867–71	36.5	24.7	35.8	0.56	0.54	0.27	40.7	39.2	12.5
1872–76	42.0	29.2	55.5	0.60	0.61	0.27	41.9	42.6	12.2
1877–81	47.5	35.5	56.8	0.63	0.63	0.24	42.8	42.8	11.4
1882–86	52.1	42.0	43.7	0.60	0.58	0.31	44.7	47.3	10.9
1887–91	60.3	49.5	43.2	*0.67*	*0.62*	*0.48*	43.0	42.8	9.2
1892–96	65.1	55.2	45.0	–	–	–	43.3	42.8	9.5
1897–01	64.7	57.7	34.6	–	–	–	43.8	42.9	8.6
1902–07	69.3	63.5	34.0	–	–	–	44.6	44.9	8.0
1908–13	65.2	61.0	35.7	–	–	–	50	–	–
	(35.1)	*(34.6)*	*(11.0)*				*(100)*		
1914–18	33.0	25.9	24.8	–	–	–	50	–	–
	(28.5)	*(25.0)*	*(15.3)*				*(100)*		
1919–20	17.3	15.7	11.5	–	–	–	50	–	–
	(24.5)	*(23.5)*	*(9.7)*				*(100)*		

Notes: "Med." is "median." Before 1906/07, the contribution payment refers to the compound-insurance period and, thereafter, to the pension-insurance section, with information on the sickness-insurance section in brackets. Regarding the period 1887–1891, the ratio of unestablished miners' and established ones' contributions only refers to years 1887 and 1888; after 1888, unestablished miners contributions were no longer reported in the KV statistics.

Sources: See Figure 3.2.

The table above also reports information on the ratio of the unestablished miners' average contribution to the established miners' per capita payment. Unfortunately, the Prussian KV statistics do not report the relevant numbers for the years after 1888. So, the impression we get might not hold for the second half of the pre-reform period. However, both the mean and median ratios were quite constant over time, implying that unestablished members, indeed, paid around 40 percent less. This holds for, at least, 75 percent of KVs observed in the respective year (not displayed in the table); in addition, at look at the maximum values (likewise not displayed) points to the fact that a few had sometimes burdened unestablished miners more heavily (up to 400 percent!) than their established co-workers and particularly heavily with regard to the lower generosity they generally experienced. As the bottom line, we can, nonetheless, confirm that for the period 1861–1888, unestablished miners paid less per capita in absolute terms.

Beyond that, if employers did not necessarily contribute to the same extent as employees, exactly how big was the difference? Table 3.27 answers this question, too. As can be seen, the mean financing share – employers' contributions referenced against total contributions – amounted to 40 percent in 1861/66, and this share climbed somewhat thereafter and then remained at a level of about 43 or 44 percent until close to the reform. The series of annual minimum and maximum values (not displayed in the table) implies that there was some variation, which was, however, always induced by few KVs that might have been in some extraordinary state of operations. So, according to the data, the *Cottenheim KV* and the *Mayen KV*, for example, two KVs responsible for workers in stone pits, saw a mean employer's share in total contributions of four percent and respectively seven percent over 1861–1884. In the opposite direction, annual social insurance contributions to the *Neusalzwerker KV* consisted, on average, of almost 85 percent employer's contributions over 1869–1907.

Although the majority of KVs is said to have usually specified a flat-rate payment per member, sometimes being differentiated in different classes according to length of service or occupation (see Chapter 2), it might be of interest how small or large the average annual contribution payment was compared to annual wage – thus, how much of the gross wage was effectively consumed by the miners' social insurance. At this point, we face the same problem of choice as with income-replacement rates: Which wage series should we apply? In the following, I display evidence derived on both the basis of Hoffmann's series on the mining sector and steel production and of subsector-specific wages. This is done for two concepts of contribution rates. The first concept is "historical contribution rates," which are simply derived by forming the ratio of per capita contributions – to be estimated from the KV statistics – and gross wage. The second concept is slightly different and is inspired by Disney's approach: "equilibrium contribution rates."[113] While the former approach simply asks how we can express the *observable* average contribution as a percentage of wage, the latter asks of what magnitude the contribution rate would have had to be if a KV had targeted a perfect financing equilibrium – or, put slightly differently, if KVs had not accumulated reserves.

Calculating such equilibrium contribution rates requires some straightforward modeling of the PAYG equation that KVs' used. In the absence of reserves, let the budget equation that KV i was to maintain in period t be of the following form:

$$(3.12)\ CR_{it} * AW_{it} * N_{it} + MR_{it} = P_{it} * ATP_{it} + N_{it} * ASPD_{it} * OSD + MC_{it}\,,$$

with: CR = contribution rate; AW = average annual wage; N = contributors; MR = miscellaneous revenues; P = pensioners (as the sum over invalids, widows, and orphans); ATP = average total pension (i.e., pension expenditure divided by pensioners); $ASPD$ = average sick pay per day; OSD = sick days; MC = miscellaneous costs.

[113] See Disney (2004), p. 274.

The budget equation implies full balance of revenues and expenditures. For the sake of reducing complexity, there are only two revenue sources: contributions and miscellaneous revenues, capturing, among other things, fees and capital income. The right-hand side accounts explicitly for expenditures on pensions and sick pay, and miscellaneous costs include residual cost items such as healthcare benefits, miscellaneous benefits and operating costs. Re-arranging expression (3.12) for CR_{it} yields:

$$(3.13)\ CR_{it} = \frac{P_{it}}{N_{it}} * \frac{ATP_{it}}{AW_{it}} + \frac{SD_{it}}{N_{it}} * \frac{ASPD_{it}}{AW_{it}} + \frac{1}{N_{it}} * \frac{(MC_{it} - MR_{it})}{AW_{it}}.$$

The first three right-hand-side quantities are already known: the PCR, the pension level, and the sick days-to-contributors ratio. The fourth quantity reflects the income-replacement standard of sick pay as referenced against *annual* wage. Finally, the product of the last two quantities accounts for net miscellaneous costs. If miscellaneous revenues exceeded miscellaneous costs, the contribution rate would be lowered. All right-hand-side quantities can be calculated from the KV statistics' data in combination with wage data, so that the equilibrium contribution rate results, as usual, in a residual.

Let us focus first on the historical contribution rate. Figure 3.12 depicts, on the one hand, the unweighted mean average contribution among Prussian KVs and, on the other hand, the size-weighted mean.

Figure 3.12: Historical contribution rates among Prussian KVs

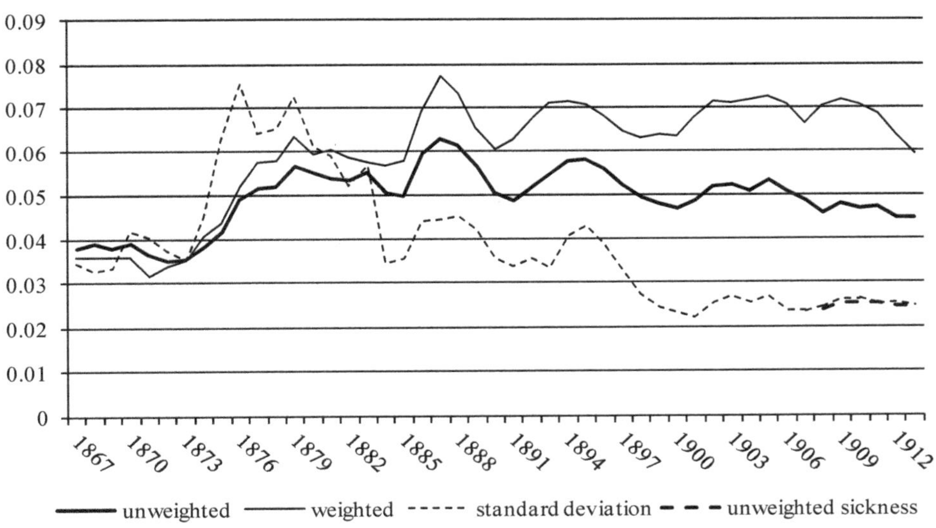

Notes: Hoffmann's wage series on the mining sector and steel production are used. The series simply named "unweighted" and "weighted" refers, after 1907, exclusively to the pension-insurance section. It is weighted with KV size.

Sources: My own calculations; for the data, see Figure 3.2 and Hoffmann (1965), pp. 599–601.

The wage series applied relates to the mining sector as a whole. Evidently, the unweighted historical mean contribution rate was close to four percent as of 1867, and it climbed to approximately 6.3 percent in 1887. Thereafter, it declined towards 4.8 percent in 1907 and remained, for the pension insurance section, quite constant until 1913; the mean contribution rate among KVs with respect to the sickness-insurance section was likewise constant at around 2.5 percent. The weighted mean contribution rate was higher as of 1887 and fluctuated between six and 7.8 percent; however, the trend was constant. In comparison to KVs and according to Hentschel (1978), the total contribution rate in Bismarckian social insurance amounted to seven percent in 1913; of that seven percent, two percent were linked with invalidity and old-age insurance and 3.5 percent with health insurance.[114]

Table 3.28, in addition, depicts the historical mean contribution rate for the post-reform period using subsector-specific annual wages from the *Statistische Mitteilungen über die beim Bergbau Preußens gezahlten Arbeitslöhne und erzielten Arbeitsleistungen*. An impression of the difference in the average contribution rates due to choice of wage scenario is given through the value for 1908. Using subsector-specific wages, the mean contribution rate in that year was about 0.5 percent higher for the pension-insurance section, but similar for the sickness-insurance section; the annual mean difference between the two wage scenarios for all years for which observations are available, however, was roughly that 0.5 percent.

Table 3.28: Mean contribution rates for the pension- and sickness-insurance sections using subsector-specific wages (Prussian KVs)

Periods	Pension insurance section		Sickness insurance section	
	Mean	(SD)	Mean	(SD)
1908	4.99 %	(2.77)	2.57 %	(0.86)
1908–1913	5.14 %	(3.96)	2.65 %	(0.84)
1914–1918	3.55 %	(1.84)	2.87 %	(0.97)
1919–1920	2.23 %	(2.20)	2.93 %	(0.90)

Notes: Displayed is the unweighted mean. "SD" abbreviates "standard deviation."
Sources: See Figure 3.2.

Furthermore, while we see a decline in the mean contribution rate in the KVs' pension-insurance section, we also see that the mean contribution rate stayed, overall, constant in the KVs' sickness insurance. We cannot verify that the overall mean contribution rate charged on the average working miner and his employer was visibly lower at the time than for many of today's systems. For this has to be seen against the background that the replacement standards of typical insurance benefits, pensions and daily sick pay

[114] See Hentschel (1978), p. 348. The rest was linked with accident insurance.

leading the way, were *de facto* lower. Conducting a simple thought experiment, if KVs had granted an average pension level twice as high as they historically did (≈ 50–60 percent) and 100-percent daily sick pay, the mean contribution rate could well have approached levels of 25 percent and higher. Thus, in this hypothetical scenario, KVs would have likely matched today's German social insurance system's parameters. Besides, we would have to consider the contribution rates linked with Bismarckian health and invalidity and old-age insurance in addition to the KVs' rates.

Referring now to equilibrium contribution rates, Table 3.29 reports, separately for the pre- and post-reform periods, information on the deviation of what was charged according to the business reports and should have been charged to ensure equilibrium. Let CR^H and CR^E denote the historical contribution rate and the equilibrium rate. Column two refers to the case in which CR^H exceeded CR^E.

Table 3.29: Historical and equilibrium contribution rates compared (Prussian KVs)

Year	Highest deviation given $CR^H > CR^E$	Median deviation	Highest deviation given $CR^H < CR^E$	Number of KVs displaying a deviation of $-0.1\% < (CR^H - CR^E) < 0.1\%$
A) Compound insurance period				
1867	-16.50 %	0.50 %	5.67 %	10 (out of 85)
1871	-2.40 %	0.40 %	5.41 %	13 (out of 91)
1876	-9.61 %	0.38 %	4.79 %	13 (out of 87)
1881	-1.84 %	0.67 %	13.75 %	4 (out of 83)
1886	-3.57 %	0.66 %	7.89 %	10 (out of 75)
1891	-2.46 %	0.54 %	9.06 %	9 (out of 74)
1896	-0.63 %	0.72 %	9.90 %	5 (out of 73)
1901	-0.80 %	0.81 %	6.10 %	8 (out of 73)
1906	-1.79 %	0.78 %	6.00 %	8 (out of 72)
B) Pension insurance section				
1908	-6.02 %	1.83 %	7.40 %	5 (out of 68)
1913	-0.27 %	2.15 %	7.98 %	3 (out of 62)
C) Sickness insurance section				
1908	-2.28 %	0.02 %	1.24 %	18 (out of 68)
1913	-15.43 %	0.13 %	2.05 %	9 (out of 62)

Sources: See Figure 3.2.

As we can see, the highest deviations in that direction occurred in the first decade. This deviation could have occurred only because KVs charged a higher contribution rate than was technically required to maintain equilibrium. Conversely, CR^E had, in some cases, definitely exceeded CR^H. This observation highlights that KVs drew on financial reserves accumulated in the past to keep contributions constant – at least for some time. So, besides having been an actuarial buffer, reserves accumulated in the past installed some room for fiscal policy – some room that was obviously used. Evaluated at the median, this gave many KVs in some years the opportunity to charge a contribution rate that was between roughly 0.3 and one percent lower than it would have had to be without the backing of reserves. However, it should be clear that such a strategy was discretionary in nature depending on the state of reserves. In fact, looking at the underlying data on the deviation of CR^H from CR^E reveals that it was positive in the majority of observed KV years, meaning that reserves were more regularly accumulated than they were reduced. Thus, the dominant strategy among funds, whether small or large, was to fuel reserves and to implicitly specify more-than-sufficient contribution rates.

This strategy becomes exemplarily clear in Figure 3.13, which shows the development of the historical contribution rate, the equilibrium contribution rate and real monetary reserves for the *Märkische KV* and, respectively, the *Allgemeine KV Bochum*, into which the former was merged in 1890 with two other funds; both funds were the largest of their time.

Figure 3.13: Historical and equilibrium contribution rates compared: the *Märkischer KV* (1867–1889) and the *Allgemeiner KV Bochum* (1890–1913)

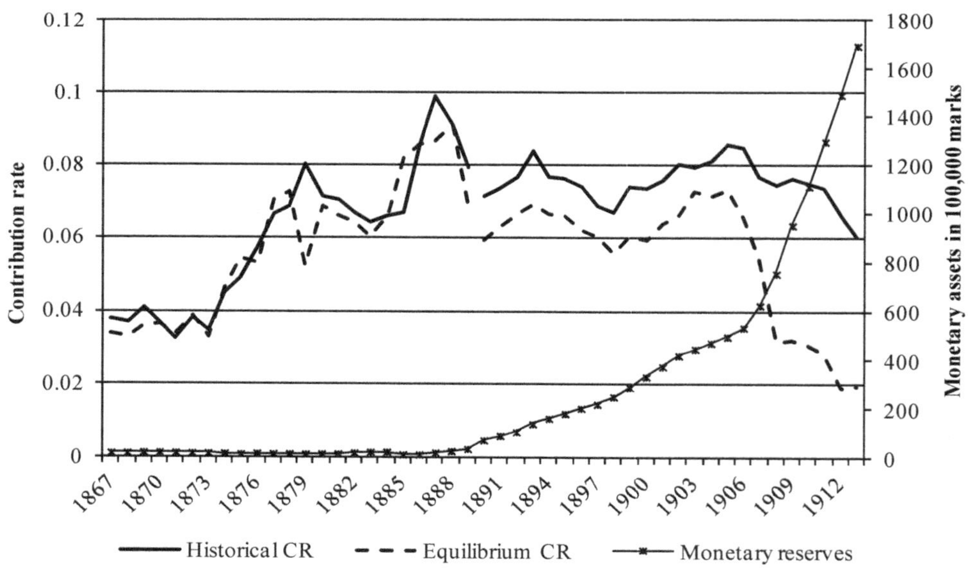

Notes: Monetary reserves are displayed in 1913 prices and exclude immovable and movable property.

Sources: See Figure 3.2.

While the *Märkische KV* apparently did not specify the average contribution payment in a way that fueled reserves to a great extent, the *Allgemeine KV* did. It could have actually charged a rate that was at least one percent lower. Turning to the years after 1907, the primary difference between CR^H and CR^E was due to the reform law of 1906.

In order to explore the KVs' reserves or, respectively, assets, Table 3.30 depicts some descriptive statistics on them and, thereby, illustrates asset structure and the magnitude of assets ready to settle unpredicted extra costs. Cash reserves and bank deposits (the latter category was introduced by 1908) made up only a minor fraction of reserves at the mean; their mean share in all assets usually was between five and six percent before 1907, and even lower, around two to three percent, thereafter in the pension-insurance section. Interest-bearing assets – or interest-bearing "papers," as this item became known in 1908 – accounted for the highest percentage. While the mean amount of such papers amounted to about 127 000 marks (in 1913 prices), it climbed enormously to over two million marks in 1907, and even beyond that sum during the post-reform period.

What is called miscellaneous assets comprises different sorts of items, such as immovable property, movables, interest-free claims and outstandings. Note that these items were not always reported by all KVs; many, for example, did not report immovable property.[115] Starting in 1908, the KV statistics also reported deposits in the reinsurance fund, the *Knappschaftliche Rückversicherungsanstalt Charlottenburg a. G.* created in 1907 and intended to back a larger amount of pension liabilities. While immediately after 1906/1907, the percentage share of reinsurance deposits was close to one quarter, the share climbed to a remarkable 50 percent in 1920. However, one should bear in mind that the reinsurance fund did essentially the same with the KVs' money as the KVs would have done – namely, to invest in interest-bearing papers of various kind (see Chapter 2). The increasing percentage rather indicates the extent of out-sourcing of investment-decisions more than it does the extent to which a different investment object was addressed. Two observations must, furthermore, be highlighted: First, the mean-median difference – or, alternatively, the standard deviation – underlines the very unequal asset distribution among Prussian KVs; and, second, a KV's assets were overwhelmingly bound in the pension-insurance section.

As of 1913, mean cash reserves in the sickness-insurance section amounted to about 45 000 marks in 1913 prices; the mean amount of interest-bearing papers was only 444 000 marks, thus much lower than for the pension-insurance section. Of course, this difference stems from the facts that pension insurance was likely to see larger fluctuations in expenditures, and part of implicit pension liabilities had to be pre-funded.

115 KVs that did not own their own buildings rented office space or must have had some at their disposable for free at firms' facilities.

Table 3.30: Descriptive statistics of the Prussian KVs' real assets (in marks)

Year	Mean (in marks)	Mean share in total assets	Median (in marks)	Standard deviation
A) Cash reserves and bank deposits				
1867	6 187	5.3 %	581	20 263
1887	37.145	6.2 %	2 018	129 478
1907	172 892	5.1 %	6 586	815 729
1913 (pension section)	67 800	1.8 %	2 366	188 367
1920 (pension section)	31 479	2.9 %	725	104 952
B) Interest-bearing assets (papers)				
1867	127 112	88.8 %	39 352	306 462
1887	379.487	89.6 %	128.880	784.456
1907	2 076 722	89.7 %	247 423	7 554 674
1913 (pension section)	3 911 081	73.8 %	278 963	16 174 854
1920 (pension section)	831 840	43.8 %	67 055	2 817 916
C) Deposits at re-insurance fund				
1913 (pension section)	1 606 092	23.9 %	13 911	8 024 084
1920 (pension section)	629 800	50.9 %	142 145	1 638 987
D) Miscellaneous assets				
1867	34 237	5.8 %	721	166 557
1887	82 595	4.2 %	654	377 774
1907	298 509	5.1 %	1 090	1 239 380
1913 (pension section)	3 257	0.1 %	0	25 412
1920 (pension section)	476	0.1 %	0	2 861

Notes: Miscellaneous assets include immovable property, movables , interest-free claims and outstandings. Figures are in current prices.

Sources: See Figure 3.2.

Contemporary observers of the KVs believed that long-term implicit liabilities arising from pension provision were not adequately covered by the funds' PAYG operations. How could the "ability of monetary reserves to cover liabilities" be quantitatively assessed? One illustrative way is to calculate a counterfactual asking: The pension expenditure of how many years would the capital reserves at the beginning of year *t* have helped to finance if no other revenues occurred? Figure 3.14 shows the results of the calculation of that hypothetical range.

Figure 3.14: Average counterfactual range of capital reserves to finance pension liabilities (Prussian KVs)

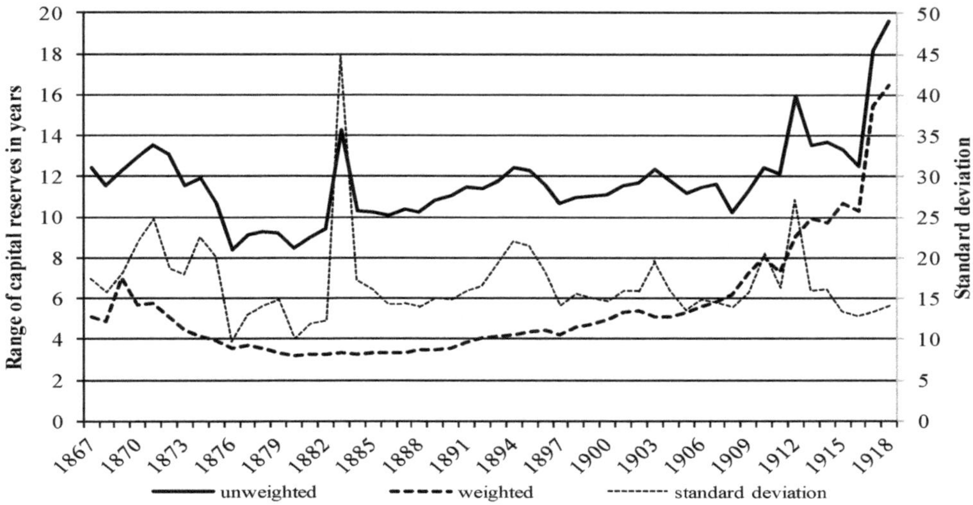

Notes: The range is calculated as (capital reserves at the end of year t/overall pension liabilities in year t). Weighted is with KV size.
Sources: See Figure 3.2.

Figure 3.15: Average counterfactual range of capital reserves by growth pattern (Prussian KVs)

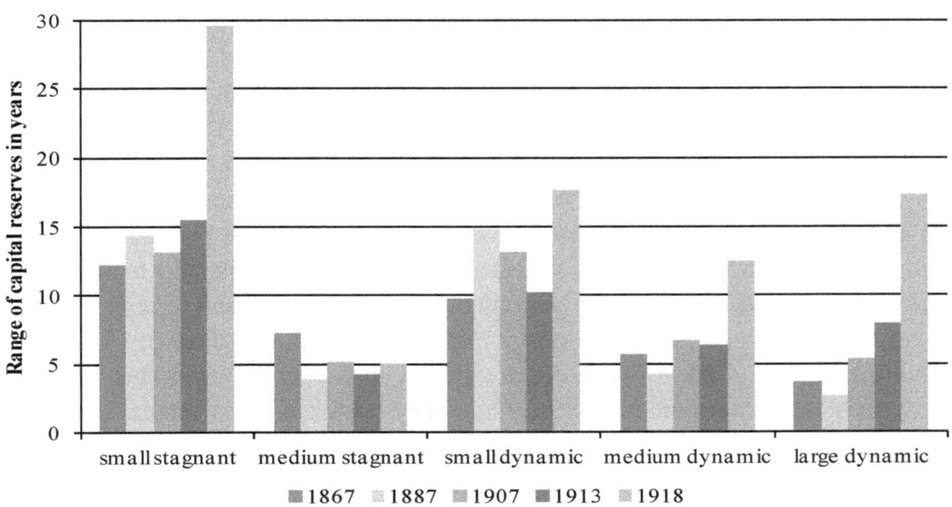

Notes: The range is calculated as [(capital reserves at the end of year t)/(overall pension liabilities in year t)]. The range is KV size-weighted.
Sources: See Figure 3.2.

Capital reserves were assumingly used exclusively to pre-fund constant future pension liabilities in the absence of regular revenues. The figure displays that range in the form of an unweighted and size-weighted mean. The difference between both series underlines that the smaller funds in particular, but also the medium-sized ones, showed a higher range than the large KVs. It is astonishing – especially in light of the contemporary critiques of the small KVs' financial instability – that those funds' reserves were still relatively large. In the unweighted scenario, the financial reserves at the end of 1867 (1913) would still have covered 12 years (16 years) of equal pension expenditures. From 1908 on, the range increased steadily due to the new pre-funding requirement introduced by the New *Knappschaft* Law.

Figure 3.15, moreover, re-arranges estimates on the hypothetical range of capital reserves by growth pattern. A KV subject to structural decline, as opposed to those that were prospering, probably was under more pressure to maintain a capital reserve capable of pre-funding future liabilities to some extent. In fact, the picture is surprising, as the small, stagnant, not the small, dynamic, KVs exhibited the higher mean range.

3.8.3. The financing equilibrium

The last issue with which this subsection is concerned is the "financing equilibrium." Here, I intend to briefly illustrate the regularity with which a net deficit at the end of a year occurred, as well as the average magnitude of such deficits. Recall that an insurer produces insurance coverage by making conditional promises to pay a certain benefit if a certain event occurs – e.g., if the insurant becomes sick or permanently incapacitated for work. KVs operated according to the principle of collective equivalence, which demands that total annual contributions (plus other revenues and assets) had to settle total annual claims costs. Yet, there might be a difference between actual (or historical) claims costs per year to be observed in retrospect and what the KV management had expected ex-ante. Since KVs specified contributions to be paid in a given time interval in advance, they were subject to the actuarial risk that costs to be covered during that time interval had been underestimated ex ante. If, at the end of a period, revenues did not match costs, a KV had to draw on reserves accumulated in the past to close the financing gap. Indeed, KVs had the discretionary alternative at hand to compensate for any reserve loss in the previous year by raising per capita contributions for the next period. The degree of the KVs' actuarial expertise in the period under study is hard to specify from available sources. Precisely how they made decisions and calculated their expectations of claims costs to be settled is, as mentioned above, still a black box.

In fact, although KVs operated according to the PAYG method, they did not always manage to exactly balance revenues and expenditures. From the KV statistics, we can calculate that, astonishingly, the relative frequency of a net deficit among all small KVs was zero only in 1909. In all remaining years, the relative frequency fluctuated between five and 30 percent before 1907 and, respectively, five and 42 percent between 1908 and 1920. At the mean, two of out of ten small funds experienced a net deficit at the end of whatever year we pick out. The ex-post probability with which a medium-sized KV wrote a net deficit was quite similar until the middle of the 1890s (with relative frequencies covering a range of between three and 28 percent), but notably lower after that (between four and 18 percent). Large KVs reflected a larger volatility in the relative frequency, especially before 1890 (with relative frequencies covering a range of between zero and 60 percent). Except for during the First World War (with relative frequencies covering a range of between 10 and 25 percent), the occurrence of net deficits in this size class declined essentially to zero thereafter. Although it is not clear whether deficits can doubtlessly be traced back to inadequate forecasting, data nonetheless underline the tendency among KVs' to have underestimated financial needs.

Table 3.31 extends the view on the occurrence of deficits by concentrating on whether a liquidation or an absorption might have been conducted as a direct consequence of a net deficit or sequence of deficits just experienced. The table reports data separately for the two time periods 1861–1907 and 1908–1920.

Regarding mergers in the former period, just 17 percent of the KVs absorbed reported a net deficit in the pre-merger year; only one KV experienced a sequence of three deficits before the merger took place. The coincidence of exit by merger and having reported a net deficit was notably higher in the latter period. Of thirteen KVs absorbed, 31 and 46 percent, respectively, experienced a deficit in the pension- and sickness-insurance sections. Regarding liquidations, half of the KVs that ceased operation before 1908 wrote a net deficit in their final year of operation. Quantities again increased in the second period. 58 percent of closed KVs show a deficit in the pension-insurance section and 66 percent in the health-insurance section; only three KVs were not closed after having reported a deficit in either section. From this perspective, historical evidence suggests that liquidations were, indeed, correlated with an immediate financial shock due to false predictions of claim costs.

However, the mere fact that net deficits occurred is not, in itself, indicative of deep financial problems occurring as a consequence. Rather, the relative size of occurring net deficits should be taken into account. This is done with evidence presented in Table 3.32. Reported are two measures of a deficit's severity: first, the deficit in relation to total expenditures in the respective year; and, second, the deficit in relation to capital reserves existing at the beginning of the year and ready to even out costs in excess of revenues. Regarding both measures, the median ratio is, in turn, depicted separately for the subsets of net deficits and, for comparative reasons, of net surpluses.

Table 3.31: Coincidence of net deficits and the occurrence of absorptions and liquidations (Prussian KVs)

	Exit by absorption	Exit by liquidation
A) 1861–1907: Compound insurance		
Number of KVs involved	24	10
Thereof deficit in the last year of operation (% of KVs involved)	4 (17 %)	5 (50 %)
Thereof deficit in the last two or more years of operation (number of consecutive years on average)	1 (3 years)	2 (3 years)
Exit without deficit in the final year of operation	20	5
B) 1908–1920: Pension and sickness section		
Number of KVs involved	13	12
a. Pension section		
Thereof deficit in the last year of operation (% of KVs involved)	4 (31 %)	7 (58 %)
Thereof deficit in the last two or more years of operation (number of consecutive years on average)	3 (3 years)	5 (2 years)
b. Sickness section		
Thereof deficit in the last year of operation (% of KVs involved)	6 (46 %)	8 (66 %)
Thereof deficit in the last two or more years of operation (number of consecutive years on average)	2 (2 years)	4 (3 years)
Exit without deficit in any section in the final year of Operation	7	3

Sources: See Figure 3.2.

If referenced against actual expenditures, net surpluses usually ranged, before 1908, between 16 and 22 percent at the median. Net surpluses were considerably higher thereafter in the pension-insurance section. What is more, the median net deficit fluctuated more over time and usually did not exceed ten percent. Whether a net deficit of ten percent was threatening after all is a matter of available buffer. So, referenced against capital reserves existing at year-start, the median net deficit usually ranged somewhere between one and five percent. This is, all in all, quite low. Assuming a regular five-percent deficit for the median KV, it would have had to experience such a deficit for, at least, 20 consecutive years, interest income aside, before finally being ruined. No KV actually reported a net deficit for more than few consecutive years. Thus, the risk of financial ruin was probably not that high for the majority of KVs.

Table 3.32: The relative size of net surpluses and deficits among Prussian KVs

Year	Measured against total expenditures at year end		Measured against capital reserves existent at year start	
	Median ratio in % if net surplus	Median ratio in % if net deficit	Median ratio in % if net surplus	Median ratio in % if net deficit
A) Compound insurance period				
1867	+21.8 %	-7.7 %	9.1 % (1868)	-4.6 % (1868)
1871	+18.5 %	-3.4 %	6.7 %	-2.7 %
1876	+16.1 %	-5.1 %	6.5 %	-2.2 %
1881	+16.0 %	-7.2 %	7.4 %	-3.9 %
1886	+20.0 %	-0.9 %	7.9 %	-3.9 %
1891	+16.0 %	-5.2 %	7.4 %	-0.9 %
1896	+19.1 %	-6.1 %	5.4 %	-3.2 %
1901	+19.7 %	-8.6 %	6.4 %	-2.8 %
1906	+20.6 %	-3.0 %	7.7 %	-1.2 %
B) Pension insurance section				
1908	60.4 %	-10.7 %	10.6 %	-2.0 %
1913	56.3 %	-3.3 %	9.2 %	-1.5 %
1918	45.4 %	-26.0 %	4.2 %	-3.0 %

Sources: See Figure 3.2.

3.9. The average *Knappschaft* fund at a glance

Finally, to summarize the previous descriptive overview, the membership and financial characteristics of the "average Prussian KV" are reported for three selected years: 1867, 1890 and 1913. I constructed the average KV per year by displaying, for each variable, the non-weighted average over all KVs operating in the particular year. Table 3.33 reports the set of basic variables on membership structure, costs, and revenues. Presented are absolute values in persons or marks, or percent. All financial quantities are displayed in current prices this time.

In 1867, the average Prussian KV had 2 435 members – 15.8 percent of who were pensioners – and provided insurance benefits in a total nominal amount of about 47 900 marks (or about 64 000 marks in 1913 prices). Reported expenditure data enable us to derive the functional division of claims costs of the early average KV's compound-insurance scheme. Of those costs, about 39 percent were spent on pensions and 33.4 percent on sickness-related benefits (e.g., sick pay, physician costs, medicine); the remaining costs relate to miscellaneous benefits to be further aggregated into funeral pay and such benefits that did not belong to the set the regulator prescribed by law, such as education subsidies for the miners' children, charitable payments to members, and operating costs. The two single most important benefits were the invalidity pension, granted if a miner was judged permanently incapacitated for work for fifty percent or more (i.e., in his physical state, the miner would no longer be able to earn half the wage he had earned before) and sick pay per day on leave.

Reported data, on the whole, draw a picture of long-term growth in almost every aspect. The average KV's contributor base increased enormously – from 2 050 in 1867, over 11 221 in 1907, to 12 102 in 1913 –, but the number of financially-dependent pensioners increased even faster – from 384 in 1867, over 2 705 in 1907, to 3 457 in 1913. In addition, the number of paid sick days peaked in 1907 – with about 140 600 (or 12.5 per contributor) – and decreased to about 123 400 (or 10 per contributor), probably due to successful efforts to reduce simulation. The growth of claims costs, in turn, outperformed membership growth, thus indicating rising per capita costs. The average KV's expenditure on invalidity pensions increased enormously from about 12 300 marks nominal to 434 000 marks. In comparison, expenditures on survivorship pensions increased from an 1867 amount roughly equal to invalidity pensions to only 35 000 marks. Since there always were more survivors than invalids, this clearly indicates a relative stronger increase in the generosity of the invalidity benefit. Moreover, nominal sick pay per sick day nearly doubled between 1867 and 1913 with 2.2 marks. In line with the secular increase of benefit expenditures, was the increase in revenues, of course.

Table 3.33: The average Prussian KV's characteristics, 1867, 1907 and 1913

Variables	1867	1907	1913
	Mean (SD)	Mean (SD)	Mean (SD)
Number of contributors	2 050 (4 934)	11 221 (42 049)	12 102 (45 621)
Share of contributors aged 16-35 in all contributors (%)	42.7 (19.6)	47.5 (18.1)	55.4 (15.4)
Number of contributors in the sickness section	–	–	15 335 (57 619)
Number of invalids	83 (232)	1 083 (4 065)	1 362 (4 781)
Number of widows	117 (276)	884 (2 671)	1 157 (3 430)
Number of orphans	184 (515)	738 (2 597)	938 (3 366)
Invalids per 100 contributors	4 (–)	10 (–)	11 (–)
Survivors per 100 contributors	15 (–)	14 (–)	17 (–)
Number of paid sick days	15 937 (37 911)	140 574 (651 581)	123 435 (460 752)
Sick days per 100 contributors	751 (–)	1 253 (–)	1 020 (–)
Total expenditure on invalidity pensions (marks)	12 255 (35 573)	270 893 (1 043 609)	434 020 (1 598 349)

Table 3.33 continued

Variables	1867	1907	1913
	Mean (SD)	Mean (SD)	Mean (SD)
Average invalidity pension (marks/invalid)	148 (–)	250 (–)	319 (–)
Total expenditure on survivorship pensions (marks)	11 031 (34 580)	47 442 (263 586)	34 697 (117 974)
Total expenditure on sick pay (marks)	9 689 (25 303)	201 995 (980 074)	269 342 (1 203 708)
Average sick pay (marks/day on leave)	0.6 (–)	1.4 (–)	2.2 (–)
Total expenditure on medical treatment (marks)	10 451 (27 541)	191 304 (731 539)	323 456 (1 145 325)
Total expenditure on miscellaneous benefits (marks)	8 235 (42 626)	11 869 (56 837)	17 293 (73 180)
Total expenditure on miscellaneous benefits in the sickness section (marks)	–	–	9 740 (53 265)
Operating costs (marks)	8 645 (49 746)	48 332 (238 543)	22 630 (97 270)
Operating costs in the sickness section (marks)	–	–	20 419 (97 199)
Payments of contribution by miners (marks)	29 961 (78 716)	577 621 (2 463 473)	536 865 (2 073 988)
Payments of contribution by employers (marks)	23 975 (64 595)	489 902 (1 941 495)	536 865 (2 073 988)
Average annual payment of contribution (marks/ contributor)	26.3 (–)	95.1 (–)	88.7 (–)
Average annual payment of contribution in the sickness section (marks/contr.)	–	–	47.3
Entrance fees (marks)	489 (1 270)	3 172 (12 591)	1 694 (5 909)
Average entrance fee per entrant (marks)	3.4 (–)	3.0 (–)	0.4 (–)
Interest on assets (marks)	3 811 (9 379)	72 139 (262 023)	162 243 (582 471)
Interest on assets in the sickness section (marks)	–	–	13 168 (54 895)
Miscellaneous revenues (marks)	8 962 (46 194)	38 976 8187 386)	39 239 (260 074)
Miscellaneous revenues in the sickness section (marks)	–	–	14 085 (77 460)

Notes: All values in 1913 that do not explicitly relate to the sickness-insurance section relate to the pension-insurance section only. Monetary figures are in current prices.

Sources: See Figure 3.2.

4. Detecting economies of scale:Did growth reduce actuarial risk and operating costs?[1]

Chapter 1.2 introduced in short what nineteenth- and early twentieth-century observers thought about the KVs' size – i.e., that their size was of utmost significance for their financial well-being, as well as for the survival of the whole institution. They argued, in particular, that large funds would be more actuarially stable. With this chapter, I explore the issue in more depth and answer the question of whether increasing memberships empirically led to reductions in the KVs' exposure to actuarial risk. In addition, I address a related question with which contemporary observers did not seem very concerned: Were larger memberships associated with fewer administration expenses per capita?

4.1. Making the case for economies of scale inherent in *Knappschaft* insurance

Asking whether it is reasonable to consider KV operations as reflecting the potential for economies of scale has brought about a qualified "yes." Two problem areas – actuarial risk and operating costs – have been identified as being most valuable to investigate. However, before using some basic econometric methodology to explore whether economies of scale existed at any size or whether there was an MES or even an optimal size, I will start with descriptive quantitative evidence. This will give a general idea of what is to follow.

In the next subsection, I ask whether there is evidence in the data that the law of large numbers, as the baseline concept in actuarial science, empirically worked. The key to understanding *why* external concentration was promoted – and, as far as we know, conducted – lies in the contemporaries' educated guesses on actuarial relationships. However, the key to understanding whether KVs did well to follow this practice lies in empirical tests, which are long overdue.

---

[1] An earlier version of this chapter (specifically, subsection 4.2), is published as part of the article "Insurance, size, and exposure to actuarial risk: Empirical evidence from nineteenth- and early twentieth-century German Knappschaften;" see Jopp (2012a).

4.1.1. Actuarial fundamentals

Recall that, in its empirical formulation, the law of large numbers states that the relative frequency with which an event occurs stabilizes around the true, but unknown, probability of occurrence, if the number of observations rises. Put another way, the variance of a series of, say, annual relative frequencies relating to an event such as "becoming invalid over the course of a year" diminishes if the collective of insurants increases. If the law in this formulation had significance for the KVs' operations, this should be reflected in the way actuarial fundamentals developed over time. From the perspective of a KV, such fundamentals arguably concerned, first and foremost, the demographics. Let us, therefore, look at a set of important relative frequencies that are already known from Chapter 3:

- the ratio of the number of active miners having become invalid over the course of a year to the person-years exposed to that risk (i.e., the probability of becoming invalid);
- the ratio of the number of active miners (or invalids and widows) having died over the course of a year to the person-years exposed to that risk (i.e., the probability of dying);
- the ratio of the number of active miners having "out-migrated" (to another KV or another sector of the economy) over the course of a year to the person-years exposed to that risk (i.e., the probability of out-migrating);
- and, conceptually different from the aforementioned ratios, the number of miners joining the KV over the course of a year as a proportion of a reference quantity, such as the number of miners existent at the beginning of the year.[2]

From a KV's perspective, it is clear that if those series showed considerable variation in the past, the ability to plan or predict future financial needs would be negatively affected. A way to deal with this risk, of course, would be to accumulate reserves to draw upon if necessary, which is essentially the same as putting a so-called risk-loading on top of the per capita contribution.[3]

There is a straightforward way to create a first impression of whether the law of large numbers, in practice, worked for the KVs. We have to compare a measure of the varia-

---

[2] This variable is conceptually different since the denominator must not be interpreted as person-years exposed to a risk, thus not as a "probability of occurrence." Besides, other ratios, on which I do not focus, include the ratio of the number of widows having remarried over the course of the year to widow-years exposed to that "risk;" the ratio of the number of sick days having occurred over the course of the year to the person-days exposed to that risk (i.e., the probability of claiming a sick day); or the ratio to the number of invalids reactivated over the course of the year to the invalid-years exposed to that risk.

[3] I will not deal with this point here in detail. For a more detailed discussion of the problem area of risk loadings, see the long version of this study, Jopp (2012b), and footnote 33.

tion in the series of relative frequencies over KVs. With Table 4.1, I assess this cross-section comparison. The coefficients of variation for six single fundamental variables – "prob (1)" to "prob (6)" (see notes on the table) – are reported for each of the 103 Prussian KVs. Calculating the coefficient of variation as the ratio of the standard deviation to the mean of a statistical series is a simple way to assess comparative variability. KVs are, therefore, put into seven categories according to whether or not they consistently fit, regarding the whole observation period, into one of the size classes: 1–199, 200–999, 1 000–4 999, 5 000–9 999, 10 000–49 999, and 50 000+ contributors. If a KV did grow through at least two classes, it was categorized "not clearly assignable;" column two would then report through which size classes the KV had grown. The size information enables us to identify whether small KVs were especially exposed to variation. Since I am interested in what the empirical rule was among KVs, I exclude the war and inflationary years.

Without going into detail on every KV, one fact stands out: The coefficient of variation among small KVs, which never operated with more than 199 contributors at once, was, by tendency, higher than it was among the rest. This definitely holds for the five proper probabilities of occurrence displayed in columns three through seven. So, for example, take the *Münster am Stein KV*. Thanks to the coefficient of variation being unit-free, we can interpret a value of 1.84 for the probability of becoming invalid as stating that the dispersion in the series over the observation period (here, 1867 to 1913) was 184 percent, so apparently quite high. The mean coefficient of variation regarding the probability of becoming invalid for the smallest size class was a little lower, at 148 percent. It seems consistent with the data to say that the mean dispersion in all five probability series declined, as a rule, towards a KV size of about 5 000 to 10 000 contributors.[4]

Now that we see statistical evidence about comparative dispersion, two questions arise: First, besides relative differences in the coefficient of variation values, do the values indicate plausible dispersion? Second, what do the results tell us, so far, about KV size and actuarial risk? Addressing the first question, it is rather a subjective matter where to draw the line between "low dispersion" and "high dispersion," or, put differently, between dispersion threatening the continuation of a KV's operation and dispersion not threatening a KV's operation at all. In my opinion, variability of more than 100 percent was definitely likely to threaten a KV's operation, in that outcomes – such as the number of working members that would die over the course of a year – were hardly predictable. Whether a dispersion of even 50 percent was too high is open to argument. Regarding the second question, I interpret a high dispersion among small KVs, on the one hand, and a low dispersion among large KVs, on the other hand, as indicating economies of scale potential. Since a KV's finances ultimately depended on how many contributing and how many economically dependent insurants existed, it was, arguably, important to

---

[4] There are, indeed, exceptions within each size class to the size-class baseline.

be able to predict future numbers based on the numbers observed in the past. The higher the variation in the past, the less certain could a KV be about its future. Contemporary observers of the KVs are proven right here in their assumption that smaller KVs were exposed to more variability in baseline data that formed the basis for operative decisions.

Table 4.1: The coefficient of variation for a set of actuarial fundamentals (Prussian KVs, 1867–1913)

KV	Relevant size classes (in case KV assigned to "G")	Prob (1)	Prob (2)	Prob (3)	Prob (4)	Prob (5)	Prob (6)
A) KV size < 200							
Goffontaine KV	–	1.41	0.36	1.81	–	2.44	2.17
Hohenzollern'sche Lande KV	–	1.25	2.40	1.54	3.03	1.48	1.61
Krupp KV	–	–	0.78	–	–	0.66	2.20
Münster am Stein KV	–	1.84	2.65	1.93	2.56	2.85	2.28
Theodorshalle KV	–	1.17	1.61	1.35	2.17	2.74	2.08
Thommer KV	–	1.44	2.23	2.71	2.80	1.82	2.50
Werl KV	–	1.04	1.35	1.07	1.65	2.24	1.75
Westernkottener KV	–	1.95	2.28	1.73	2.65	2.66	2.02
Hohensteinischer KV	–	1.07	1.10	1.04	1.39	1.43	3.04
Schmalkaldener KV	–	1.15	0.70	0.97	0.66	1.89	3.39
Altenbeken KV	–	2.56	2.95	1.80	2.50	2.29	3.24
Gottesgabener KV	–	2.05	3.64	1.67	3.10	3.39	9.16
Königsborner KV	–	1.37	0.64	0.72	1.05	1.10	3.14
Neusalzwerker KV	–	1.39	2.17	0.89	1.22	2.01	3.18
Rothenfelder KV	–	1.81	1.86	1.43	2.11	1.18	2.90
Salzkottener KV	–	2.24	1.52	2.42	1.90	2.40	2.72
Sassendorfer KV	–	2.24	4.23	1.61	3.24	1.25	2.71
Artern'scher KV	–	0.68	0.81	0.32	1.38	1.73	2.74
Berliner KV	–	1.41	0.62	0.94	0.55	1.61	2.37
Erfurter KV	–	0.92	1.55	1.58	2.45	1.76	2.00
Halle Saline KV	–	1.08	0.92	1.06	0.82	1.48	1.76
Henneberger KV	–	1.41	–	-	1.41	1.41	1.41
Kamsdorfer KV	–	1.41	1.41	0.85	0.45	1.41	1.35
Stolberg KV	–	1.25	1.71	0.96	1.55	0.89	2.18
Mean coefficient of variation		*1.48*	*1.72*	*1.38*	*1.85*	*1.84*	*2.66*
B) 199 < KV size < 1 000							
Arnsberger KV	–	0.87	0.56	0.64	0.87	0.90	1.34

Table 4.1 continued

Knappschaft	Relevant size classes (in case KV assigned to "G")	Prob (1)	Prob (2)	Prob (3)	Prob (4)	Prob (5)	Prob (6)
Eschweiler Pümpchen KV	–	0.67	0.94	0.65	1.00	0.87	1.53
Ichenberger KV	–	1.53	0.77	1.44	1.71	0.89	1.11
Olpe KV	–	0.65	0.61	0.81	0.92	0.64	1.37
Rheinböller Hütte KV	–	0.82	0.97	1.43	1.02	0.83	1.40
St. Goar KV	–	1.26	0.97	0.70	1.27	1.18	3.14
Wied KV	–	0.84	0.42	0.45	1.09	0.74	1.30
Muskauer KV	–	0.77	0.74	0.63	1.26	0.93	4.11
Dürrenberger KV	–	0.70	0.79	0.38	0.62	1.15	1.26
Schönebecker KV	–	0.35	0.49	0.55	0.37	0.71	4.63
Wernigeroder KV	–	0.53	0.62	0.45	0.46	1.44	0.83
Mean coefficient of variation		*0.82*	*0.72*	*0.9*	*0.96*	*0.93*	*2.00*
C) 999 < KV size < 5 000							
Deutz KV	–	0.43	0.50	0.55	0.39	0.49	0.64
Eschweiler KV	–	0.55	0.53	0.30	0.30	0.99	1.18
Meinerzhagener KV	–	0.45	0.38	0.29	0.58	0.64	1.47
Oberbergischer KV	–	0.70	0.43	0.85	0.78	0.52	1.46
Wetzlarer KV	–	0.51	0.50	1.17	5.46	0.73	3.28
Schaumburger KV	–	0.32	0.39	0.31	0.41	1.21	–
Georgs-Marien-Hütte KV	–	0.86	0.42	1.03	1.06	1.29	2.20
Mülheimer KV	–	0.32	0.47	0.75	0.49	0.32	2.80
Brandenburg-Pommern KV	–	0.30	0.65	0.86	1.06	0.76	–
Mean coefficient of variation		*0.49*	*0.47*	*0.68*	*1.16*	*0.77*	*1.86*
D) 4 999 < KV size < 10 000							
Lahn KV	–	0.09	0.04	0.07	0.13	0.04	–
Nassau Allgemeiner KV	–	0.46	0.24	0.24	0.62	0.29	0.36
Mean coefficient of variation		*0.27*	*0.14*	*0.15*	*0.37*	*0.16*	*0.36*
E) 9 999 < KV size < 50 000							
Essen-Werdenscher KV	–	0.26	0.24	0.18	0.23	0.48	3.88
Hallescher KV	–	0.22	0.35	0.20	0.40	0.45	–
Mean coefficient of variation		*0.24*	*0.29*	*0.19*	*0.31*	*0.46*	*3.88*
F) KV size > 49 999							
Allgemeiner KV Bochum	–	0.32	0.28	0.14	0.11	0.92	1.58

Table 4.1 continued

Knappschaft	Relevant size classes (in case KV assigned to "G")	Prob (1)	Prob (2)	Prob (3)	Prob (4)	Prob (5)	Prob (6)
G) Not clearly assignable							
Briloner KV	B, *C*	0.46	0.45	0.30	0.34	0.82	0.79
Brühler KV	B, C, D	1.02	1.08	0.51	0.67	0.49	2.32
Burbacher Hütte KV	C, *D*	0.72	0.79	0.58	0.98	1.61	1.29
Cottenheimer KV	A, *B*	0.85	0.67	1.16	1.52	1.49	1.46
Dillinger Hütten KV	*C*, D	0.80	0.52	0.31	0.41	1.85	3.78
Eifel KV	*B*, C	0.77	0.68	0.74	0.89	0.88	1.63
Emser KV	B, *C*	0.78	0.58	0.47	0.64	1.38	1.31
Günnersdorfer KV	A, *B*	1.30	0.48	0.75	1.39	3.52	2.10
HalbergerKV	B, C	1.55	0.81	0.76	0.88	1.30	2.23
Heller KV	C, *D*	0.37	0.52	0.30	0.31	0.86	0.93
Holzappeler KV	*B*, C	0.58	0.66	0.48	0.57	0.74	2.13
Hostenbach KV	*B*, C	0.60	0.74	0.54	0.46	1.28	1.99
Lendersdorfer KV	A, B, *C*	1.44	0.28	0.53	0.83	2.18	2.82
Mariahütte KV	A, *B*	0.86	1.11	1.10	2.45	1.38	1.44
Mayener KV	B, C	0.56	0.62	0.77	1.08	1.32	1.84
Mosel KV	A, B	1.43	0.73	0.72	0.91	1.10	2.66
Müsener KV	B, *C*	0.51	0.52	0.49	0.52	0.91	1.92
Neunkirchener KV	*C*, D	0.40	0.62	0.34	0.53	1.44	1.33
Niedermendinger KV	A, B	1.27	0.97	1.33	1.35	1.03	1.05
Quinter KV	*B*, C	0.66	0.77	0.74	0.52	1.93	1.49
Rheinischer KV	C, D	0.51	0.45	0.31	0.36	0.73	1.93
Rheinpreußen KV	A, B, C, D	0.91	1.45	0.96	2.48	1.08	5.75
Saarbrücker KV	*E*, F	0.39	0.44	0.17	0.17	0.91	1.47
Siegener KV	C, *D*	0.35	0.40	0.37	0.24	0.97	0.91
Stolberger KV	*C*, D	0.54	0.53	0.33	0.55	1.41	0.95
Stromberger Hütte KV	A, *B*	1.40	0.84	1.45	3.29	1.12	3.04
St. Wendel KV	A, B	2.37	2.03	1.80	1.55	3.73	–
Wittgensteiner KV	A, *B*	1.81	0.91	1.30	2.33	0.70	1.03

Notes: Italic print in column two indicates the size class in which the respective KV mostly operated. No such print indicates that it, by and large, operated an equal amount of years in the stated classes. "Prob (1) is the "probability of becoming invalid (established miners)";"prob (2)" is the „probability of dying (established miners, across all ages)"; "prob (3)" is the "probability of dying (invalids, across all ages)"; "prob (4)" is the "probability of dying (widows, across all ages)"; "prob (5)" is the "probability of out-migrating (established miners)"; and "prob (6)" is "inflowing contributors as a proportion of miners existent at the beginning of the year", which is not a true probability.

4.1.2. Operating costs

This subsection provides a baseline assessment of economies of scale potential regarding the KVs' administration. Subchapter 3.7 only touched on the costs associated with operating KVs. Here, I provide an overview.

First, I focus on "operating costs" and not "administrative costs" in the more narrow sense. These additionally include expenses on building maintenance, arbitration trials, and interest on debt.[5] Figure 4.1 reports descriptive statistics on real operating costs per working member.

Figure 4.1. Descriptive statistics of real average operating costs per working member (in marks) related to the Prussian KVs' compound insurance, 1861–1907, and pension-insurance section, 1908–1920

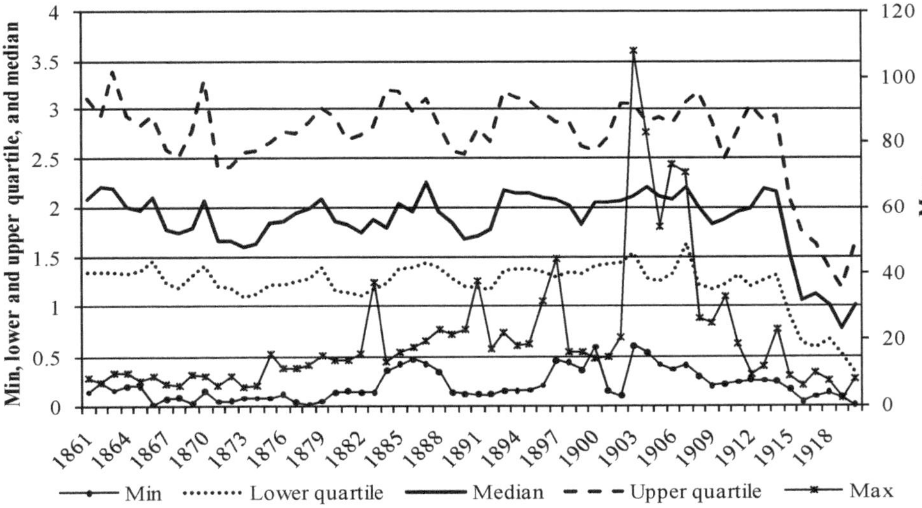

Notes: The maximum values for 1867 (384.5) and 1868 (500 marks), belonging to the *Kamsdorfer KV*, which operated with only one contributor in those years, have once more been dropped from the figure.

Sources: See Table 4.1.

---

[5] Investigating Dutch pension funds, Bikker and De Dreu (2009), pp. 67–82, in particular, differentiate administration costs from investment costs, with the latter category accounting for costs linked with asset management. Since the KVs had assets not merely in the form of cash or bank deposits, but especially in the form of interest-bearing papers, they certainly had to bear investment costs as well. As, however, the percentage distribution of costs in Subsection 3.7 made clear, there was no cost category besides "administration costs" that could be captured as reflecting that sort of additional operating costs. Not even with the post-reform accounting framework are we able to clearly identify such costs. The only cost category implying a link to investment is "interest on debt."

Note, first, that the maximum values are depicted *on the right axis* because of their magnitude and, second, that figures on the post-reform period relate to the pension-insurance section exclusively. Median operating costs fluctuated, in fact, between around 1.8 and two marks; for a given year, 75 percent of KVs usually produced at costs of no more than 2.8 marks per contributor.

Figure 4.2 displays the same information for the post-reform sickness-insurance section. Until the middle of the First World War, the median operating costs per capita ranged a little bit below that obtainable for the pension-insurance section, but this difference disappeared thereafter.

Figure 4.2: Descriptive statistics of real average operating costs per working member (in marks) related to the Prussian KVs' sickness-insurance section

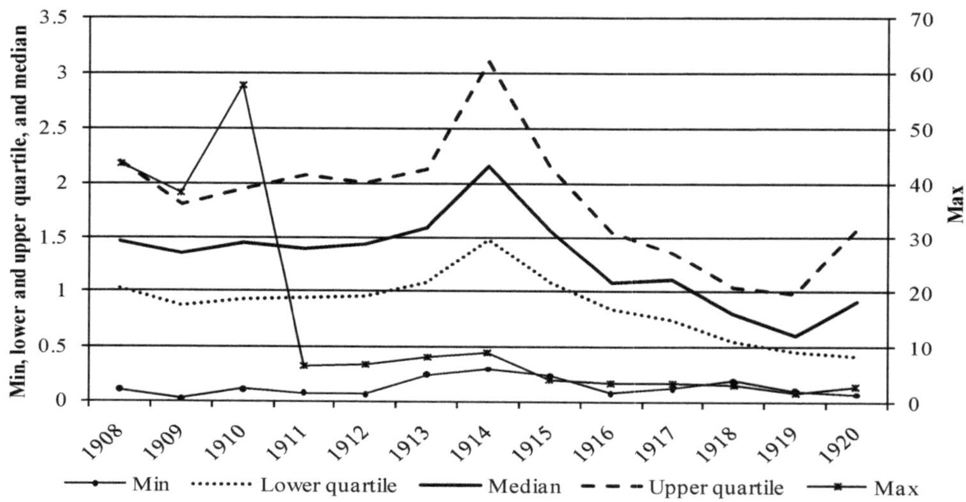

Sources: See Table 4.1.

In order to assess whether there were economies of scale regarding operating costs stemming from fixed-cost degression, I now consider average operating costs per size class and operating costs as a percentage of total insurance benefit expenditures. This approach is in line with, for example, Bikker, Steenbeek and Torracchi (2010), Bikker and De Dreu (2009), Mitchell (1998), and Estrin (1988).[6] Tables 4.2 and 4.3 contain results on both measures, organized by combinations of five-year periods and size classes.

Beginning with Table 4.2 on the compound-insurance period, data indicate, for example, that operating costs as a percentage of total insurance benefits usually decreased

6 See Bikker, Steenbeek and Torracchi (2010), p. 13, Bikker and De Dreu (2009), p. 69, Estrin (1988), p. 29, and Mitchell (1998), p. 406.

across size classes; regarding the period 1861 to 1865, the percentage amounted to 12.3 percent for the smallest size class, slightly increased to 13.2 for KVs sized between 200 and 999 working members, but then decreased to only 4.8 percent for KVs beyond a size of 10 000. This picture is broadly consistent with what happened in the following five-year period, although the differences in the advantage of the next-largest size class became smaller over time. It seems as if, after the 1890s, a considerable difference in operating costs as a percentage of total insurance benefits persisted only between the smallest and next-smallest KVs, implying positive, but marginally decreasing returns to scale.

Table 4.2: Operating costs among Prussian KVs by size class, 1861–1907

Year	Real operating costs…	Size classes					
		1 – 199	200 – 999	1 000 – 4 999	5 000 – 9 999	10 000 – 49 999	50 000 +
1861–1865	per contributor (marks)	1.8	1.6	1.1	2.2	1.4	–
	in % of total benefits	12.3 %	13.2 %	9.4 %	7.5 %	4.8 %	–
1866–1870	per contributor (marks)	6.9	1.4	1.3	1.6	1.3	–
	in % of total benefits	7.6 %	8.4 %	7.7 %	5.4 %	4.2 %	–
1871–1875	per contributor (marks)	2.0	1.7	1.3	2.1	1.6	1.3
	in % of total benefits	6.6 %	8.1 %	6.6 %	6.8 %	3.5 %	3.1 %
1876–1880	per contributor (marks)	2.5	1.9	1.5	2.2	1.7	1.3
	in % of total benefits	7.3 %	7.4 %	4.9 %	6.5 %	3.1 %	2.7 %
1881–1885	per contributor (marks)	2.5	2.0	1.4	1.7	1.5	1.2
	in % of total benefits	7.3 %	6.8 %	4.2 %	4.8 %	3.1 %	2.6 %
1886–1890	per contributor (marks)	3.3	1.8	1.4	1.5	1.8	1.4
	in % of total benefits	8.6%	5.2 %	4.0 %	4.2 %	3.1 %	2.9 %
1891–1895	per contributor (marks)	4.6	1.9	1.4	1.6	2.0	1.3
	in % of total benefits	7.8%	4.5 %	3.8 %	4.0 %	3.1 %	2.6 %
1896–1900	per contributor (marks)	4.8	1.8	1.4	1.4	2.0	1.4
	in % of total benefits	7.3 %	4.2 %	3.8 %	4.0 %	3.5 %	2.6 %
1901–1905	per contributor (marks)	8.0	1.8	1.6	1.6	2.1	2.6
	in % of total benefits	9.4 %	4.4 %	3.5 %	4.1 %	3.5 %	4.0 %
1906–1907	per contributor (marks)	10.3	1.9	1.7	1.5	2.1	2.2
	in % of total benefits	8.9 %	4.6 %	3.3 %	3.0 %	3.9 %	2.5 %

Sources: See Table 4.1.

While there obviously was potential for economies of scale with regard to operating costs as a percentage of total insurance benefits, the picture for real operating costs per capita is less clear. In almost all periods, the smallest KVs accounted for the relatively highest per capita costs. However, data on the remaining five size classes do not suggest a clear negative relationship, where the largest KVs would show up as the ones that reflected the lowest costs. Indeed, within each period, operating costs per capita appear to follow a U-shaped pattern over size classes, with the descending branch falling more steeply than the ascending branch rises. This observation implies that there very likely was an optimal KV size. The econometric analysis below is aimed at determining the nature of this relationship more clearly.

Finally, regarding Table 4.3 and the separate insurance sections in the period 1908 to 1920, we find that real operating costs per capita, or as a percentage of total insurance benefits, generally increased towards 1920 and do not suggest a clear negative linear relationship, but either a U-shaped pattern (pension-insurance) or even a positive one (e.g., sickness-insurance section, 1914–1918). In all, a descriptive analysis of KV size and operating costs makes us doubt that the slogan "larger is better" accurately describes KV's historical experience.

Table 4.3: Operating costs among Prussian KVs by size class, 1908–1920

Year	Real operating costs...	Size classes					
		1 – 199	200 – 999	1 000 – 4 999	5 000 – 9 999	10 000 – 49 999	50 000 +
A. Pension insurance section							
1908–1913	per contributor (marks)	4.8	1.7	2.0	1.5	2.2	2.8
	in % of total benefits	7.4 %	8.0 %	8.8 %	8.4 %	6.7 %	4.3 %
1914–1918	per contributor (marks)	4.8	1.9	2.6	1.8	2.5	3.6
	in % of total benefits	5.3 %	6.1 %	6.8 %	6.7 %	6.1 %	4.1 %
1919–1920	per contributor (marks)	10.5	4.4	4.5	9.1	10.1	12.9
	in % of total benefits	5.5 %	9.1 %	6.0 %	16.6 %	14.4 %	15.1 %
B. Sickness insurance section							
1908–1913	per contributor (marks)	4.1	1.4	1.4	1.0	1.8	1.6
	in % of total benefits	14.0 %	5.2 %	4.9 %	3.3 %	4.8 %	4.2 %
1914–1918	per contributor (marks)	2.7	2.0	2.1	2.3	2.7	3.2
	in % of total benefits	13.3 %	4.7 %	4.5 %	4.4 %	4.8 %	5.6 %
1919–1920	per contributor (marks)	5.5	2.6	4.8	7.7	8.6	12.4
	in % of total benefits	8.4 %	2.4 %	4.2 %	4.9 %	5.6 %	12.0

Sources: See Table 4.1.

One might ask how the figures on KVs relate to modern pension and sickness systems' parameters. We can compare operating costs as a percentage of total benefit expenditures. In this respect, it seems sufficient to look at Mitchell's (1998) and Estrin's (1988) collections of data for a set of developing and developed countries' social security systems. Focusing first on the former source, as of the early 1990s, the mean percentage of administrative costs against social security benefit expenditures among OECD countries was 3.1 percent, while the mean percentage for Latin American and Caribbean countries, to which they were compared, was far higher, at roughly 28 percent. Germany ranged slightly below the OECD average (2.84 percent), and Greece showed the highest percentage among OECD countries at the time (6.7 percent). Estimates provided by Estrin cover the years 1966 to 1983 for a set of developed countries (including the U.S., Canada, and Japan). In Germany, the mean percentage, taking all social security programs into account, was 5.1 percent in 1966 and almost half that (2.7 percent) in 1983. A decreasing percentage was a general phenomenon among almost all countries studied.[7] So, given these figures, the historical mean percentages among KVs are not outstanding in the one direction or the other.

4.2. Searching for the minimum efficient size of a *Knappschaft* – Findings on the empirical relationship between size and actuarial risk

4.2.1. Empirical strategy

At its very core, this chapter is about an idea that has developed into an important issue in modern insurance economics: That is, the size of an insurance pool matters for the efficiency of producing insurance coverage. As argued above, many studies have already focused on the empirical question of whether returns-to-scale potential exists for insurance providers, especially with respect to administration costs. An answer to this question would provide policymakers and businessmen alike with helpful information to decide how to shape an insurance system – which is basically a regulatory question – or how to develop the scale of one's own insurance company to become or remain competitive. In addition, we have also seen above that contemporary observers of the KVs' operations dealt with that matter, but focused more on the question of whether or not larger insurance funds were less exposed to variability in claims. They were already aware of the fact that production of insurance coverage is inevitably connected with actuarial risk, which exists simply due to the stochastic foundation of insurance, and, consequently, they asked how to reduce it.

To begin with, the applied definition of actuarial risk needs to be clarified. According to the literature, there seems not to be simply *the* actuarial risk of an insurer, but rather

[7] See Estrin (1988), pp. 30–31, and Mitchell (1998), p. 406. Engaging in such a comparison across the time-line, we have to consider, however, that quantities may not be measured equally. As Mitchell points out, percentages compiled refer, first of all, to pension provision, leaving health care and other types of social security expenditures aside.

several dimensions to be addressed. An insurance provider is fundamentally exposed to at least three dimensions of actuarial risk: first, the "relative actuarial risk" of an insurer, which is the risk that an insurant's *actual* claim size to be settled during a period will exceed the *expected* claim size;[8] second, the "absolute actuarial risk," which is the risk that the *actual* aggregate claim size of a period, relating to the entire collective of insurants, exceeds the *expected* aggregate claim size. These two dimensions obviously focus on an insurer's risk from an expenditure perspective, which is the traditional view of actuarial risk. A third and broader concept, the probability of ruin, also internalizes the revenue side. Actuarial risk might then be considered as the risk that generated revenues (plus financial reserves, if existing) will not suffice to settle all claims costs occurring in a given period.[9] Strictly speaking, analysis of actuarial risk according to one of the above definitions only makes sense if we consider a multi-period context: In such a context, the fluctuations in the risk – i.e., in the deviation of actual from expected values – are under focus. Then, regarding "relative actuarial risk," the variance of the average claim and its change over time due to fluctuations in the number of insurants, more precisely, reflects the risk.

In this study, I concentrate on the exposure to relative actuarial risk – i.e., the risk that the variance of the average claim (to be financed by one contributor) is large. This is for three reasons. First, this dimension is consistent with contemporaries' thinking (see Subsection 1.2). Second, it is straightforward to argue that absolute actuarial risk must necessarily rise alongside a growing collective of insurants, which does not leave much room for empirical surprises.[10] Third, as the institutional overview has shown, KVs had to form expectations about the amounts of claims that might have to be settled

---

8 The risk is termed "relative" because the sum of all individual claims per period is, so to say, expressed in relation to the average insurant; this procedure reflects the assumption of independent and identical (homogenous) insurants. Such a deviation is only possible if we assume that there exist a bandwidth of possible outcomes (i.e., claims) and if we assume that we cannot say in advance for sure which outcome will appear for the average insurant.

9 See Zweifel and Eisen (2003), pp. 240–243, for a standard textbook description. Precisely, Zweifel and Eisen define the "relative actuarial risk" as the risk that the actual average insured loss exceeds its expected value by more than *k*-times its standard error. The probability that this happens, however, tends towards zero given the number of insurants rises. Furthermore, they define the "absolute actuarial risk" as the risk that the actual aggregate insured loss exceeds its expected value by more than *k*-times its standard error.

10 See Albrecht (1982), pp. 504–515. The logic behind is that each (potential) insurant is assumed to reflect a non-zero variance. *Ex-ante* it is unclear how often an insurant would make a claim or how large a claim would be. So, principally, there is a menu of possible outcomes, to which the insurer is prone to ascribe probabilities of occurrence and which is captured by the variance. In the ideal case, each insurant and, thus, each variance is independent of others (covariance is equal to zero). If independence between insurants can be assumed, the variance of the aggregate claim, related to the entire collective, is equal to the sum over individual insurants' variances. It follows that, given a variance greater than zero, adding additional insurants to the collective inevitably means adding amounts to the variance of the aggregate claim such that it rises.

at the end of the "insurance period" in order to specify per capita revenues such that they match expected claims costs. Thus, the deviation of the actual average claims costs to be financed by one contributor from the expected average claims costs was relevant. This holds all the more since it was not customary to use the instrument of charging immediate additional cover to contributors.[11]

I do not concentrate on the probability of ruin here explicitly since the answer to the question of whether KVs could really adjust their revenues – especially contributions – upwards immediately after a need of additional resources to cover deficits had been determined, is not, at all, clear. With "immediate adjustment" I mean an adjustment in the form of a discretionary ex-post assessment – i.e., a discretionary, but temporary, ex-post increase in per capita contributions or a discretionary permanent increase in per capita contributions not visible in the KV's statute. As mentioned in Chapter 2, the non-discretionary way towards a (permanent) increase of the per capita-level of contributions was to officially change the KV's statutes. Although I cannot trace the time-sequence of statute changes for *every* KV over the whole observation period, available evidence suggests that statutes were, by far, not changed every year or even several times over the course of one year. Thus, once a KV had determined the contribution to be charged to an insurant according to its expectations about the insurant's future claims behavior *ex-ante*, at the beginning of a period, contribution revenues to flow in and claims costs likely to occur are kind of delinked over the rest of the period. So if *ex-post* all insurants modestly exceeded expectations by having claimed larger amounts, the KV would have come under financial pressure. The question of whether or not ruin would have occurred is, then, mainly a matter of reserve availability or the ability to borrow money via the capital market. Since revenues were formally fixed over some longer time-period, the risk of ruin arguably played some role. However, as long as we cannot rule out that KVs could also rely on the "discretionary channel," the question of the importance of the risk of ruin cannot, at all, be answered satisfactorily. Besides, quantitative evidence provided in Subsection 3.8, shows that a KV's existing reserves, if a deficit occurred, were almost always more than sufficient to compensate for it. What could have brought a KV near ultimate insolvency was a sequence of deficits in more than just a few consecutive periods, or strong shocks to the size of the contributor base. Such shocks, however, were usually related to the long-term decline of a mining area and, thus, were not really stochastic. My conclusion is that focusing on the probability of ruin would not add much surplus value to the investigation as it stands.

My main measure of a KV's exposure to actuarial risk is the *variance of claims costs per contributor*, or

---

[11] A note on the concept of the "insurance period" is necessary: As I said in Chapter 2, the official way to change expenditure and revenue parameters was by a revision of the KV's statute. Thus, accurately, an insurance period was the period between two statutes. However, although official revisions were not done year-by year, the use of the KV statistics' data implies an insurance period of exactly one year.

$$(4.1)\quad Var[\frac{Z_{eit}}{N_{it}}]=\frac{\{Z_{eit}-E[Z_{eit}]\}^2}{N_{it}^2}\ .$$

In this expression, Z denotes the *actual* aggregate claims costs of category e that KV i had to settle in year t; the data situation does not allow for modeling periods shorter than a year. I distinguish between two categories of claims, namely all sorts of pensions ($e = 1$) and all sorts of sickness benefits ($e = 2$). $E[Z]$ denotes *expected* (or *predicted*) aggregate claims costs. N is the number of contributing miners, not of the entire membership. It is, therefore, important to note that Z_1/N must not be interpreted as the average claim size in the sense of an average pension that a pensioner received, but as the fraction of the total pension cost that the average contributor had to finance with his payments.

The insurance literature predicts an inverse relationship between an insurer's size and the variance of the average claim. To an extent, the KVs' contemporaries have put the right issue on their agenda; however, they did not attempt to explore the issue quantitatively. This seems an important task since it is unclear whether or not the exposure to actuarial risk could have been reduced strictly monotonically with growth instead. There might even have been a minimum efficient size beyond which no more reductions in the exposure could have been realized, or even an optimal size implying that beyond some point, diseconomies of scale would be realized. This chapter's baseline model is not intended to explicitly explore whether KVs generated internal or external growth because they believed in contemporary actuarial considerations. Rather, the model is intended to explore whether KVs had effectively realized economies of scale by growing, such that increasing the pool size was empirically (*ex-post*) – not only theoretically (*ex-ante*) – a useful strategy.

The claim that the larger a KV was, the better it could capitalize on the law of large numbers in predicting future costs is consistent with the implications of both the empirical and mathematical formulations of the law of large numbers: The variance of the average claim size must converge towards zero if the size of the insurants' collective converges towards infinity.[12] Thus, it seems reasonable to test the following hypothesis for the universe of Prussian KVs:

$$(4.2)\quad Var[\frac{Z_{eit}}{N_{it}}]\xrightarrow{N\to\infty}0.$$

Given that predictions of $E[Z]$ are available, I fit the following panel regression specification separately for categories $e = 1$ and $e = 2$:

[12] See Albrecht (1982), p. 504.

$$(4.3) \quad Var[\frac{Z_{eit}}{N_{it}}] = \delta_i + \delta_1 D_{1906} + \delta_2 SIZE_{it} + \delta_3 D_{1906}*SIZE_{it} + u_{it}\,.$$

Variables are in logs. The dependent variable is the category-specific variance, and *SIZE* is a placeholder for two different measures of KV size. The first and main measure is the total number of contributors (*total contributors*), and the second measure is the number of contributors per works divided by the total number of contributors (*correlated claims*). Besides, a dummy variable controls for the new 1906 regulations (D_{1906}); it takes on the value one for all years 1908 to 1913 (when the regulations came statistically into effect) and zero otherwise. In fact, I will estimate equation (4.3) in two alternative ways: The one alternative is the baseline model, which may be called "piece-wise constant" since I estimate the relationship between the variance and KV size for several size classes separately; this procedure does not require non-linearities. The other alternative is a model in which the several size classes are put together; in order to allow for a non-linear relationship between the variance and KV size, I include a squared term of *SIZE*, which makes the model a "quadratic" one. The reason why a "piece-wise constant" model seems to be appropriate is outlined below, in the part on results. Estimating a "quadratic model" at once, in addition, seems to be in order to test the baseline model's sensitivity to specifying several size classes.

While the main measure of KV size is simply intended to display the true financing power of the KVs, the second measure – *correlated claims* – needs some more substantiation. Regarding this, recall that a functioning balancing-of-risks-in-the-collective (*kollektiver Risikoausgleich*) requires insureds' risks to be stochastically independent of each other. This means that miner Friedrich's sickness or invalidity claim and miner Wilhelm's sickness or invalidity claim did not originate from the same event. If, however, an epidemic had spread or a mass accident had occurred, individual risks would not have been independent of each other, which clearly would have destroyed the stochastic basis of insurance. Such a state of affairs would have arguably resulted in a serious actuarial problem – insolvency – since a fund would have been immediately overwhelmed with claims that could not be settled in a timely manner. Whether a fund was incapable of immediate claims settlement ultimately depended on the state of its financial reserves.[13]

Since mining inevitably meant a high risk of massive accidents probably involving a lot of mineworkers at the same time and, thereby, immediately overwhelming a KV with sickness or invalidity claims, claims might have been to some extent correlated. If a KV was in charge of exactly one mine, the measure *correlated risks* equals one, meaning that the *potential* for correlation of risks due to a mass accident was high, simply because the entire working membership worked together in close proximity and

---

13 We might call this sort of risk to the insurer the "accumulation risk," a hopefully workable translation of the German technical term "Kumulrisiko."

could easily have been affected by an accident at the same time. The more mines a KV was in charge of, the smaller was the ratio and the lower arguably was the potential for immediate financial disaster from correlated risks. It has already been made clear that underground mining operations were connected with a high *ex-ante* accident risk. Data from the *Knappschafts-Berufsgenossenschaft* (the employer-based carrier of *Reich* accident insurance) introduced in Subsection 2.2 show that the accident rate among miners markedly exceeded the national average. Explaining the variance of the average claim according to my definition by the variable *correlated claims* means to quantify what happens to the variance if the potential for dependent risks changes when we consider (i) relative average mine size and (ii) that an accident in one mine does not automatically cause an accident in another mine. For the reason that we cannot measure the dependence of individual risks among the collective of workers of a particular mine, we can only consider ourselves satisfied with modeling more or less accurately the dependence of sub-collectives of insurants – i.e., dependence across single mines.[14]

In order to calculate the variance measure, I need to form expectations or predictions of a KV's historical costs, Z_e, in the observed years. Yet, data on the KVs' expectations – in the form of an historical document, for example, that says "we expect costs of x marks in year t" – are not available.[15] So, basically, there are three ways to compensate for this lack of historical information: (i) I could model static expectations – i.e., expecting for year t the costs of the preceding year; (ii) I could model adaptive expectations – i.e., expecting for year t the costs of year t-1, corrected in some way for the experience in the years before t-1; or (iii) I could model rational expectations based on more information than just past expenditure series. Option three is what I use hereafter. To be precise, I specify a pooled regression model based on data on all 103 Prussian KVs in order to explain their historical pension and sickness costs in each year under observation. From the regression model, I recover the predicted values that I use as input into equation (4.1). According to Emery and Emery (1999) and Broten (2010), this regression model might be called an "aggregate claims distribution."[16]

---

14 There is one important assumption implicit in the insurance model: insurants forming a KV's membership and representing claims are identical and independent. For the reason that I have data on the KV-level, I cannot check whether, or model that, individual miners were not independent of each other. For the reason mentioned in the text, they were likely not independent. Creating a variable *correlated risks* is a way to indirectly account for miners having been independent insurants. This way is all but perfect, but in my opinion the best procedure given the limits of the data. Note that I do not try to prove the existence of dependence among insurants here, but take its existence for granted. In particular, I do not focus on the true extent of the accumulation risk, but simply on a concept that I call the *potential* for correlated claims.

15 Sniegs (1998), in comparison, has such "formulated expectations" at hand for Bismarckian invalidity and old-age insurance. In her study, she compared those expectations or, respectively, projections on invalidity and old-age insurance's finances with historical developments.

16 See Emery and Emery (1999), p. 144, and Broten (2010), p. 33. Emery and Emery model the claim-producing process as one that is driven, in fact, by two stochastic sub-processes – namely, the process that brings out the number of claims and the process that brings out the claim size per

I use the generalized linear-models (GLM) approach according to Nelder and Wedderburn (1972) to estimate a baseline model, in which the aggregate claim is Poisson distributed. The GLM approach allows – instead of using maximum likelihood estimation – using iterated reweighted least squares optimization, which is said to improve the robustness of the conditional mean estimates in case distributional misspecification might be suspected.[17] The dependent variables are the natural logarithms of total pension costs (*pension costs*) and of total sickness costs (*sickness costs*) of KV i in year t, which are calculated as the number of claims – i.e., the number of pensioners or the number of sick days – times the average payout. For both combinations of claim category e and the period 1867–1913, I estimate the conditional mean according to equation (4.4). The independent variables are intended to substantiate a KV's risk structure. Except for the last regressor, subscripts e and t are omitted for convenience:

$$\begin{aligned}(4.4)\quad Costs = {} & \beta_0 + \beta_1 Age(1625) + \beta_2 Age(2635) + \beta_3 Age(3645) + \beta_4 Age(4655) \\ & + \beta_5 Age(>55) + \beta_6 Membership + \beta_7 Averagepayout + \beta_8 Burden \\ & + \beta_9 Burden^2 + \beta_{10} Firmshare + \beta_{11} Firmshare^2 + \beta_{12} Estab \\ & + \beta_{13} Estab^2 + \beta_{14} Hardcoal + \beta_{15} Browncoal + \beta_{16} Ironore \\ & + \beta_{17} Otherores + \beta_{18} Halite + \beta_{19} Stone + \beta_{20} Related + \beta_{21} Salt \\ & + \beta_{22} FirmKV + \beta_{23} Laggedcosts + \beta_{24} Breslau + \beta_{25} Clausthal \\ & + \beta_{26} Dortmund + \beta_{27} Halle + \sum_{1868}^{1913} \beta_t Year + \text{v}.\end{aligned}$$

First of all, controls include the five age-group shares with respect to established contributors that are recoverable from the KV statistics. The variable is constructed as the ratio of age group w's size to all established contributors. The w denote five age groups of established miners: those aged between 16 and 25, 26 and 35, 36 and 45, 46 and 55, and 56 and older. The age-group shares are intended to capture a KV's age structure as well as possible. While the KV statistics do provide age-group data on established mineworkers, they do not, unfortunately, provide such data on unestablished ones. Recent empirical evidence indeed suggests that the incidence and severity of sickness or inability to work rise with age, such that we may hypothesize that aggregate claims increased with an increase in the age-group share.[18]

Moreover, since a larger KV is likely to have generated more claims than a smaller KV, membership size – i.e., in this case, the sum of all working members and all pen-

---

incidence. Since both processes are described by a random variable, exponentially distributed regarding the claim number process and Poisson distributed regarding the average claim size process, the resulting aggregate claim size is modeled as gamma-distributed. As in Broten (2010), I model the distribution as a single stochastic process.

17 See Nelder and Wedderburn (1972). In the case of the Poisson distribution, misspecification might occur because the assumption of the variance-mean-equality is violated by the data due to under- or overdispersion. For the panel version of the GLM approach, see Zeger et al. (1988).

18 See Gorsky et al. (2006).

sioners – is incorporated in the model as well (*membership*). In addition, the variable *average payout* – either overall pension costs per pension or sick pay per day – is intended to measure generosity, or what I addressed in Subsection 3.7 as intensive growth of social insurance expenditures; this variable refers to the current year. The variable *burden* is a placeholder, denoting either the PCR (*pensioner burden*) if pension costs are addressed, or the sick-days-to-contributors ratio (*sick day burden*) if sickness costs are addressed. Following Guinnane and Streb (2011), the variable *firm* is included, as well, and defined as the ratio between employers' contributions and total costs. The mentioned authors find that part of an increase in the number of sick days can be explained by, among other things, an increase in the employers' financing share. So, they conclude that miners were presumably more prone to abuse funds if they knew they had come from employers rather than from their fellow contributors.[19]

Furthermore, the variable *established ratio* measures the share of established contributors among all contributors. Since established miners were said to be costlier than unestablished ones, the proportion is assumed to matter in explaining the aggregate claim. Here, as in the other cases, the non-linearities notably improve the fit. I also take into account the production structure of the mining area or firm to which a KV was tied by incorporating variables that measure how many of a KV's insurants were employed in the different sub-sectors. According to the KV statistics, I once more concentrate on eight sub-sectors: hard-coal, brown-coal, iron-ore, miscellaneous ores, halite, stone, salt, and steel; the variable *related* is equal to the share of KV members employed in steelworks and ore-processing plants. As outlined above, contemporaries believed that the different subsectors reflected different occupational hazards because of variation in the production processes.[20] If that were so, we would expect to find significant level differences across dummy variables.

Finally, to complete specification (4.4), *firm KV* is a dummy taking on the value one if a KV was a firm-related fund. As mentioned previously, the KV statistics *do not* provide information on whether a KV was firm- or area-related. So, I used information I was able to collect from the literature or concluded from the KV's name that it must have been firm-related.[21] The variable *lagged costs* measures the pension or sickness costs of the preceding period, and four dummy variables control for the mining-administration regions (*Oberbergamtsbezirke*) in which the KVs were located: Breslau, Clausthal, Dortmund, and Halle; the dummy for the fifth region, Bonn, is omitted in order to avoid multicollinearity. Year dummies are included for all years but 1867 to allow for effects common to all KVs (e.g., macroeconomic effects, legislative effects).

A final point has to be made regarding independent variables' selection: While the selection of those variables introduced above is intended to explain as well as possible

---

[19] See Guinnane and Streb (2011), p. 87.

[20] See Karwehl (1907), p. 75, for arguments on mixed hazard classes.

[21] So, for example, the name *Dillinger Hütten KV* or *Burbacher Hütte KV* doubtlessly imply the company linkage.

the heterogeneity among KVs and the KVs' aggregate sickness and pension claims, I cannot exclude that I omitted important explanatory variables. In this respect, the model certainly is not perfect. There especially is one factor of which I think that it very probably leads to such an "omitted variable" bias, namely the incidence of accidents per KV per year. In fact, we might measure that factor by either the variable *number of mass accidents among all mines belonging to a KV* (per KV per year) or, respectively, the average *number of miners involved per mass accident* (per KV per year) or, respectively, the *amount of pension or health expenditures directly caused by a mass accident relative to non-mass accident-caused expenditures*. Two things should be born in mind: (i) all three alternatives cannot be measured with the data available from the KV statistics; and (ii) all three alternatives directly link up with the issue of correlated claims and, thus, the assumption of independent insurants mentioned when equation (4.3) was introduced. Although my collective risk model of the KVs' insurance operations implicitly rests on the assumption of identical and independent insurants, I am aware that especially the latter assumption is not realistic. In order to account in some way for the violation of this assumption hidden in the KV-level data, I consider the additional variable *correlated claims* as explained above. Alternative to this variable, a better measure of the extent to which a KV faced *dependent* claims might be one of the three alternatives measuring the incidence and severity of mass accidents. If I had these variables, I could incorporate them either in equation (4.4) – the "aggregate claims distribution"-equation – or in equation (4.3) which explains the variance in a straightforward manner. Due to data unavailability, however, I do not have those variables to model the extent to which dependence of risks might be relevant in the context of KVs.

In order to check for the model's robustness to alternative distributional assumptions, I also perform a Gaussian and a negative binomial regression of equation (4.4) (not displayed below).[22] Moreover, the regression equation is estimated for five size classes separately because I suspect that the influence of the explanatory variables on costs might have been different for different KV sizes. In addition, this approach improved the fit since it reduced the strong influence of observations on coefficient estimates for the very large KVs.

4.2.2. Empirical results

This section begins with a discussion of the regressions explaining total pension and total sickness costs. Tables 4.4 and 4.5 display the baseline estimation results based on the assumption of a single Poisson aggregate claim process. What do estimation results tell us about the KVs' scheme? Obviously, not all explanatory variables have significant

---

22 See Jopp (2012a) for the negative binomial and Gaussian estimates. The negative binomial distribution is an alternative to the Poisson distribution, especially if suspecting overdispersion (mean < variance).

explanatory power, and only a few variables are significant in all ten models. Because elasticities are displayed, a coefficient can be interpreted as indicating the percentage change in the dependent variable given a one-percent change in the independent variable. Let us look at the variable *membership* first. Claims costs significantly increase if membership increases;[23] this holds for all models. Like *membership*, the variables *average payout*, *burden* (*pensioner burden* or *sick-day burden*), and *firm-related KV* are important, too. In all cases but one, raising the generosity per pension or per sick day leads to an increase in costs; this fact is straightforward, of course. Furthermore, a rising PCR drives costs up. We can interpret this effect as a direct consequence of an ongoing aging process among the KVs' members. However, the effect is not linear, implying that in a collective that had already aged, costs were increasingly controlled and kept down. The significance of the dummy that indicates firm-related KVs shows that those had, in the majority of cases, lower minimum costs than area-related KVs, which highlights the power that the "monopolistic" mine owner had in the administration.

With regard to total pension costs, the variable *firm share* is important only for the smallest KVs, implying that an increase in the employers' financing share increased costs. This also holds regarding sickness costs – except for the third size class – and leads to the conclusion that mineworkers were occasionally more prone to claim invalidity or sickness if they knew their fellows' pocketbooks would be relatively less burdened. It would, moreover, be consistent with the view that established miners were more costly per capita than unestablished mineworkers if the coefficients of the variable *established ratio* were statistically significant. What we can find, in fact, is that the structure of the contributor base with respect to this insurant characteristic helps to explain total costs in half the cases. The message is ambivalent: Regarding sickness and pension costs of the largest KVs, the elasticity is negative, meaning that an increase in the proportion of unestablished miners is connected with an increase in costs. This implies that those miners were, on average, not less costly per capita. This, in turn, might point to a higher claim number per year, possibly because they were less experienced in their jobs – as they were often part-time workers – and, thus, more endangered, or they reacted to a comparatively lower per capita benefit by claiming (and simulating) more.

Moreover, contrary to my expectations, the age structure of established contributors is not always important. Regarding pension costs, an increase in the share of older contributors, to the detriment of younger contributors, increases costs (except for the largest KVs). Regarding sickness costs, however, an increase in the same share reduces costs. This holds for the three largest size classes. An explanation might be that older contributors, who were past their productivity peak, were often re-allocated to less perilous occupations at the surface resulting in less sickness. Regarding pension costs, the potential for becoming disabled nonetheless remained. This is the reason why a positive coefficient would make sense.

---

23 This is in line with Emery and Emery (1999), p. 79, and with Broten (2010), p. 33.

Table 4.4: Explaining total pension costs (baseline Poisson model, dependent variable is the log of *pensions costs*, elasticities displayed)

	Model 1 (Size: 1–199)	Model 2 (Size: 200–999)	Model 3 (Size:1 000–4 999)	Model 4 (Size: 5 000–9 999)	Model 5 (Size: > 10 000)
Age(1625)	-0.020**	0.021	0.260**	-0.038	0.323
	(0.026)	(0.077)	(0.132)	(0.146)	(0.540)
Age(2635)	0.084	0.194	0.567**	0.099	0.612
	(0.053)	(0.146)	(0.232)	(0.163)	(0.983)
Age(3645)	0.099	0.060	0.453**	-0.056	0.585
	(0.069)	(0.137)	(0.189)	(0.151)	(0.662)
Age(4655)	0.168***	0.130	0.338***	-0.118	-0.029
	(0.061)	(0.086)	(0.106)	(0.085)	(0.288)
Age(>55)	-0.028	0.125***	0.120***	0.056	-0.223***
	(0.048)	(0.038)	(0.039)	(0.041)	(0.077)
Membership	0.554***	1.081***	0.850***	0.972***	0.391***
	(0.041)	(0.023)	(0.024)	(0.080)	(0.030)
Average payout	0.813***	1.100***	0.655***	0.697***	1.104***
	(0.054)	(0.024)	(0.021)	(0.043)	(0.069)
Pensioner burden	0.189***	0.939***	0.845***	2.394***	3.090***
	(0.019)	(0.022)	(0.033)	(0.151)	(0.222)
Pensioner burden^2	-0.038***	-0.138***	-0.136***	-0.711***	-0.872***
	(0.004)	(0.005)	(0.008)	(0.066)	(0.107)
Firm share	0.295***	0.243***	-0.014	0.006	0.046
	(0.075)	(0.090)	(0.031)	(0.078)	(0.041)
Firm share^2	-0.063**	-0.080**	0.005	0.001	-0.006
	(0.028)	(0.039)	(0.004)	(0.024)	(0.005)
Established ratio	0.355	0.825***	1.206***	-0.055	-0.701**
	(0.520)	(0.113)	(0.164)	(0.319)	(0.295)
Established ratio^2	-0.008	-0.367***	-0.522***	-0.046	0.293
	(0.310)	(0.067)	(0.091)	(0.172)	(0.179)
Hard-coal share	-0.008	0.344**	-0.091***	-0.113	-2.281***
	(0.013)	(0.136)	(0.012)	(0.169)	(0.265)
Brown-coal share	-0.002	0.157**	-0.053***	-0.206	-0.571***
	(0.001)	(0.076)	(0.009)	(0.208)	(0.079)
Iron-ore share	-0.004*	0.207**	-0.085***	-0.341	0.009
	(0.009)	(0.091)	(0.011)	(0.355)	(0.012)
Miscellaneous ore share	0.019	0.511**	-0.183***	-0.171	-0.459***
	(0.010)	(0.202)	(0.021)	(0.170)	(0.072)
Halite share	0.007	–	-0.004***	-0.024	-0.203***
	(0.013)		(0.001)	(0.021)	(0.032)
Stone share	-0.039**	0.244**	-0.054***	–	-0.000
	(0.019)	(0.001)	(0.005)		(0.005)

Table 4.4 continued

	Model 1 (Size: 1–199)	Model 2 (Size: 200–999)	Model 3 (Size:1 000–4 999)	Model 4 (Size: 5 000–9 999)	Model 5 (Size: > 10 000)
Steel share	0.000	0.746**	-0.192***	-0.127	-0.064***
	(0.012)	(0.301)	(0.025)	(0.159)	(0.005)
Salt share	0.134	0.018**	–	–	0.001
	(0.111)	(0.007)			(0.002)
Firm-related KV	-0.233***	-0.027***	-0.028***	0.003	0.068***
	(0.082)	(0.007)	(0.007)	(0.004)	(0.012)
Lagged pension	0.158***	-0.286***	0.005	-0.007	-0.065***
costs	(0.023)	(0.014)	(0.016)	(0.041)	(0.027)
Number of observations	756	1 168	1 031	292	347
Deviance	280 075	557 459	1 775 659	386 693	4 552 489
Pearson	261 501	528 302	1 644 436	339 672	4 550 141
Residual degrees of freedom	683	1 095	958	211	273
BIC	275 548	549 725	1 769 013	385 438	4 550 892

Notes: Standard errors are in brackets. Mining administration region and year effects are not displayed. *, **, and *** denote significance at the one-, five- and ten-percent levels.

Let us now turn to the empirical findings on the existence of scale economies regarding the KVs' actuarial risk as defined in the previous section. The contemporary observers of the KVs' operations believed that there was a positive relationship between KV size and financial stability. A reduction in the average costs' variance due to an increase in the insurance collective was linked with the stability issue, in that this effect was said to improve the predictability of costs; this would have been due to a reduction of the bandwidth of possible economic outcomes. Besides, as Albert Caron stated, a reduction in the variance would have been the necessary precondition to lower the implicit degree of risk-loading on top of the premium and, thus, to charge a premium on the representative mineworker that was closer to the expected value of cost; this would, then, have been an actuarially fairer premium. However, the question of whether or not KVs actually passed on such scale economies to their insurants is different from the question of whether or not there was the *potential* for such economies of scale at all.

According to equation (4.2), did the variance diminish if the collective of insurants increased? The empirical message of Table 4.6 with regard to the main size measure – *total contributors* – is that the variance did indeed diminish in both the pension- and sickness-insurance sections, but only up to a certain size.

Table 4.5: Explaining total sickness costs (baseline Poisson model, dependent variable is the log of *sickness costs*, elasticities displayed)

	Model 1 (Size: 1–199)	Model 2 (Size: 200–999)	Model 3 (Size:1 000–4 999)	Model 4 (Size: 5 000–9 999)	Model 5 (Size: > 10 000)
Age(1625)	0.087***	-0.085	-0.139**	-0.093	1.241
	(0.031)	(0.075)	(0.061)	(0.110)	(0.882)
Age(2635)	-0.074	-0.087	-0.241**	0.173	1.650
	(0.079)	(0.144)	(0.103)	(0.111)	(1.608)
Age(3645)	0.050	-0.020	0.032	0.015	1.670
	(0.096)	(0.135)	(0.089)	(0.108)	(1.085)
Age(4655)	0.034	-0.113	-0.194***	-0.171***	0.699
	(0.084)	(0.085)	(0.054)	(0.059)	(0.473)
Age(>55)	-0.077	-0.048	-0.076***	-0.073**	-0.233*
	(0.060)	(0.039)	(0.023)	(0.032)	(0.121)
Membership	0.504***	0.385***	0.749**	0.523***	0.611***
	(0.044)	(0.026)	(0.018)	(0.042)	(0.028)
Average payout	0.075***	0.049***	0.122***	0.238***	-0.044
	(0.023)	(0.007)	(0.004)	(0.042)	(0.053)
Sick day burden	0.173***	0.735***	0.592***	0.428***	0.909***
	(0.024)	(0.088)	(0.109)	(0.131)	(0.217)
Sick day burden^2	-0.017***	-0.194***	-0.077*	-0.015	-0.124
	(0.003)	(0.036)	(0.047)	(0.056)	(0.086)
Firm share	0.073	0.533***	-0.135***	0.100	0.150**
	(0.080)	(0.111)	(0.036)	(0.071)	(0.067)
Firm share^2	-0.020	-0.238***	0.010*	-0.033	-0.020**
	(0.022)	(0.050)	(0.006)	(0.022)	(0.009)
Established ratio	0.809	-0.010	0.478***	-0.145**	-0.303
	(0.508)	(0.123)	(0.149)	(0.210)	(0.386)
Established ratio^2	-0.843***	0.043	-0.206**	0.251**	0.302
	(0.312)	(0.075)	(0.081)	(0.119)	(0.259)
Hard-coal share	-0.013	-0.122	-0.023*	-0.150	-2.109***
	(0.019)	(0.167)	(0.013)	(0.147)	(0.359)
Brown-coal share	-0.000	-0.095	0.002	-0.165	-0.565***
	(0.001)	(0.084)	(0.007)	(0.166)	(0.096)
Iron-ore share	-0.029**	-0.110	-0.047***	-0.379	-0.010
	(0.013)	(0.113)	(0.110)	(0.309)	(0.020)
Miscellaneous ore share	-0.003	-0.230	-0.064***	-0.087	-0.409***
	(0.015)	(0.249)	(0.022)	(0.125)	(0.090)
Halite share	-0.009	–	0.001	-0.009	-0.173***
	(0.019)		(0.001)	(0.018)	(0.041)
Stone share	0.021	-0.115	-0.011**	–	0.014**
	(0.025)	(0.121)	(0.005)		(0.007)

Table 4.5 continued

	Model 1 (Size: 1–199)	Model 2 (Size: 200–999)	Model 3 (Size:1 000–4 999)	Model 4 (Size: 5 000–9 999)	Model 5 (Size: > 10 000)
Steel share	-0.028*	-0.308	-0.041	-0.121	-0.060***
	(0.016)	(0.345)	(0.026)	(0.137)	(0.008)
Salt share	0.039	-0.001	–	–	0.001
	(0.128)	(0.001)			(0.003)
Firm-related KV	0.201***	-0.023***	0.045***	-0.006**	0.108***
	(0.062)	(0.008)	(0.007)	(0.003)	(0.013)
Lagged sickness costs	0.210***	0.329***	0.102***	0.162	-0.184***
	(0.019)	(0.018)	(0.008)	(0.020	(0.012)
Number of observations	676	1 118	983	280	347
Deviance	132 037	571 426	1 570 583	171 087	6 202 664
Pearson	119 681	493 755	1 548 053	171 885	6 240 473
Residual degrees of freedom	603	1 045	910	209	273
BIC	128 108	564 091	1 564 312	169 910	6 201 068

Notes: Standard errors are in brackets. Mining administration region and year effects are not displayed. *, **, and *** denote significance at the one-, five- and ten- percent levels.

With regard to the variance of the average pension costs that a representative mineworker had to finance, the following findings need to be highlighted: At first, the variance diminished by around 2.5 percent if the number of contributors increased. Between a size of 200 and 999 contributors, the variance even diminished by 3.4 percent, given a one-percent increase in size. The variance decreased further – by 0.62 percent between 1867 and 1907 and by 0.96 percent between 1908 and 1913 – up to the size of about 5 000 contributors. Beyond that size, the statistical models suggest that there were no more scale economies. With regard to the variance of the average sickness costs, the picture is slightly different: A statistically significant effect cannot be detected for the smallest KVs, but model 2 suggests that the variance significantly diminished, by 1.5 percent (1867–1907) or 1.3 percent (1908–13), between 200 and 999 contributors, given a one-percent increase in size. While there is no significant effect in the third size class, there is one in the fourth, but in the opposite direction. Interestingly, the conditional increase of the variance is 1.4 percent between 5 000 and 9 999 contributors. There is, again, no significant effect for the largest KVs.[24]

[24] The estimates depicted in Table 4.6 are different from those I displayed in Jopp (2010). The difference is mainly a matter of using sets of controls that are not identical. However, the implications of both models are more or less equal, although my earlier – I would now say preliminary – piecewise linear model suggests for both insurance sections an optimal KV size of 5 000.

What does the second measure of size – *correlated risks* – tell us? This alternative measure is based on the assumption that a single workplace – i.e., a mine, a steelworks or a stone pit – can be treated as a closed system without being physically interrelated with another workplace. The KV statistics report how many of those workplaces existed in a KV's area or belonged to the company a KV was responsible for. The smaller the percentage of the average mine's (steelworks', stone pit's) size of *total contributors* was, the lower was the probability that a high number of correlated claims occurred at the same time given that we can measure the dependence of risks (the "accumulation risk" of an insurer) only very inaccurately – i.e., on the KV-level, and there only across mines, but not within mines. With regard to pension insurance, Table 4.6 suggests that an increase in the potential for correlated risks is significantly associated with an increase in the variance for the smallest KVs. Beyond that, regarding sickness insurance, the effect exists between 200 and 9 999 contributors. However, the finding for the largest KVs is counter-intuitive: In both our (hypothetical) pension- and sickness-insurance sections, a decrease in the ratio of *contributors per mine* to *total contributors* leads to an increase in the variance. How can this finding be explained? Assume that it is possible to assign to each mine a probability greater than zero but smaller than one that an accident will occur at a particular point in time. Let us consider that there is a second probability that we should account for – namely the probability that for a given number of different work places, for which a KV was responsible, an accident always occurred. Assumingly, this probability was the higher, for the more workplaces the KV had at the same time to provide insurance coverage. Now, if we take into account that the largest KVs were, indeed, responsible for more workplaces than one of the smaller KVs, the probability that always one, two, or even more than two mass accident(s) occurred – involving many miners at the same time – might have become much higher for larger KVs as the number of mines linked with them increased.

Depending on whether or not the issue of correlated claims is considered crucial, the findings make the case for sickness funds that should even be larger than pension funds. This is because, up to a size of approximately 10 000 contributors, the variance decreased if the potential for correlated claims decreased. This is particularly important with regard to the issue of infectious or parasitic diseases triggering epidemics. Bluma (2009) told a story about the hookworm (*anchylostomiasis*) epidemic in the Ruhr area at the time of the *Kaiserreich*, which was based on parasitic disease due to the lack of hygiene at the surface and especially underground.[25] In addition, Lauf (2006c) reviewed the influenza pandemic of 1918.[26] Findings imply that a *large* sickness fund covering *many* workplaces (i.e., separated mines) often was – in contrast to our thought experiment above – superior to a small KV in controlling for correlated risks.[27]

[25] See Bluma (2009a) and (2009b).

[26] See Lauf (2006c).

[27] This seems to have been especially true in dealing with epidemics. However, this implies that a large KV covering only few different mines was not superior.

Table 4.6: The relationship between KV size and actuarial risk (dependent variable is the variance of the average claim, elasticities displayed)

	Model 1 (Size: 1–199)	Model 2 (Size: 200–999)	Model 3 (Size:1 000–4 999)	Model 4 (Size: 5 000–9 999)	Model 5 (Size: > 10 000)
	Pensions	Pensions	Pensions	Pensions	Pensions
(1a) Contributors	-2.469***	-3.391***	-0.622**	0.880	-0.077
	(0.625)	(0.873)	(0.262)	(0.793)	(0.310)
(2a) Slope-dummy 1908–1913	-0.044	-0.050	-0.334***	-0.435	0.021
	(0.095)	(0.142)	(0.053)	(0.361)	(0.047)
R-squared (overall)	0.011	0.002	0.216	0.024	0.014
F-statistic	9.040	9.820	88.310	2.060	0.07
Prob > F	0.000	0.000	0.000	0.106	0.976
Number of observations	756	1,168	1,031	292	347
(3a) Correlated risks	3.770***	0.272	-0.122	-0.819	-2.342***
	(1.260)	(0.555)	(0.184)	(0.701)	(0.631)
(4a) Slope-dummy 1908–1913	-0.149	0.110	0.175***	-0.010	-0.233***
	(0.203)	(0.072)	(0.029)	(0.091)	(0.077)
R-squared (overall)	0.003	0.010	0.215	0.035	0.010
F-statistic	4.06	1.510	77.580	7.790	6.970
Prob > F	0.007	0.210	0.000	0.051	0.000
Number of observations	756	1 168	1 031	292	347
	Model 1 (Size: 1–199)	Model 2 (Size: 200–999)	Model 3 (Size:1 000–4 999)	Model 4 (Size: 5 000–9 999)	Model 5 (Size: > 10 000)
	Sickness	Sickness	Sickness	Sickness	Sickness
(1b) Contributors	-0.788	-1.530***	-0.404	1.429*	0.072
	(1.910)	(0.195)	(0.468)	(0.814)	(0.254)
(2b) Slope-dummy 1908-1913	-0.015	0.206***	(-0.163	-0.043	-0.032
	(0.335)	(0.043)	(0.102)	(0.289)	(0.045)
R-squared (overall)	0.003	0.061	0.046	0.133	0.010
F-statistic	0.070	25.54	9.960	4.00	4.11
Prob > F	0.978	0.000	0.000	0.008	0.007
Number of observations	676	1 118	983	280	347

Table 4.6 continued

	Model 1 (Size: 1–199)	Model 2 (Size: 200–999)	Model 3 (Size:1 000–4 999)	Model 4 (Size: 5 000–9 999)	Model 5 (Size: > 10 000)
	Sickness	Sickness	Sickness	Sickness	Sickness
(3b) Correlated risks	2.153	0.436***	1.466***	1.379**	-1.269***
	(4.204)	(0.153)	(0.367)	(0.680)	(0.466)
(4b) Slope-dummy 1908–1913	-0.077	-0.029	0.100**	0.120	-0.002
	(0.707)	(0.020)	(0.048)	(0.090)	(0.057)
R-squared (overall)	0.003	0.011	0.026	0.192	0.545
F-statistic	0.10	3.290	17.640	5.710	7.020
Prob > F	0.960	0.020	0.000	0.001	0.000
Number of observations	676	1 118	983	280	347

Notes: All variables in natural logarithms. Standard errors are in brackets. *, **, and *** denote significance at the one-, five- and ten-percent levels. All models are estimated with fixed effects as the Hausman-test suggests doing.

I now add a quadratic model to the piecewise linear one estimating the relationship for the entire addressed period, 1867–1913, and the full data set at once. Results using the fixed effects estimator are reported in Table 4.7. The coefficient of *contributors* is significantly different from zero and implies the expected inverse relationship between KV size and the variance of the average claim. An expansion of the collective of insurants appears to have reduced actuarial risk. That the signs of the incorporated squared terms are positive, however, suggest that actuarial risk, at some KV size, increased. This, in turn, clearly suggests not merely the existence of a minimum efficient size, beyond which neither further economies of scale nor diseconomies of scale were to be expected, but an optimal size. This does not hold if we focus on the alternative measure *correlated risks,* which is statistically significant in the context of sickness insurance only.

Table 4.7: Non-linear model of the relationship between KV size and the variance of the average claim

	Pensions	Sickness
(1) Contributors	-2.466*** (0.456)	-3.946*** (0.680)
(2) Contributors squared	0.146*** (0.037)	0.277*** (0.059)
(3) Constant	11.235*** (1.507)	14.033*** (1.838)
R-squared overall	0.283	0.049
R-squared between	0.674	0.345
R-squared within	0.024	0.031
F-statistic	19.35	28.17
Prob > F	0.000	0.000
Number of observations	3 594	3 403
(4) Correlated risks	0.402 *(0.438)*	0.627** (0.296)
(5) Correlated risks squared	-0.120 (0.091)	0.030 (0.091)
(6) Constant	3.371*** (0.497)	2.010*** (0.343)
R-squared overall	0.113	0.108
R-squared between	0.155	0.240
R-squared within	0.020	0.005
F-statistic	5.87	7.64
Prob > F	0.004	0.000
Number of observations	3 594	3 403

Notes: All variables in natural logs. Standard errors in brackets. *, **, and *** denote significance at the one-, five- and ten-percent levels. All models are estimated with fixed effects as the Hausman-test suggests doing.

While it is helpful to have estimation results of the non-linear model on hand, we cannot immediately derive an estimate of optimal KV size from them, as it was possible from the piecewise linear model. It actually requires some simply math to recover an estimate. Therefore, we have to create the first derivation of the regression equation underlying Table 4.7. Let us denote the coefficients of *contributors* and *contributors squared* with δ_1 and δ_2. At any point, the elasticity of the variance of the average claim with respect to KV size is

$$\frac{\partial \ln(var)}{\partial \ln(contributors)} = \delta_1 + 2 * \delta_2 * \ln(contributors). \tag{4.5}$$

Recall, also, that an elasticity of smaller than -1.0 indicates, in our particular case, a more-than-proportionate decrease in actuarial risk. An elasticity of between -1.0 and zero implies a less-than-proportionate decrease. So, as long as the elasticity ranges below zero, a KV would have experienced economies of scale regarding the reduction of actuarial risk. Moreover, consider an elasticity of between zero and +1.0 and an elasticity of beyond +1.0 as reflecting, respectively, a less-than-proportionate increase and a more-than-proportionate increase in the variance if size is raised; as long as the elasticity ranges above zero, a KV would have experienced diseconomies of scale accordingly. From the viewpoint of economics, operating on either side of an elasticity of zero implies an inefficient scale of production.

Solving equation (4.5) for the values of KV size associated with elasticities of -1.0, zero, and +1.0, given the model parameters and KV size data, enables us to define the ranges of increasing and decreasing returns to scale concretely. Table 4.8 reports those values for the pension- and the sickness-insurance sections and compares them to what we gathered from the piecewise linear model. While the latter suggests a MES in the pension-insurance section of about 5 000 contributors and of somewhere between 1 000 and 5 000 contributors in the sickness-insurance section, the quadratic model suggests an optimal KV size of about 4 750 contributors for the pension-insurance section and about 1 240 contributors in the sickness-insurance section. Complementing Table 4.8, Figure 4.3. gives a stylized visual impression of the optimal KV sizes recovered from the quadratic model, the estimates of which are a little more precise than those from the piecewise linear model.

Table 4.8: Summary of evidence – was there an optimal KV size?

Model	Actuarial risk elasticity with respect to size				Judgement
	$\beta < -1$	$-1 < \beta < 0$	$0 < \beta < 1$	$\beta > 1$	
A) Piecewise linear (size class) model					
Variance of the average pension claim	*SIZE* <1 000	*999* <*SIZE*< 5 000	–	–	MES
Variance of the average sickness claim	*199* <*SIZE*< 1 000	–	–	*4 999* <*SIZE*< 10 000	MES
B) Non-linear (quadratic) model					
Variance of the average pension claim	*SIZE* <153	153 <*SIZE*< 4 746	4 746 <*SIZE*< 146 890	*SIZE* >146 890	OS
Variance of the average sickness claim	*SIZE* <204	204 <*SIZE*< 1 240	1 240 <*SIZE*< 7 539	*SIZE* >7 539	OS

Notes: KV size values recovered from equation (6.5).

Figure 4.3: A graphical summary of the non-linear model

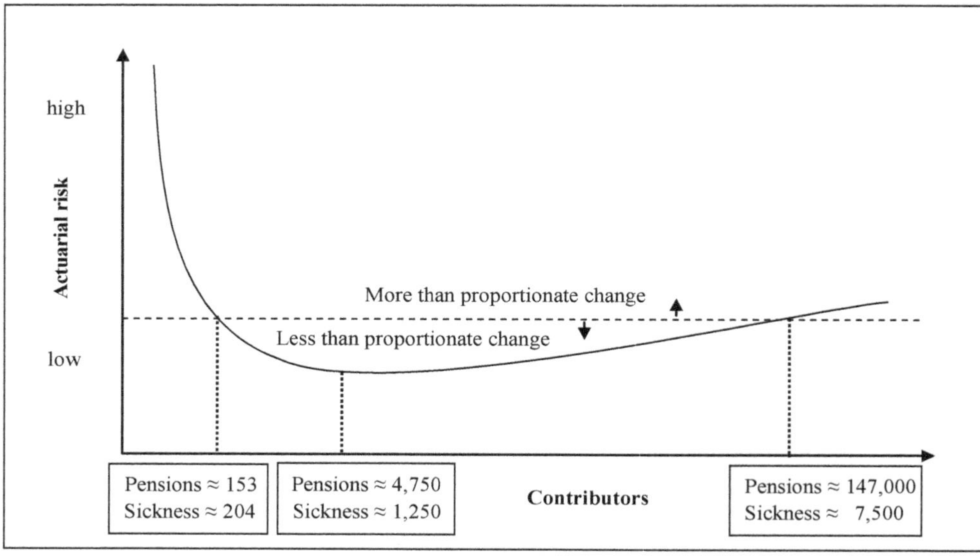

Notes: Actuarial risk is measured by the variance of the average claim.
Sources: My own depiction.

First, estimates suggest the existence of economies of scale with respect to lowering the actuarial risk inherent in providing insurance; thus, contemporary observers were on the right track when they claimed "the larger the better." Second, however, the optimal KV sizes regarding pension insurance, on the one hand, and sickness insurance, on the other hand, were definitely much smaller than any contemporary observer would pro-bably have postulated if forced into making a definitive statement.

Thus, contemporary observers were wrong to assume a strictly monotonic, negative relationship between KV size and actuarial risk up to infinity. The smallest KVs – all the ones that were perceived to be most ineffective in providing insurance to the miners – could have actually benefited from more-than-proportionate returns to scale for up to 150 or 200 contributors if they had thrived. By the same token, they could have experienced over-proportional diseconomies of scale if they had been subject to shrinkage. Growing towards the optimal scale would have meant realizing less-than-proportionate returns to scale but remaining positive. Beyond the optimal scale, actuarial risk had increased, either less-than-proportionately up to a KV size of 147 000 for pensions or up to 7 500 for sickness-related benefits, or over-proportionally once having surpassed even those scales – meaning that a one-percent increase in KV size had pushed actuarial risk up again by more than one percent.[28]

28 In the context of the quadratic model, as in the context of the piecewise linear model, controls beyond that for size are lacking. One might consider this a deficiency, in particular because the R-

Third, we can conclude that it obviously mattered for KVs' exposure to actuarial risk that they provided pension and sickness insurance in one scheme for a long period of time. Indeed, contemporary observers, from the perspective of the quadratic model, seem to have been definitely right to claim that not having separate funds for different life risks was a basic deficiency. KVs seem to have faced a trade-off between mutually exclusive scales as long as the bundling existed.

The fact that the optimal size of the sickness-insurance section, with respect to reducing actuarial risk, was notably lower than that of the pension insurance section is consistent with the case that Guinnane and Streb (2011) and Guinane et al. (2012b) made regarding small KVs' sickness-insurance section: Smaller entities might have had some advantage in controlling costs through social sanction mechanisms, for example.[29]

In light of the empirical results presented in this subsection, the merger wave among KVs seems to be not justified – not entirely, at least – by economic reasoning since (i) many KVs remained operational at a scale below the suggested MES/OS and were not subject to what I called above, "rescue absorptions;" and (ii) many mergers exceeded the suggested OS and, thus, might have experienced rising actuarial risk. How can the insufficient quality of the observable concentration process be explained? From the viewpoint of the whole insurance system, it must have been of the utmost priority to ensure the security of supply with insurance coverage. In particular, this means to ensure every member's legal entitlements to (future) benefits. In this respect, it would have been collectively rational to conduct mergers that eliminated all small KVs and to integrate those memberships into larger funds. In fact, the growth opportunities of many KVs were naturally bounded from above by the dimension of their area or their company (i.e., the capacity of the underlying resource deposits). This limited the number of miners that were employable and, thus, the possibility of approaching the OS exclusively by internal growth. However, KVs could break through this obstacle by either absorbing another KV or by being absorbed. Chapter 6 will extend this issue.

Departing from the purely technical implications of the model concerning which KV size is optimal – or whether there simply was a MES –, it might be, in fact, appropriate to downgrade the identified optimal scales to minimum efficient scales. This is for the reason that the actuarial risk increased, after passing the OS, a lot more slowly than it decreased in the case a small KV grew towards the OS. If we can accept the MES-scenario – KVs needed to have a certain size to operate efficient, but this size was not

---

squared values are rather low. My main argument for not including further controls is that I estimated the variance of the average claim in a two-step procedure, in which almost all controls one might think of are used to estimate the expected values of the total claims. Having said this, I believe that it would be inconsistent to use them again to explain the variance. What I have actually done is to re-estimate the quadratic model with the PCR as an additional control (not displayed), such that the effect of KV size is measured at the mean of the PCR. In fact, optimal KV size does not change dramatically when derived with the alternative coefficient estimates.

29 See Guinnane and Streb (2011) and Guinnane et al. (2012b).

that large –, what remains, in this case, still is the insight that having been larger than those scales would not have given a KV a significant advantage: The KV could have been larger without decreasing efficiency considerably, but it would not have needed to be.

So, after having interpreted the implications of the models, I still owe a substantive explanation for why the KVs' optimal size should have been, in fact, so low. I think that we can explain this counterintuitive result against the background of accumulation risks inherent in mining and, thus, the KVs' operation. As I said above, I cannot prove the relevance of correlation between claims. Rather, I take it for granted that accidents involving more than just one miner and epidemics constituted relevant accumulation risks. These risks itself could not be eliminated or substantially reduced unless miners were pooled with other occupations that generally reflected a lower degree of accumulation risk, or none at all. Actually, those accumulation risks can be thought of as being prevalent at every level of size. Hence, minimization of actuarial risk by diversification among an increasing number of insured became optimal at a quite low number of insurants; if there had been no correlation among claims, actuarial risk would have likely been minimized further and further by growing, as theory suggests. The obvious implication is that contemporaries had the wrong idea of how accumulation risks and KV size were interrelated.[30] Let me put it this way: From an actuarial risk perspective, a necessary precondition for the creation of large KVs – as a socially desired aim, so to say – would not have been, in the first instance, the separation of the KVs' pension- and sickness-insurance sections, but the pooling of the mining sector with a lot of other sectors. However, as long as contemporaries of the KVs supported the idea of maintaining a strictly specific social insurance for miners, they implicitly voted – for actuarial reasons and not that they knew, obviously – for a lot of small(er) funds.[31]

Finally, if a KV did not need to be larger – or, for technical reasons, could not be larger! – than about 5 000 contributors or, depending on the emphasis, than 1 000 contributors to adequately deal with actuarial risk, would there still be arguments in favor of larger organizations? In my view, there are at least two relevant arguments. On the one hand, one could debate whether a large KV size well beyond 5 000 or 10 000 mem-

---

30 It would be wrong to say contemporaries neglected the importance of accumulation risks at all. As Subsection 1.2 showed, they incorporated those risks explicitly in their argumentation.

31 There is an interesting point arising if we consider the *Knappschaftliche Rückversicherungsanstalt a. G.* of 1907 (see Chapter 2). Recall that taking reinsurance is a way to spread locally very important accumulation risks (like earthquakes and tsunamis) over geographical areas where the respective accumulation risks are not that important. Against this background, the KVs' reinsurance arrangement is an example of spreading risks on a small geographical scale, namely over localities within Prussia. This spread, however, seems to be strange in that risks – let us assume the *Knappschafliche Rückversicherungsanstalt* really spread risks (like the longevity risk) – were distributed over localities that reflect the same accumulation risks. Besides the fact that this is a consequence of the job-relatedness of the KVs' insurance, this seems not to be the best foundation for reinsurance to function.

bers would have been helpful in lowering the relative administrative overhead of the system. This question is highly relevant with regard to, for example, the current German social security system. It might turn out in an empirical analysis that larger KVs had serious advantages in that respect. The subsequent subsection will deal with precisely this question. On the other hand, one could argue that economies of scale existed with respect to holdings of financial reserves: A large size might have enabled a KV to reduce the financial reserves necessary to maintain a desired security level as measured by the probability of ruin.[32] Exploring this relationship, however, is left out here, in particular because it seems that we cannot take for granted that a KV's main financial aim was to ensure a specific security level with least possible reserves.[33]

32 Jopp (2010a) experimented with a kind of event-history model, which is based on assuming two basic structural breaks – one occurring between the transition from the pre-merger to the post-merger period and the other occurring when an assumed period of a few years after the merger have passed. The idea is to assess whether the exposure of newly merged KVs to actuarial risk was significantly different for a small period immediately after the merger. Results do not strictly imply a significant shock to actuarial risk in the one direction or the other. However, I do not consider that model here.

33 Left out at this point is an extension to the basic actuarial risk framework serving to broaden the perspective on economies of scale. This extension establishes a relationship between the pricing of the average insurance contract and actuarial risk by assuming that KVs specified risk loadings as part of the average contribution payment. Consider the average contribution to consist of three parts: the part that covers expected claims; the part that covers operating costs; and the part that covers unexpected financial shocks. The latter part is substantiated by the risk loading (e.g., as a percentage of expected claims). Contemporary observers thought about the effects that a growing or shrinking size would have on the pricing of the average insurance contract. They argued that risk-loading on top of the net contribution payment had to be higher for smaller KVs, reflecting more variation and a higher actuarial risk to be covered. Analogously, less actuarial risk associated with larger collectives of insurants would have enabled KVs to reduce the risk-loading included in the average contribution. As a consequence, the contribution would have been more actuarially fair, which would have benefited insurants. This extension sets out to test the hypothesis that KVs passed cost advantages due to growing size on to miners by lowering the risk loading of the average contribution. What we find, in fact, is the opposite: As a bottom-line, KVs did not systematically reduce risk loadings, provided they grew; rather, if any, they accumulated "unused" risk loadings to build up reserves. The ultimate reason for this policy surely is the lacking pressure by competition. Although, as pointed out earlier, we cannot exclude that members' turn-over from one KV to another forced KVs into some competition, it is clear that KVs did not compete each other directly via establishing branches everywhere; see the long version of this study (i.e., Jopp 2012b) for empirical strategy and results.

4.3. Searching for the minimum efficient size of a *Knappschaft* – Findings on the empirical relationship between size and operating costs

4.3.1. Empirical strategy

In formulating a statistical model to assess the relationship between non-benefit costs and KV size, I choose an approach in which it is not necessary to properly specify input prices and output quantities. Measuring both quantities is exactly the problem I face with KVs. There is no doubt that their administrative fundament was, to some extent, based on honorary work, especially if we consider the Elders, who are said to have taken over certain monitoring activities or maintaining membership list to keep the board up-to-date; this saved costs *ceteris paribus*. KVs, as compulsory insurers, did not even face the sort of operating cost that other funds, based on voluntary membership, would have been likely to spend – namely, acquisition or advertising costs. However, to produce insurance coverage, KVs needed inputs such as professional personnel – if not for small memberships, then certainly, from a certain size on, for larger memberships (e.g., an actuary, hospital staff, legal staff, clerks); some office space and related maintenance; depreciation of tangible assets; possibly debts; and some kind of basic "data management system." Prices to be associated with such inputs include corresponding wages, rents and corresponding interest rates if debts existed. To the extent that we can assume that not all board members – especially not the employers' representatives – would work for free, we even have to consider some wage-equivalence for doing KV board work.[34] All these input prices are rather difficult to measure. Nevertheless, had I done all of that, I would likely have a set of KV-independent input prices on hand that would not be useful, in any sense, to identify between-KV-heterogeneity.

Furthermore, measuring the output of insurance production – the coverage, so to speak – seems to be linked with certain problems as well. The basic problem is that the theoretically correct measure of insurance output – the sum of conditional promises produced over a certain period of time expressed in monetary units – is rather difficult to determine.[35] As a proxy for output, the empirical literature, thus, proposes the effectively settled amount of claims; in the case of KVs, this is total insurance-benefit expenditures as annually reported in the KV statistics. As Subsection 4.1.2 highlighted, operating costs per produced unit of insurance benefit seem to have decline over size classes, even though not over the full range of KV size. However, I use as a proxy of output the number of contributors and pensioners, on the one hand, and paid sick days

---

[34] The KV statistics principally allows incorporating the numbers of different personnel.

[35] This definition of the output of insurance follows Zweifel an Eisen (2003), pp. 185–187.

per contributors, on the other hand, avoiding possible shortcomings of total insurance benefits as a monetary measure.

In order to test for the determinants of the KVs' operating costs (*OP*), I specify a cost function in which no input prices appear, but a set of controls that assumingly influenced the amount of generated operating costs.[36] The cost function is specified according to Bateman and Mitchell (2004) as

$$(4.6) \quad \ln(OP_{it}) = \alpha_0 + \alpha_1 \ln(SIZE_{it}) + \alpha_2 \ln(X_{it}) + \upsilon_{it}$$

and is intended to explain the *i*-th KV's real total operating costs in period *t*, suggesting a panel-data specification.[37] *SIZE* represents a linear and a quadratic term of the variable *participants* as the sum over working members and pensioners. *X* represents controls that capture part of the heterogeneity on the fund level. Included is a set of continuous variables that are partly known from above: the percentage of pensioners, the percentage of survivors, the percentage of established members among all working members, the employers' financing share in costs, the young ratio as a proxy for a KV's age structure, the number of paid sick days per working member, and real assets per member in marks net of immovable and movable property to account for the possibility that a KV had to engage in asset management. The dependent variable and all continuous right-hand side controls are in natural logarithms.

A set of dummy variables complements the specification. I controll for (i) the mining administration region in which a fund was situated; (ii) the subsector in which the majority of a fund's members were employed; (iii) the link to a company; (iv) the First World War; and (v) the existence of immovable property. In the first three cases, one dummy was omitted to avoid multicollinearity (*Bonn*, *hard-coal*, and *area-related fund*). The dummy variables would, of course, cancel out if the fixed-effects estimator were found superior to the random-effects estimator. Rather than to hope that the latter would be the one suggested by, for example, a Hausman-type specification test, so that the dummy variables' coefficients are reported, I additionally perform a pooled OLS regression of (4.6).

I estimate equation (4.6) separately for the compound-insurance period (1861–1907), the post-reform pension-insurance section, and the post-reform sickness-insurance section (both 1908–1918). Finally, I drop the years of disguised inflation once more.

4.3.2. Emprical results

To begin with, Table 4.9 displays coefficient values for the three specified models if equation (4.6) is estimated using pooled OLS. Displayed are cost elasticities indicating by how many percent *real total operating costs* increase or decrease if the explanatory

---

36 See, for example, Bikker and De Dreu (2009) and Mühlenkamp (1995a) for similar approaches.

37 See Bateman and Mitchell (2004) for the pooled approach toward equation (4.6).

variable changes by one percent. The basic idea behind the pooled specification is that it allows me to figure out the influence of certain dummy variables, which is, in the panel data model presented hereafter, cleaned out by the fixed effects. Note that, in a pooled OLS regression framework, observations on KVs are treated as independent of each other – especially within cross-sections. Since this assumption might be unrealistic, I report clustered robust standard errors, which allow for intra-KV correlation.

Let us begin with model 1 on the compound-insurance period, which provides baseline evidence on the empirical relationship we are interested in here. As expected, the variable *participants* is highly significant, implying a *ceteris paribus* less-than-proportionate percent increase in *real total operating costs* of only 0.64 percent if KV size increased by one percent. The squared term is significant, too, suggesting that the cost elasticity with respect to participants increased as the number of participants did. This clearly implies the existence of an optimal KV size, which can be recovered as done above: A KV reached constant returns to scale at about 42 000 participants, evaluated at the mean of all other variables, so, in particular, at a mean percentage of pensioners of about 22 percent.

Furthermore, operating costs are significantly positively related to the percentage of pensioners, the young ratio, sick days per working member, and real assets per member. An increase in these variables led to a less-than-proportionate increase in operating costs. If a KV had immovable property, fixed costs were likewise raised. KVs associated with the brown-coal subsector apparently had higher fixed costs. In contrast, KVs linked with related industries and stone and halite production had lower fixed costs. Finally, an increase in the percentage of established members decreased costs.

The results on the post-reform period tell a different story with respect to size. First, regarding the pension-insurance section, only the variable *participants* is statistically significant. This implies increasing returns to scale over the full size range compatible with the view "the larger the better." Every increase in participants led to a less-than-proportionate increase in operating costs. In contrast, regarding the sickness-insurance section, the significant coefficients on *participants* and *participants squared* suggest that there existed scales associated with increasing, constant *and* decreasing returns to scale. The optimal scale is recoverable from the estimates and amounts to about 5 750 working members, which tells of a clearly different situation in both insurance sections after the institutional separation. Beyond that, the youth ratio is significant in both cases but is positively related to costs for pension insurance and negatively related for sickness insurance. The fact that KVs had assets to be managed was another significant cost driver, in the pension-insurance section more than in the sickness-insurance section.

Table 4.9: The determinants of the KVs' operating costs in a pooled model (dependent variable is *real total operating costs*, cost elasticities displayed)

Variables	Model 1: Compound insurance (1861–1907)	Model 2: Pension insurance section (1908–1918)	Model 3: Sickness insurance section (1908–1918)
(1) Participants	0.638***	0.813***	0.394**
	(0.146)	(0.175)	(0.180)
(2) Participants squared	0.017*	0.006	0.035***
	(0.009)	(0.010)	(0.010)
(3) Percentage of pensioners	0.124*	-0.035	–
	(0.065)	(0.168)	
(4) Percentage of established members	-0.193**	–	–
	(0.081)		
(5) Employers' share in costs	0.059	–	–
	(0.065)		
(6) Young ratio	0.181**	0.294	-0.271*
	(0.071)	(0.183)	(0.140)
(7) Sick days per 100 working members	0.161***	–	0.079
	(0.050)		(0.093)
(8) Real assets per member	0.178***	0.421***	0.201***
	(0.045)	(0.070)	(0.062)
(90) Immovables dummy (1 = yes, 0 = no)	0.150*	0.293**	-0.067
	(0.084)	(0.131)	(0.132)
(10) Firm-related fund dummy	-0.064	-0.114	0.188
	(0.200)	(0.176)	(0.170)
(11) OBAB Breslau dummy	-0.127	0.422*	0.302
	(0.279)	(0.215)	(0.227)
(12) OBAB Clausthal dummy	0.053	-0.133	-0.757***
	(0.207)	(0.291)	(0.279)
(13) OBAB Dortmund dummy	-0.262	0.150	-0.110
	(0.159)	(0.212)	(0.171)
(14) OBAB Halle dummy	0.159	0.218	0.161
	(0.115)	(0.174)	(0.233)
(15) Hard coal dummy	-0.078	-0.135	-0.076
	(0.151)	(0.240)	(0.317)
(16) Brown coal dummy	0.428***	-0.048	0.516*
	(0.145)	(0.269)	(0.303)
(17) Salt dummy	-0.413	-0.280	-0.992*
	(0.341)	(0.280)	(0.532)
(18) Stone and halite dummy	-0.291*	-0.051	-0.339
	(0.152)	(0.213)	(0.259)
(19) Related industries dummy	-0.653**	-0.787***	-0.582**
	(0.258)	(0.196)	(0.276)

Table 4.9 continued

Variables	Model 1: Compound insurance (1861–1907)	Model 2: Pension insurance section (1908–1918)	Model 3: Sickness insurance section (1908–1918)
(20) Mixed dummy	-0.205	-0.209	0.141
	(0.148)	(0.202)	(0.247)
(21) WWI dummy	–	-0.364***	-0.191***
		(0.051)	(0.062)
(22) Constant	1.362**	-0.699	1.908**
	(0.675)	(1.021)	(0.921)
R-squared	0.914	0.952	0.9329
F-statistic	264.17	193.20	469.56
Prob > F	0.000	0.000	0.000
Number of observations	3 508	653	627

Notes: Clustered robust standard errors are in brackets (it is clustered by KV). *, **, and *** denote significance at the one-, five- and ten-percent levels.

Now that we have evidence from pooled models on hand, what do the panel-data models suggest? Actually, the picture these draw is quite different, as Table 4.10 shows. Addressing the compound-insurance period, it is only the squared term that is statistically significant. This means, however, that the cost elasticity with respect to the number of overall participants depends on scale. To be precise, the panel model of the period 1861–1907 suggests that small KVs experienced the highest economies of scale. The threshold number of participants, evaluated at the mean percentage of pensioners of 22 percent and reflecting constant returns to scale, was, though, extremely high – namely, about 3.8 million participants. This is way beyond the size that KVs reached in the observation period and even well beyond the size realized by the foundation of the *Reichsknappschaft* in 1923 (even beyond the size of the carriers of Bismarckian invalidity and old-age insurance). So, it seems reasonable to assume that KVs over the full size range realized increasing returns to scale during the compound-insurance period. This finding is, in turn, compatible with the slogan "the larger, the better" – in contrast to what has been found concerning the exposure to actuarial risk.

As in the pooled case, real operating costs increased as sick days per contributors and real assets per member did. Relying upon immovable property also can be associated with higher baseline operating costs. Interestingly, however, neither the youth ratio nor the percentage of established members plays a major role in explaining costs. Besides, the employers' share in costs is significant and suggests a modestly positive relationship.

Table 4.10: The determinants of the KVs' operating costs in a fixed-effects model (dependent variable is *real total operating costs*, cost elasticities displayed).

Variables	Model 4: Compound insursance (1861–1907)	Model 5: Pension insurance section (1908–1918)	Model 6: Sickness insurance section (1908–1918)
(1) Participants	0.224	1.493***	0.833***
	(0.203)	(0.518)	(0.270)
(2) Participants squared	0.033***	-0.051	-0.030
	(0.012)	(0.033)	(0.021)
(3) Percentage of pensioners	0.042	-0.215*	–
	(0.040)	(0.120)	
(4) Percentage of established members	0.024	–	–
	(0.058)		
(5) Employers' share in costs	0.085**	–	–
	(0.042)		
(6) Young ratio	0.062	0.017	-0.099
	(0.037)	(0.048)	(0.098)
(7) Sick days per 100 working members	0.065**	–	0.007
	(0.026)		(0.048)
(8) Real assets per member	0.174***	0.391***	0.141***
	(0.037)	(0.077)	(0.038)
(9) Immovables dummy (1 = yes, 0 = no)	0.133**	0.134	-0.037
	(0.060)	(0.157)	(0.070)
(11) WWI dummy	–	-0.281***	-0.162**
		(0.046)	(0.065)
(12) Constant	3.264***	-2.786	2.765**
	(0.863)	(1.990)	(1.299)
R-squared overall	0.869	0.877	0.826
R-squared between	0.900	0.886	0.827
R-squared within	0.519	0.337	0.153
F-statistic	34.4	29.97	10.18
Prob > F	0.000	0.000	0.000
No. of observations	3 508	653	627

Notes: Clustered robust standard errors are in brackets (it is clustered by KV). *, **, and *** denote significance at the one-, five- and ten-percent levels. A Hausman-type specification test suggested using fixed-effects.

The models on the post-reform period indicate, on the one hand, that KVs operated with decreasing returns to scale, regardless of scale, in the case of pensions (coefficient on *participants* > 1.0) and, on the other hand, with increasing returns to scale, regardless of scale, in the case of sickness insurance (coefficient on *participants* < 1.0). So, from the perspective of sickness benefit provision, the unbundling of the pension and the sickness insurance sections after 1906/07 eliminated some synergy. Otherwise, the coefficient of about 1.5 is not reasonably explainable.

To finish this subsection, I ask: What do we do with all the evidence pointing in different directions, implying economies of scale, on the one hand, but diseconomies, on the other hand? First of all, I am not convinced that one should take the models on the post-reform period too seriously. I come to this conclusion for the following reason: We cannot be completely certain that KVs as a whole ran two entirely de-linked administrations after 1906/07. The separation of the pension- and sickness-insurance sections was, as reflected by the KV statistics, a separation with respect to accounting, which renders the whole issue of reporting revenues and expenditures an issue of distributing costs to sources. While expenditures on pensions, for example, were easily assignable to the pension section from an accounting point of view, I doubt that this was that easy for KVs concerning operating costs. If KVs had not duplicated their administrative unit, how can we be sure that the operating costs reported for pension insurance and for sickness insurance give a true picture? The observation that KVs apparently operated with decreasing returns to scale in the pension-insurance section might, thus, be an artifact of the way KVs somehow distributed costs that were not clearly assignable to cost centers. That would make the statistical evidence not very informative at all. It seems, thus, reasonable to put less emphasis on the implications that the post-reform models have regarding KV size.

What remains, then, are the implications of the models for the period 1861 to 1907. Using the pooled-model, I have estimated an optimal KV size of about 42 000 members (or approximately 33 000 contributors, given a mean percentage of pensioners of 22 percent). Using the fixed-effects model, the estimated OS lies beyond any scale realized in the observation period (so, factually, it implies increasing returns to scale). In the former case, the cost elasticity becomes larger than 1.0 after surpassing optimal scale, but remains well below 1.08 until an overall size of about 442 000 members (or 345 000 contributors) is reached. Such a cost-elasticity might still be economically seen as "close to constant returns to scale."

5. The hazard of exit: Explaining the concentration process[1]

This chapter investigates the determinants of mergers and liquidations among the Prussian KVs, drawing on the results from the previous chapters. As shown above, most contemporary sources conveyed the impression that KVs *were to be* merged – and actually *were* merged – to stabilize the entire insurance scheme by sorting out actuarially unviable and financially distressed funds. In contrast, the statistical evidence presented below suggests that the KVs were absorbed over time primarily because they offered advantages to the absorbing KVs and, hence, were quite attractive targets. This contradicts the rescue hypothesis and the idea of cross-KV solidarity.

5.1. To merge or not to merge … or to liquidate?

Regarding internal and, especially, external concentration, the findings from the previous chapter suggest that a merger was not necessarily efficiency-enhancing. Insofar as it pushed the absorbing KV (the "mother," so to speak) way beyond the MES/OS, actuarial risk was likely to stay constant, or to increase. For a small KV to be merged into a larger one could have resulted in quite an essential release of pressure for the newly-formed KV since the variance of the average claim it faced was very probably larger pre-merger than post-merger. In addition, mergers presumably did not affect operating costs that negatively. In light of what the previous chapter presented, whether or not a particular merger – evaluated from an ex-post perspective – was useful, KVs were indeed merged with others or, alternatively, liquidated. This begs for explanation, which I will give in this chapter using mass data on the KVs that may be captured as observable realizations of unobservable decisions.

According to the basic regulations of 1854 and 1865, the mining administration lost its function to run KVs on the government's behalf. As Chapter 2 outlined, KVs were, instead, to form their own managing boards (*Knappschaftsvorstände*) consisting of miners' and employers' representatives. Those individuals decided on all business mat-

[1] An earlier version of this chapter is published as a part of the article "The hazard of merger by absorption: Why some Knappschaften merged and others did not, 1861–1920;" see Jopp (2011a).

ters – thus also on mergers and liquidations. Contemporary observers such as Heinrich Imbusch or Ferdinand Bertrams expressed their worries about the firms having the real power on the managing boards. They were concerned that firms could make decisions according to their own interests, thanks to the Elders, who often elected representatives more friendly to employers than to their official principals.[2] Lauf recently stressed that it was the mine owners or salaried managers, respectively, who had a clear intellectual advantage over Elders, which gave them "opinion-leadership."[3] Besides, employers could credibly threaten to fire uncooperative elder miners (as long as they were active, of course) or influence the loyalty to their boss by giving them some kind of administrative tasks in the KVs (control the sick, for example).[4] As Geyer argues, the question of who really ruled KVs is decisive to understand KV business policy and who profited most from decisions.[5]

Based on the presumption that employers and employees had different interests, much scholarly attention has been devoted to the modes of social and economic interaction among miners, owner-managers or salaried managers, on one side, and the state as provider of the relevant regulatory framework, especially regarding conflict regulation, on the other.[6] The industrial relations historiography has one important implication for this study: Given that conflicts of interest between miners and their employers existed regarding contractual issues, working conditions, safety issues and mine owners' paternalistic attitudes, and given that those conflicts deepened as miners organized their opposition in the form of trade unions, those conflicts must also have carried over to the KV boards and must have influenced the decision-making process to a certain extent.[7]

It is still unclear which role the different actors played. It also remains unclear why a particular alternative – continuation of operations, merger or liquidation – was chosen in a particular period. This part of the investigation explores indirectly whether one can assume economic, and especially actuarial, motives. In general, the economic literature proposes a variety of motives for firms to merge – for example, to increase market share, diversify risks, exploit economies of scale and scope, reach minimum efficient size, and also to avoid insolvency.[8] As discussed in the introduction, it may have been in the interest of a KV's management to view a merger as an insolvency alternative in the face of financial distress.[9] In contrast, it might have been in the interest of a KV to

---

2 See Imbusch (1910), p. 63, and Bertrams (1912).

3 See Lauf (2006a), p. 272.

4 See Imbusch (1910), p. 63, and Geyer (1992), p. 1 050.

5 See Geyer (1992), pp. 1 046–1 047.

6 See, for example, Adelmann (1963), pp. 413–427, Tenfelde (1977), Ullmann (1979), Weisbrod (1989), and Wisotzky (1992).

7 See Lauf (2006a), p. 272.

8 See, for example, Shrieves and Stevens (1979), Patena and Ruland (1986), Jensen (1988), BarNiv and Hathorn (1997), Andrade (2001), and Focarelli et al. (2002).

9 In a general economic context, BarNiv and Hathorn elaborate on the "insolvency alternative" motive; see BarNiv and Hathorn (1990).

absorb another fund because there were perceived opportunities to stabilize one's own growth and reduce one's own costs; members of the absorbed KV may have profited, too, or may not have profited at all.

From the opinion in the literature that the firms dominated KV boards, it follows that business policy very likely reflected mine owners' interests. What were their interests? In my opinion, their main interest was economic at its core: being able to maintain a healthy, satisfied and productive workforce at the lowest possible cost. Several findings in the statistical section of this chapter will be in line with this view. Since employers' contributions to KVs were a true cost item – even more so after implementing additional labor costs due to co-financing Bismarckian insurance – they were interested in keeping costs down. The point is that a KV encountering financial trouble, whatever the cause – business-cycle fluctuations, structural decline, long-term demographic changes, or erratic expenditure shocks due to accidents –, led to additional costs for firms. If the PCR had risen, this inevitably would have required raising contributions, which firms also had to pay. Hence, it was in the economic interest of firms to operate a financially healthy KV. In fact, the contemporaries' discussion on actuarial deficiencies of KVs gave firms attractive arguments for how to improve a fund's financial soundness.

The following empirical exercise will test two straightforward hypotheses against each other: According to the "rescue" hypothesis, KVs absorbed other funds to rescue them from financial distress – for example, in the form of too-high operating costs per member, too-high actuarial risk or a too-high burden from pensioners. According to the "self-interest" hypothesis, KVs absorbed other funds only if absorptions helped to improve their own growth prospects. This issue is assessed from the perspective of the KVs that exited, due either to absorption or to liquidation. Before turning to the empirical results, the subsequent subsection clarifies the methodological approach towards KV exit (or, interchangeably, KV survival).

5.2. Empirical strategy

5.2.1. A Competing risk approach to *Knappschaft* survival

This chapter makes use of the information from the number of years that a KV was in operation from 1861 until one of the following three events occurred: first, a KV ceased operation due to liquidation; second, a KV was merged into another fund; or third, a KV reached the formal end-year of the quantitative analysis, 1920; the latter event reflects what the literature calls a case of censoring. Analysis of time-to-event data has become known as "survival analysis" and the relevant body of literature has become increasingly rich over the three decades since Cox's seminal contributions.[10]

---

[10] See Cox (1972, 1975).

To start with, let index i denote the observed KVs, $i = 1, \ldots, D$. Furthermore, let the survival time of a KV be defined as the time elapsed from its foundation to its market exit. KVs exited the market due to either the event "merger" or the event "liquidation." Chapter 3 showed that most KVs were absorbed by an already established fund, rather than being merged into a new one (type A and type B mergers). What I call "absorption" is actually synonymous with the more commonly used term "acquisition." Furthermore, liquidation of a firm could principally be due to voluntary withdrawal from business or to bankruptcy. While this distinction is often made in empirical studies on firm survival, KV data do not allow making it.

Thus, let T^M and T^L denote two nonnegative random variables representing uncensored survival time until a merger or liquidation occurred. Furthermore, let t denote each of the years in which KV_i is factually observed, $t \in [t^{beg}, t^{end}]$, where t^{beg} and t^{end} mark the beginning and end of the study, and t^M and t^L denote the periods in which the event of interest took place. Indeed, it is ideal to observe uncensored survival times for each KV. This means that both a KV's true founding date and exit date do not lie outside the observation period. Observing a KV from the known start-up year, t^S, onwards until the particular event of interest, it would, thus, hold that $t^{beg} \leq t^S \leq t^M$ $(t^L) \leq t^{end}$; the data would be neither left-censored nor right-censored. Nevertheless, the presence of censoring obviously would imply methodological problems.

In fact, many duration data studies are based on incomplete observations because of censoring, and this problem matters for this analysis, too. One should bear in mind that its presence leads to deviations from the ideal model: If censoring occurs, survival time is incorrectly measured, implying a potential error in measurement of the dependent variable. Right-censoring occurs if KV_i had not exited in 1920, the end-year of the quantitative analysis, and the event of interest potentially occurred sometime thereafter, t^M $(t^L) > t^{end}$. Given that $t^S \geq t^{beg}$, the observed duration of the KV's operational time is the minimum duration until merger or liquidation and right-censoring. A particular advantage of duration data models is that they explicitly allow for right-censored durations, so that this form of censoring is no greater problem. Left-censoring, in contrast, is a problem for which, to the best of my knowledge, no systematic correction procedure exists. Thus, true survival time is unknown, and the survival time observed is best labeled as "minimum survival time."[11] To summarize, of 103 Prussian KVs 15 represent a full uncensored case. Right-censoring occurs for 17 and (potential) left-censoring for 44 KVs. Finally, 27 KV are subject to double-censoring meaning that they came into being sometime before 1861 and survived beyond 1920.

Moreover, regarding competing risk regression design, there basically are three ways to construct firm survival models given that firms may exit through liquidation (volun-

[11] See Kiefer (1988), p. 647, Wooldridge (2002), pp. 695–696, and Hosmer, Lemeshow and May (2008), pp. 6–9. Unfortunately, the KV statistics themselves do not allow inferring a KV's precise start-up year, provided that the KV did not enter the scene as early as 1862 or later.

tary or forced) or through merger by absorption: i) Ignore that different modes of exit potentially imply different circumstances affecting the decision to exit; in this case, all data are pooled, and what is evaluated is simply the relationship between explanatory variables and the probability of exiting the market at some point in time, without taking care of the "how;"[12] ii) account for the potential difference by separately evaluating the relationship between explanatory variables and the probability of merger, on the one hand, and explanatory variables and the probability of liquidation on the other. However, separate estimation is based on the assumption that the modes of exit are statistically independent. Estimating the covariates' influence on the probability to be subject to merger is done by treating liquidations as censorings, and vice versa;[13] and iii) estimate the relationship between covariates and the probability of merger by simultaneously accounting for the competing risk of liquidation, and vice versa. This approach allows for statistical dependence of the modes of exit and, thus, may better reflect reality. Statistical dependence could mean that the event "liquidation" prevented the event "merger" from happening since it had happened first, or that liquidation as an alternative mode of exit had at least altered the ex-ante probability of occurrence of merger.[14]

In the presence of multiple modes of KV exit, it seems appropriate to apply a competing risk model of the third type, thereby allowing KV_i to cease operation principally due to the occurrence of one of $j = 1, \ldots, k$ different events. The random variable C_j then indicates the event by numerical value where right-censoring is indicated by $C_j = 0$. In the following, I use the proportional subhazards model proposed by Fine and Gray (1999), allowing for correlation between the competing risks of absorption and closure by modeling the joint distribution of survival times T_j. Their model is based on Cox's (1972) well-known semi-parametric proportional hazards model.[15]

Starting from a single-exit mode setting, assume that T has a cumulative probability distribution $F(t)$ with corresponding probability density $f(t)$, and the survivor function $S(t)$ with $S(t) = 1 - F(t)$. $F(t)$ states the probability that a KV exits before t, or $F(t) = \Pr(T \leq t)$. Accordingly, $S(t)$ represents the unconditional probability that a KV reaching t would not exit. Let the instantaneous hazard rate – i.e. the conditional probability that a KV will exit shortly after t given it survived until t – be defined as $h(t) = f(t) / S(t)$, which can be also expressed in terms of equation (5.1):[16]

---

12 See, for example, Fotopoulos and Louri (2000) and Van Kranenburg et al. (2002).

13 See, for example, Buehler et al. (2006), Isachenkova and Weeks (2008) and Esteve-Pérez et al. (2010). Although he does not apply survival analysis, Schary (1991) may be mentioned as well, since he estimates the influence of certain covariates on exit routes (merger, voluntary liquidation, and bankruptcy).

14 See Berg et al. (2008), who conduct a competing risk analysis with dependent risks for unemployment durations, and Jopp (2011a).

15 See Cox (1972) and Fine and Gray (1999). Pintile (2007) and Putter (2007) provide useful introductions from the field of medical statistics.

16 See Wooldridge (2002), p. 692.

$$(5.1)\qquad h[t; X(t)] = \lim_{\Delta t \to \infty} \frac{\Pr\left[t \le T < t + \Delta t, \middle| T > t, X(t)\right]}{\Delta t}.$$

The widely applied Cox model to assess the effect of covariates on the hazard of exit, regardless by which mode, is then given by

$$(5.2)\qquad h[t; X(t)] = h_0(t) * \exp[X(t)'\beta]\,,$$

where $h_0(t)$ is a baseline hazard that depends only on time. It is a basic feature of semi-parametric duration models that they do not provide estimates of the baseline hazard explicitly. Yet, given some additional assumptions, the baseline hazard may be recovered from the estimates. Maximization is done via maximum partial likelihood.[17]

In the bivariate case with $j = 2$ mutually exclusive exit modes ($j = 1$: merger; $j = 2$: closure), basic quantities have to be generalized to the multiple-exit setting and the two alternative approaches. According to Hosmer et al. (2008), generalization of the hazard function according to (5.1) leads us to

$$(5.3)\qquad h_j[t; X(t)] = \lim_{\Delta t \to \infty} \frac{\Pr[t \le T < t + \Delta t, C = j | T > t, X(t)]}{\Delta t}.$$

The cause-specific effect of covariates on either exit mode can be assessed by fitting the regression equation

$$(5.4)\qquad h_j[t; X(t)] = h_{0j}(t) * \exp[X(t)'\beta_j]\,.$$

Event-specific effects can be derived either by separately fitting for each exit mode, assuming the other mode, in turn, to be right-censored, which reduces (5.4) to (5.2), or, simultaneously, by use of a data-duplication method, where data are replicated j times and additional indicator variables must be created. Lunn and McNeil (1995) and Hosmer (2008), for example, describe this method.[18]

However, assuming independent risks might be too strong an assumption regarding KVs, in particular, or firm exit, in general. Fine and Gray's (1999) proportional subhazards model allows for a joint distribution of survival times, $F(T_1, T_2)$. Technically, this model does not estimate the event-specific hazard as if there were no other risks, but the respective hazard of the subdistribution, which is a function of both hazards.[19] The

17 See Cox (1972), Cox (1975), pp. 269–276, and Sueyoshi (1992). The model to be introduced is also called the proportional risks model in the literature; see Hosmer, Lemeshow and May (2008), p. 335.

18 See Lunn and McNeil (1995), pp. 524–525, and Hosmer, Lemeshow and May (2008), pp. 335–336.

19 See Fine and Gray (1999). Besides being of theoretical relevance, Fine and Gray's model is of practical importance since it is implemented in a statistical software package such as STATA and, therefore, renders a data-replication approach unnecessary.

subdistribution for event j is also called the cumulative incidence function. Regarding our two exit modes, the proportional subhazards model generalizes to

$$(5.5) \qquad h_j[t; X(t)] = \lim_{\Delta t \to \infty} \frac{\Pr[t \le T < t + \Delta t, C = 1 | (T > t \; or \; (T \le t \; and \; C \ne 1)), X(t)]}{\Delta t}$$

and expresses the conditional probability of exit by absorption in the presence of the risk of exit by liquidation, or vice versa. The difference between equations (5.3) and (5.5) stems from the condition in brackets concerning T. It not only says that event $j = 1$ did not happen until t, but states, in particular, that KV_i might have exited because of $j = 2$. Hence, the incidence of $j = 1$ depends on the incidence of $j = 2$. The regression equation to be estimated from the data corresponds to equation (5.4), where the competing risk is included in the likelihood function by way of a weighting technique.[20] Furthermore, equation (5.4) is estimated using a combination of time-invariant and time-varying covariates.[21]

5.2.2. Determinants of exit with explanatory potential

In the following, I briefly summarize the set of variables applied. Making a clear distinction between mergers and liquidations is important since both modes of exit are likely not to be affected by covariates in the same way. Moreover, as has been already argued above, one basic business-related decision at any point in time was to decide whether to continue operation or to exit by merger or liquidation. By including both modes of exit, this model goes beyond usual firm survival models.[22]

The nineteenth- and early twentieth-century discussion on the design flaws inherent in KVs centered on the claim that KV size would matter and that many KVs had still not reached the appropriate KV size. Indeed, Chapter 4 presented empirical evidence that shed some light on the "true" minimum efficient or, respectively, optimal size

20 There are two important issues involved in almost every empirical investigation. The first is unobserved heterogeneity. Unfortunately, the approach used, in combination with the software, does not allow me to include some KV level fixed or random effect. "Shared frailty" models, which introduce group-specific random effects into equation (5.4), can be estimated in the original Cox-model only. The second issue concerns a piecewise-constant model structure, as, e.g., in Van Kranenburg et al. (2002), which would allow coefficients to vary across some defined sub-periods while remaining constant within. Even though relaxing the assumption of constant coefficients would introduce more flexibility into the models, I do not consider a kind of structural break model further since I suppose that the number of observations and, in particular, the number of events that happened are not sufficiently high; see Han and Hausman (1990) and Van Kranenburg et al. (2002).

21 Beyersmann and Schumacher (2008) provide a technical overview of a competing risk model in combination with time-varying covariates.

22 See Manjón-Antolín and Arauzo-Carod (2008) for a recent literature survey.

among KVs. Observers of the KVs' operations proposed merging funds as an effective measure to increase the size and to improve the actuarial underpinnings of small KVs. Against this historical background, KV size is hypothesized to play a major role in explaining exit decisions. As shown above, the annual size distributions were extremely unequal, and many KVs were arguably too small to operate on a sound actuarial basis. Findings on the relationship between actuarial risk and size support this argument to a certain extent. In order to measure the importance of that relationship, I incorporate several controls with respect to size in the model. First, I measure current KV size, as before, by the sum of *contributors*. Second, I also include *start-up size* – the size of a KV in the first year for which an observation on it is available. This is in line with many empirical studies. Another time-invariant variable is the geometrically averaged mean growth rate of KV size over a fund's entire life cycle; we may call this variable the average "biological interest rate" of the KVs' PAYG schemes or, simply, *growth pattern*.[23] By measuring the long-term growth pattern, I get an indication of structural change regarding the mining areas underlying the KVs' insurance. A negative average growth rate implies, by tendency, long-term shrinkage and, correspondingly, a positive one implies long-term prosperity.

Furthermore, I employ several measures of financial distress according to the "rescue hypothesis" described in Chapter 1. First, the *ICR* and the *SCR ratio* measure the current burden from invalidity pensioners and from widows and orphans; these are the components of the pensioners-to-contributors ratio already introduced. In fact, the burden from pensioners evidently increased, as a rule, for many KVs, especially for the comparatively smaller ones and those subjected to long-term structural decline. In addition, inclusion of those ratios follows directly from the functioning of the PAYG mechanism, which the KVs applied. Depending on the development of wages – hence, productivity – the additional financing burden on contributors might, in the end, have reduced disposable income and lifetime implicit rates of return. In my view, the ICR serves as a good proxy variable for unsustainability among KVs. Second, the *start-up ICR* and the *start-up SCR* measure the initial burden from pensioners. In order to allow the effect of *start-up size* to vary with the initial burden from pensioners, I incorporate interactions with that variable and the respective ratios. Third, analogous to the ICR and SCR, the *SDCR* measures the financing burden on contributors that arose from the sickness-insurance section. The more sick days the average contributor had to finance – caused by incentives to malinger, for example – the more lasting financial pressure a KV faced. Further alternative measures of financial distress are a KV's annual (joint) net surplus (*net surplus*), operating costs per member (*operating costs*), and exposure to actuarial risk (*actuarial risk, pensions* and *actuarial risk, sickness*).[24]

---

[23] See von Weizsäcker (2010), p. 155.

[24] For the purpose of this empirical exercise, the variance of the average pension claim and of the average sickness claim have been estimated for the years after 1913, according to the framework set out in Subsection 4.2.

A few more variables are incorporated into the model to capture part of the heterogeneity among KVs. One variable is the share of established contributors among all contributors (*established ratio)*. Established contributors were, in general, more costly than non-established ones in that they often received higher benefits for equal contributions. Keeping in mind an increasing ICR or SCR, the decreasing share of established miners may have released the KV from some financial pressure. Indeed, it may also have indicated structural problems since ever fewer miners obviously had applied for established status and, thus, had decided not to enter into a long-term relationship with the KV. Following Guinnane and Streb, the firms' share in costs (*employers' share*) is included, as well.[25] Furthermore, the *young-to-old ratio* equals the ratio between established contributors aged 16 to 35 and those aged 36 and older. This is an already known proxy measure for a KV's age structure. A continuously diminishing ratio can be interpreted as a hint at structural aging of the contributor base.

The diversification of a KV's membership across the different mining subsectors is measured by a Hirschman-Herfindahl index (HHI), using subsector shares for each KV – i.e., the number of insurants employed in hard-coal mining divided by number of all insurants. Let the relevant subsectors be denoted with q, $q = 1, \ldots, Q$, and the number of insurants with N. I constructed HHI indices per year according to

$$(5.6) \qquad Diversification = \sum_{q=1}^{Q} \left(\frac{N_q}{\sum_N N_q} \right)^2 .$$

If a KV insured miners that were employed in only one subsector, the HHI equals one. If insurants were equally distributed over all subsectors, the HHI equals 0.083. Hence, the closer the HHI is to the latter quantity, the less concentrated were a KV's insurants in only one subsector. Contemporaries, such as Harry Karwehl, believed that the various subsectors reflected different degrees of occupational hazard, with hard-coal mining being the most hazardous activity.[26] A mixture of hazards might have been of actuarial advantage and might presumably have lowered dependency to the economic fate of only one subsector. While in the case of liquidations, one might assume a positive relationship between the hazard rate and the level of diversification, a negative relationship seems to make sense regarding absorptions since this would make a target more attractive.

In addition, I incorporate a variable already known from Chapter 4 – *correlated risks* – intended to measure the potential for correlated risks. As mentioned above, mining was among the economic activities that exhibited a high risk of massive accidents involving many insurants at the same time and, thereby, overwhelming a KV with many

---

25 Guinnane and Streb (2011), p. 87.

26 See Karwehl (1907), p. 74.

immediate sickness or invalidity claims. Clearly, massive accidents destroy the stochastic foundation of insurance because the risk that one miner will claim for sickness or invalidity is not independent of the risk that another miner had claimed so before; this is because both individual risks share the same origin (the accident).[27]

To measure the business cycle to which KVs were subjected, I include the variable *mining cycle*, which measures the percentage deviations of subsector-specific production (hard coal, brown coal, iron ore, zinc ore, lead ore or copper ore) from its trend. Positive deviations enter with a positive sign, negative deviations with a negative. Therefore, the trends of the production series have been estimated on the level of mining administration regions between 1861 and 1920.[28] To each KV, I ascribed the trend deviation of the single most important product that insured miners were engaged in extracting. I assumed that steelworks were subjected to the regional iron-ore trend, and salt works and stone pits to the hard-coal trend, if they were situated in Dortmund, and to the iron-ore trend, if they were situated in Bonn and Halle.

Finally, I checked explanatory variables for multicollinearity by evaluating the pairwise correlation coefficients across them.[29] As expected, there is a high correlation (beyond 0.75) between the ICR and SCR, and the SCR and the sick days-to-contributors ratio. However, all variables arguably measure different aspects and are, therefore, kept in the regressions. Table 5.1 summarizes how variables are expected to affect the conditional probability of exit. The models presented below will reveal whether there were systematic differences in how covariates and the hazard of absorption and liquidation were related.

5.3. Merge or fail? Competing risk models of exit among *Knappschaften*

Table 5.3 contains the estimation results and helps to evaluate ideas, from a statistical point of view, on why KVs might have been absorbed or, alternatively, liquidated. I focus on the statistically significant coefficients. Though, from the mere fact that in some of the presented models, certain variables apparently do not play a statistically significant role, we can already draw some striking conclusions about the process of external concentration among KVs. The regression model is formulated from the perspective of the KVs that were absorbed or liquidated. This means that the model sheds light on target characteristics and on how these affected the likelihood of being ab-

---

27 From the perspective of an insurer, the risks that individuals sharing the same occupation will make their claims seem, in general, to be correlated simply because the occupation brings about the same accident risks and diseases for everyone.

28 In almost all cases, a linear or quadratic trend yielded a reasonable fit.

29 Feinstein and Thomas (2002), p. 323.

sorbed or liquidated. It is important to note that the model does not focus directly on the characteristics of the absorbing KVs. However, it is possible to draw conclusions about the motives of the absorbing KVs by linking target characteristics with the aforementioned interests of firms as the dominant actor.

Table 5.1: Summary of expected signs

Hypotheses	Sign
H1: KVs with small start-up size face a higher conditional risk of exit than medium- or large-sized ones;	(-)
H2: Size or growth in size, respectively, inversely relates to KV exit;	(-)
H3: Both the ICR and SCR are positively related to the conditional probability of exit;	(+)
H4: KVs with a high start-up burden from pensioners face a higher conditional risk of exit;	(+)
H5: The sick days-to-contributors ratio positively affects the conditional probability of exit;	(+)
H6: The higher the actuarial risk, the higher the probability of absorption.	(+)
H7: A net surplus is inversely related to the hazard of exit;	(-)
H8: A high per capita burden with operating costs drives the hazard of exit up;	(+)
H9: A high per capita amount of monetary assets lowers the hazard of exit;	(-)
H10: The share of established contributors among all contributors is inversely related to the hazard of exit;	(+)
H11: The employers' financing share and the conditional probability of exit are inversely related, too;	(-)
H12: The lower the young-to-old ratio, the higher is the hazard of exit;	(-)
H13: A greater diversification lowers the hazard of liquidation, but might increase the probability of absorptions.	(-/+)
H14: A high potential for correlated risks drives the conditional probability of exit up.	(+)
H15: Negative deviations from the production trend increase the conditional probability of exit.	(+)

Note that coefficients, and not hazard ratios, are displayed. Since the estimated models are exponential in nature, coefficients are to be interpreted as semi-elasticities, indicating the percentage change in the log hazard rate due to a one-unit change in the respective explanatory variable. Hazard ratios reporting the percentage change in the hazard rate given a one-unit change in the independent variables may, nonetheless, be conveniently recovered from the expression exp(β). If one wants to know how a change in an explanatory variable by *m* units affects the hazard rate, one simply has to calculate exp(*m**β). Finally, a positive coefficient implies a hazard ratio larger than one and, thus, an increase in the hazard rate. Before turning to the interpretation of the competing risk regressions, have a quick view at Table 5.2, which provides, at a glance, descriptions of the set of explanatory variables applied.

Table 5.2: Variable description

Variable	Description
(1) Start-up size	Number of a fund's contributors in first observed year
(2) Start-up ICR	Number of invalids per 100 contributors in first observed year
(3) Start-up ICR	Number of survivors per 100 contributors in first observed year
(4) Start-up size × Start-up ICR	Interaction term
(5) Start-up size × Start-up SCR	Interaction term
(6) Growth pattern	Average geometric growth rate of size
(7) ICR	Number of invalids per 100 contributors per year
(8) SCR	Number of survivors per 100 contributors per year
(9) ICR × SCR	Interaction term
(10) SDCR	Number of paid sick days per contributor per year
(11) Actuarial risk pensions	Variance of the average pension claim
(12) Actuarial risk sickness	Variance of the average sickness claim
(13) Operating costs	Real operating cost per member
(14) Net deficit	Dichotomous variable taking on the value one if a net deficit occurred and zero otherwise
(15) Size	Number of contributors per year
(16) Young ratio	Number of established contributors aged 16 to 35 to those aged 36 and older per year
(17) Established members	Share of established contributors in all contributors per year
(18) Firm's cost share	Ratio of employers' contributions to total costs per year
(19) Diversification 2	Herfindahl index with respect to subsector employment shares per year ($0.9 \leq HHI < 1.0$)
(20) Diversification 3	Herfindahl index with respect to subsector employment shares per year ($0.7 \leq HHI < 0.9$)
(21) Diversification 4	Herfindahl index with respect to subsector employment shares per year ($HHI < 0.7$)
(22) Correlated risks	Average number of contributors per mine divided by all contributors
(23) Mining cycle	Deviation from production trend as percentage of trend

Let us focus, for the moment, on model 1, in which the financial-distress variables are the ICR, SCR, and SDCR. As argued above, KV size played a major role in the contemporary discussion, especially to motivate merger activity. Therefore, we would expect to find size essential in explaining absorptions. In fact, current size – i.e., size at a particular point in time (hereafter referred to simply as size) – does not play an essential a role as the insignificance of coefficient (15) implies. From a statistical viewpoint, there is no evidence that the current size of the target KV was decisive. In particular, there is no evidence in favor of absorptions as a measure intended to systematically help smaller KVs out of their actuarial trap or a measure to systematically improve the absorbing KV's actuarial position. The finding that size does explain liquidations, in contrast, supports this conclusion. The model says that the smaller the size was, the more likely it was that a KV ended up in liquidation than in absorption. Although contemporaries had perceived a small KV size as an actuarial disadvantage, which could – and, they maintained, should! – be solved by mergers of those small funds among themselves or with larger KVs, the statistical model implies that KVs had, on average, not acted according to this behavioral norm. Hence, KVs did not feel solidarity with their fellow funds in a way that would show up in the data. Members of a prospering larger fund probably did not want to share their prosperity with members of a distressed fund.

Although not related to size, the conditions at the start in 1861 mattered indeed. Start-up size has explanatory power in both models. Regarding absorptions, the model says that KVs that began with a high invalids-to-contributors ratio and a small size – coefficient (4) – were more likely to be absorbed at any point in time than were KVs with the same initial burden with invalids, but a larger size.[30] This finding implies two things: First, starting positions predetermined, to some extent, the funds' future survival process. Second, the finding supports the "rescue hypothesis." Regarding liquidations, the effect of start-up size is significantly positive.

Let us focus now on the main financial-distress variables, the ICR and SCR. While the ICR does not help to explain absorptions, especially not in the presence of alternative "liquidation," the SCR indeed does. The positive sign of coefficient (8) implies that a KV was more likely to become a target for absorption, the higher the current burden with survivors was. This finding also makes the case for rescue mergers that were fueled by – in addition to other motivations – the desire to ensure the welfare position of survivors, who would have otherwise, in case of liquidation, lost income. Yet, this finding does not yield an explanation for why invalids were not statistically relevant in absorbing another KV. When all potential targets were burdened, to some extent, with pensioners, it is possible that the targets with the least costly burden – that is, survivors – were chosen. Regarding the competing risk "liquidation," however, both the ICR and SCR, as the main financial-distress variables, have no significant explanatory power. Thus, we can conclude that a high burden from pensioners was not, in itself, a reason to shut down a fund.

30 From a technical point of view, incorporating an interaction with the initial burden from invalids has the consequence that the effect of start-up size itself is evaluated at a value of zero of the interacted variable, and not at its mean. Accordingly, a significant interaction implies a modification of the original coefficient.

Table 5.3: The determinants of mergers and liquidations assuming competing risks (dependent variable is log hazard rate, semi-elasticities displayed)

Variable	Model 1		Model 2	
	Absorption	Closure	Absorption	Closure
Time-invariant variables				
(1) Start-up size	0.00003	0.00049**	0.00002	0.00055*
	(0.00008)	(0.00022)	(0.00007)	(0.00028)
(2) Start-up ICR	0.05718	-0.18035	0.05964	-0.05297
	(0.06314)	(0.21767)	(0.06356)	(0.16458)
(3) Start-up ICR	-0.01478	0.07451	-0.01874	0.03365
	(0.02775)	(0.07822)	(0.02727)	(0.06338)
(4) Start-up size × Start-up ICR	0.00003*	0.00001	-0.00003**	-0.00009
	(0.00001)	(0.00009)	(0.00001)	(0.00022)
(5) Start-up size × Start-up SCR	6.06e-06	-0.00008	7.80e-06	-0.00006
	(7.09e-06)	(0.00007)	(7.27e-06)	(0.00008)
(6) Growth pattern	-0.01335	-0.16845**	-0.08098	-0.10032
	(0.08612)	(0.08803)	(0.04950)	(0.07763)
Time-varying variables				
(7) ICR	-0.00009	-0.00064	–	–
	(0.00044)	(0.00087)		
(8) SCR	0.00024***	-0.00002	–	–
	(0.00008)	(0.00020)		
(9) ICR × SCR	3.72e-08	-3.59e-07	–	–
	(9.02e-08)	(3.17e-06)		
(10) SDCR	-0.00031	0.00032	–	–
	(0.00078)	(0.00069)		
(11) Actuarial risk pensions	–	–	-9.62e-09**	2.72e-09
			(3.68e-09)	(3.90e-09)
(12) Actuarial risk sickness	–	–	1.35e-07	-2.29e-07
			(2.97e-07)	(2.65e-07
(13) Operating costs	–	–	–	–
(14) Net deficit	–	–	–	–
(15) Size	-1.67e-07	-0.00001**	-2.91e-07	8.68e-06*
	(3.03e-07)	(4.69e-06)	(3.70e-07)	(4.83e-06)
(16) Young ratio	0.07359**	-0.06852*	0.07877**	-0.07324**
	(0.03667)	(0.03980)	(0.03544)	(0.03053)
(17) Established members	-0.03008	0.05947	-0.01933	0.00321
	(0.02965)	(0.04674)	(0.02681)	(0.03665)
(18) Firm's cost share	0.05567**	0.07724	0.04517*	0.03432
	(0.02258)	(0.04718)	(0.02344)	(0.04456)

Table 5.3 continued

Variable	Model 1		Model 2	
	Absorption	Closure	Absorption	Closure
(19) Diversification 2	0.03970*	-0.03889	0.02855	-0.04069
	(0.02332)	(0.02668)	(0.02255)	(0.02987)
(20) Diversification 3	0.03707**	0.00479	0.02924	0.01080
	(0.01887)	(0.01485)	(0.01901)	(0.01616)
(21) Diversification 4	0.03808*	-0.03996	0.02873	-0.03856
	(0.02030)	(0.02726)	(0.01914)	(0.02817)
(22) Correlated risks	-0.01694	-0.03839***	-0.02932	-0.03151**
	(0.02350)	(0.01230)	(0.02189)	(0.02229)
(23) Mining cycle	-0.01246	-0.01204	-0.00628	-0.00797
	(0.02509)	(0.04744)	(0.02616)	(0.02608)
Number of observations	4 439	4 439	4 443	4 420
Log-pseudo-likelihood	-133.93	-72.15	-139,42	-136.79
Wald χ^2	111.34	126.50	79.94	82.78
Prob > χ^2	0.000	0.000	0.000	0.000

Variable	Model 3		Model 4	
	Absorption	Closure	Absorption	Closure
Time-invariant variables				
(1) Start-up size	0.00004	0.00044*	0.00001	0.00064**
	(0.00008)	(0.00022)	(0.00007)	(0.00028)
(2) Start-up ICR	0.05457	-0.17757	0.05340	-0.00526
	(0.05749)	(0.23493)	(0.06652)	(0.15956)
(3) Start-up ICR	-0.01003	0.07743	-0.01297	0.01491
	(0.02573)	(0.08614)	(0.02691)	(0.06237)
(4) Start-up size × Start-up ICR	-0.00003**	7.46e-06	-0.00003*	-0.00013
	(0.00001)	(0.00017)	(0.00001)	(0.00022)
(5) Start-up size × Start-up SCR	6.19e-06	-0.00006	6.94e-06	-0.00005
	(7.15e-06)	(0.00009)	(7.26e-06)	(0.00008)
(6) Growth pattern	-0.05273	-0.10471	-0.09292***	-0.04193
	(0.03210)	(0.16909)	(0.02277)	(0.04503)
Time-varying variables				
(7) ICR	–	–	–	–
(8) SCR	–	–	–	–
(9) ICR × SCR	–	–	–	–
(10) SDCR	–	–	–	–

Table 5.3 continued

Variable	Model 3		Model 4	
	Absorption	Closure	Absorption	Closure
(11) Actuarial risk pensions	–	–	–	–
(12) Actuarial risk sickness	–	–	–	–
(13) Operating costs	0.00064***	-0.00088	–	–
	(0.00021)	(0.00130)		
(14) Net deficit	–	–	0.01848*	0.01462
			(0.01085)	(0.01138)
(15) Size	-2.64e-07	-9.75e-06**	-2.67e-07	-0.00001**
	(3.64e-07)	(4.22e-06)	(3.29e-07)	(4.80e-06)
(16) Young ratio	0.03811	-0.01808	0.08926**	-0.07005**
	(0.03711)	(0.02812)	(0.04006)	(0.03439)
(17) Established members	-0.02811	0.27591***	-0.02031	-0.00539
	(0.03138)	(0.08399)	(0.02816)	(0.03682)
(18) Firm's cost share	0.05349**	-0.07501	0.05813**	0.04834
	(0.02391)	(0.07763)	(0.02250)	(0.03769)
(19) Diversification 2	0.02996	-0.03602	0.03240	-0.03492
	(0.02008)	(0.02491)	(0.02224)	(0.02984)
(20) Diversification 3	0.02851*	0.00482	0.026960	0.01346
	(0.01629)	(0.01504)	(0.01925)	(0.01803)
(21) Diversification 4	0.02827*	-0.03644	0.03055	-0.03771
	(0.01696)	(0.02740)	(0.01870)	(0.02977)
(22) Correlated risks	-0.02923	-0.02459*	-0.02888	-0.02394*
	(0.01336)	(0.01452)	(0.02238)	(0.01353)
(23) Mining cycle	-0.02541	-0.02826	-0.00557	-0.02258
	(0.05923)	(0.04887)	(0.02737)	(0.06328)
Number of observations	4 443	4 420	4 443	4 443
Log-pseudo-likelihood	-78.59	-57.81	-138.82	-78.72
Wald χ^2	90.80	58.04	64.34	90.94
Prob > χ^2	0.000	0.000	0.000	0.000

Notes: Displayed are coefficients, not hazard ratios. Standard errors in brackets are clustered robust (clustered is by KV). For ties, the Breslow method is used. ***, ** and * denote statistical significance at the one-, five- and ten-percent levels. Diversification 1 (HHI=1) is omitted as reference.

Another variable that is important – not the least because of the coefficients' magnitude in both cases – is the youth ratio, which measures a KV's age structure. Indirectly, it measures the growth opportunities embedded in the respective mining area and, thus, how attractive it was for young people to enter the mining sector as new employees.

With respect to absorptions, coefficient (16) implies that the likelihood of becoming a target for merger by absorption increased notably if the youth ratio was high. This definitely matches the idea that a KV, interested in improving its financial position by absorption, absorbed another fund only if it was really attractive. A relatively high number of young contributors relative to old contributors must have been appealing to a KV, especially the mine owners, because this is a signal of prosperity. Perhaps miners' and mine owners' representatives shared the opinion of KV board members that to ensure the fund's future, they had to ensure a steady inflow of young insurants. This was definitely in the interest of mine owners, who wanted to seek opportunities to keep costs low and to avoid dealing with "wounded" KVs. This view is supported by the finding on liquidations. If the ratio got worse – and, hence, the number of young miners declined relative to the number of old ones – KVs were liquidated. That is to say, KVs that obviously suffered from structural decline of the mining industry in their area were not attractive targets. This fact is straightforward. However, if the actors in the absorbing KVs' boards shared the contemporaries' view – namely, that KVs in trouble were to be rescued – they did not draw the same conclusion, which was to perform corresponding rescue absorptions. The growth-pattern variable, coefficient (6), clearly points in the same direction.

The financing share of the firms, coefficient (18), implies that the higher the share was, the more likely a KV was to become a target of absorption. Board members of the potential absorber KV, especially the mine owners, night have interpreted a rise in that cost share in two alternative ways: On the one hand, they might have found that it also signaled prosperity in the "targeted" mining area, which allowed the mine owners there to unburden insurants out of additional profits. Thus, it might have been attractive to incorporate those employers into one's own KV since they were obviously willing to contribute more – or, at least, they were *used to* contributing more. On the other hand, there is room to speculate about an alternative interpretation: A rise in the entrepreneurs' cost share could also have signaled growing financial distress – hence the opposite. Given that contributors were already sufficiently burdened with the financing of costs, entrepreneurs were under pressure to inject additional resources in order to prevent the KV from being underfunded. Accordingly, the absorption would not have been done because of an advantage for the absorbing KV, but would have been more of an insolvency-avoidance merger. I tend to interpret the coefficient as implying growth potential.

Finally, and not surprising, KVs were also more likely to be absorbed if they were diversified over at least two mining subsectors. This finding also supports the "self-interest" hypothesis. However, contrary to my expectations, a decreasing potential for correlation of individual risks – which is not an actuarial disadvantage at all – drove the probability of liquidation up, not down. This observation is somewhat counterintuitive. One of the following alternative explanations may apply: First, because decision makers were simply not aware of those actuarial relationships, they were not relevant to the decision. Second, such an actuarial advantage is worth nothing if a KV faced structural or actuarial problems. Hence, if, in practice, the potential for correlation of risks might have decreased, this alone was not helpful.

How do the other models fit in the picture? First of all, regarding time-invariant variables, the remaining models do not vary, except in one case: The growth pattern is, now that we incorporate the variable *net surplus* as the financial distress variable, significantly inversely related to the conditional probability of absorption. This is another counterintuitive result since it implies that a KV was the less likely to be absorbed, the larger the long-term average growth rate of KV size was. This finding, as it stands, definitely supports the "rescue hypothesis" because long-term, shrinking collectives of insurants would have been merged into more stable ones. However, I would take this way of interpreting the estimates with caution since this pattern is not consistent over all displayed models.

The additional major financial-distress variables are, except for the variance of the average sickness claim, statistically significant in relation to the hazard of absorption. Obviously, as model 2 states, a high exposure to actuarial risk resulting from pension provision was, *ceteris paribus*, an obstacle to being merged with another fund. Technically speaking, an increase in the exposure to actuarial risk was associated with a decrease in the probability of merger. This finding clearly contradicts the rescue hypothesis which implies the direct opposite of what the coefficient estimate tells. Unfortunately, the effect of the variance of the average pension claim on the hazard of exiting by absorption, though significant, is extremely weak. Evidence on the variables *operating cost* and *net deficit* tell a different story. Coefficient (13) implies that an increase in average operating costs was positively related to the probability of becoming a target for absorption, which might point to an absorption pattern that is consistent with the rescue hypothesis. Furthermore, the coefficient of the variable *net deficit* indicates that a KV was more likely to be absorbed if it experienced a deficit. In the presence of the exit alternative "liquidation," a KV displaying a net deficit was definitely a target for absorption.

With respect to the remaining explanatory variables, it seems safe to say that they correspond to what was pointed out regarding model 1. The "safest" way to slide into a situation where liquidation remained as the only exit option, thus, was a combination of small KV size and recruitment problems, certainly indicating structural problems.

In fact, referring to those findings on the merger and liquidation activities of the KVs, absorbing many of the small KVs was obviously not individually rational. Larger KVs absorbed smaller ones only if they were attractive enough, and many smaller KVs did not fulfill this condition since they were stagnating and not prospering. The question is whether the mining administration should have forced mergers on KVs even before the new regulations of 1906. The answer seems to be yes, but the evidence suggests that this was not done. 14 mergers involving 30 different KVs were effectively conducted between 1867 and 1907, and four between 1908 and 1913 involving a further nine different KVs. In addition, eight liquidations occurred before the new regulation of 1906 and even more thereafter. Regarding the median size of KVs, this number of mergers appears to have been rather insufficient to improve the financial stability of KVs. The liquidations, in particular, seem to have been the natural consequence of not being able to handle the actuarial risk in either way (internal or external growth).

6. Conclusions

The German miners' KVs represent one of the oldest collective solutions to provision against the contingencies of life, in general, and work, in particular. They grew out of local associations of mediaeval miners who wanted to bear together the costs of sickness, invalidity and survivorship arising from the hazards of their occupation. Sickness meant being temporarily unable – for some days or weeks – to work in the mines due to recovery from illness or injuries (e.g., from bruises and contusions). In contrast, invalidity meant being unable to do miners' work for the rest one's life due to disease (e.g., black lung) or loss of a limb (e.g., a leg or hand). Survivorship, meanwhile, meant that the breadwinner was dead and that his wife and children were on their own. The costs of any such contingency came as forgone wages and, thus, the loss of the ability to ensure the survival of one's family.

Cooperation in KVs was mutually beneficial, and the mechanism was, at first, simple: charity. Miners paid voluntarily into a common cash box according to their individual capability and received benefits should they need them. Soon, payments to the cash box were made mandatory, mine owners were integrated as co-sponsors, and the "solidarity contract" among actors, involving intra- as well as intergenerational transfers, was backed by mining codes enacted by the local or regional lords. During the mercantilist episode, in which the KVs' mutual character was replaced by the sovereign's paternalism, Prussia emerged as the German core-mining state. Abolishing full state control over mining, the mining reform of 1851 to 1865 set out to re-establish KVs as publicly-licensed, local or regional social insurance providers. From then on, they produced insurance coverage according to the equivalence principle: Miners paid mandatory social insurance contributions and, in return, they would gain the right to receive a proportional sick-pay or invalidity pension benefit. The early mutual associations of miners had already involved one conditional promise: your fellow miners would help you in case of need – as long as you would help in return. However, the new social insurance system re-shaped the promise in that benefits to you would be somewhat commensurate with what you had paid in.

The transformation of the *Knappschaft* in the middle of the nineteenth century laid some of the foundations on which the modern Bismarckian-type welfare state rests. Taking that as a starting point, the quantitative exercises in this study have, in different ways, produced a new account of the *Knappschaft* institution and the Prussian KVs as its major carriers. Why is that new account necessary? Prior studies have not examined in every detail the way that KVs dealt with social insurance provision between 1854 and 1923 and the outcomes that they produced. Indeed, various authors have looked at the history of the KVs as a *social*, *societal*, *political*, *technical* or *medical* history. However, in concentrating on the substance of the KVs' history as insurance providers at the time, we find that their history is, to no lesser degree, an *economic* history. Looking both at mass data on the KVs' operations and at the contemporary observers' perceptions of the *Knappschaft* institution reveals a story of the search for an insurance system's institutional design that would match all the involved actors' demands for stability, generosity, cost effectiveness, and, perhaps, economic or political power, at once.

To explain this story in economics terms: External experts, but also internal actors such as the miners, complained about the initial, post-reform "market" structure of KVs as reflecting unhealthy fragmentation, down-grading its effectiveness – if not endangering the survival of the whole institution. Obviously, whether KVs would be successful in fulfilling their tasks or would fail was judged intrinsically, depending on their size. Thus, many KVs were identified as so small that they encountered unnecessarily high operational risks in pension provision. And many KVs were, in the same breath, identified as too large to provide effective sickness insurance and to control for moral hazard, functions for which funds needed to be small. Hence, implementing the optimal size in every KV in the two separate insurance sections became the agenda. The ideal state, implied in the contemporary observers' arguments, was for each KV to operate one large pension fund for its collective of insurants, but to split its members into many small, self-standing sickness funds to limit malingering. However, a principle obstacle to implementing the optimal size in each insurance section was seen in the bundled provision of insurance coverage that was perceived as reflecting a trade-off. Putting priority on pensions, "growing" KVs to their optimal size with regard to efficient pension provision should be done through mergers. Liquidation was the ultimate way to take an unviable KV out of the market. This was the thinking reflected in the contemporary literature.

This study fills an existing research gap when focussing on the economic facts that constituted the KVs' nineteenth- and early twentieth-century history. Historians have tried to explain the observable concentration process among Prussian, as well as Bavarian and Saxon, KVs with several arguments at hand: with some actuarial considerations linked with the issue of size; with the existence of a portion of cross-KV solidarity that let important rescue absorptions become reality and successfully ensured the "securing of benefits;" with the success of the 1906 reform that legally enabled regulators to merge or liquidate KVs – thus, to facilitate the miners' insurance system's institutional change; with some KVs' will to become a powerful player on the insurance landscape, which made mergers necessary (see the *Allgemeine KV's* history); or with some KVs' will to retain autonomy, which let a number of small KVs refuse to be absorbed and thereby hinder, to an extent, desired institutional change to the advantage of the whole insurance system.

Taking it as a fact that miners and their employers, who operated the non-profit KVs altogether, faced a set of fundamental economic problems linked with insurance production, some questions naturally arise: To what extent did potential for economies of scale really exist in the KVs' insurance operations? What was the KVs' optimal size? Furthermore, did the bundling of the pension and sickness insurance sections really hinder KVs from realizing the optimal institutional design? What, in fact, *is* the optimal institutional design of the "market"? And what is the surplus value of mergers in developing the insurance system towards the perceived optimal state? In the context of all these questions, recall that neither contemporary observers nor recent research have provided definitive answers. The questions, however, are definitely latent in the historical discussion on the appropriate KV size, as briefly discussed in Subchapter 1.2. In fact, it should have become obvious that experts' and critics' opinions on optimal institutional design were far from concrete. They argued only that, generally speaking, "the larger the KV, the better" defined the agenda regarding pension insurance, and the "smaller the KV, the better" regarding sickness insurance. Therefore, it seems reasonable, at first glance, to trace the observable merger and liquidation activities back to these fundamental presumptions regarding the underlying actuarial issues.

However, to provide answers myself, I have taken the questions listed above and made them part of my central research question: Was there a minimum efficient or even optimal scale of operation that would have justified *ex-post* the historical strategy of clearing the "market" of inefficient KVs through absorption or liquidation? I have derived some hypotheses. To test them empirically means to derive some baseline evidence that is required to answer the broader question:

Concerning the null hypothesis that the actuarial risk to which a KV was exposed decreased steadily alongside a growing collective of insurants.

Based on the nineteenth-century discussion among observers of the KVs' insurance operations, I investigated the empirical relationship between the exposure to actuarial risk and KV size. The analysis yields several noteworthy findings. First, the historical example of the German miners' insurance scheme turns out to deviate from stories on prominent schemes already discussed in the literature – e.g., the English friendly societies or U.S. and Canadian fraternal lodges. The questions of the importance of scale economies and of risk-pooling plagued the KVs much more than the question of how to price the individual insurance contract. In this respect, the KVs' history allows for a different perspective on the emergence and development of nineteenth-century worker insurance schemes and, thereby, extends our knowledge.

Second, the straightforward empirical models applied suggest that the KVs' actuarial risk could have been – and partly was – reduced by increases in the size of the collective of insurants. In this respect, contemporaries put the right strategy on the agenda. In fact, three paths led to these increases in size: internal growth, external growth, or a mixture of these. A KV grew internally primarily because the mining area to which it was linked was rich in terms of the amount of resources in the ground that were – technically *and* economically – extractable. In interaction with the demand side, rich deposits offered mining companies growth opportunities over the long run, which resulted in rising local employment. In times of prosperity, when demand for labor was high – but qualified labor was particularly scarce – a KV that offered more attractive insurance conditions than others might have even facilitated the relocation of labor between two mining areas, such that local demand was matched. However, a KV linked with a mining area in which deposits were near exhaustion became, sooner or later, locked into a "structural-decline" path leading straight to a shrinking collective of insurants. In contrast, a KV grew externally by way of absorbing another fund's collective of miners, realizing an immediate jump in size. Empirical evidence shows that the absorbing KV was always larger than the absorbed KV, such that the main advantage for the absorbed members was to be integrated in a collective of insurants that had likely encountered a lower actuarial risk.

To the same degree that the empirical modeling of the KVs' exposure to actuarial risk highlights the positive effect of growth in size, it highlights the negative consequences of being locked into the "structural-decline" path and the importance of rescue absorptions as the "de-locking" event. However, the following questions arise: Were the positive effect, which is the one side of the coin, and the negative effect, which is the other side, always of importance? Thus, how vital was it for KVs to grow, and was decline in size a situation that required immediate reaction? In the logic of contemporary thinking, and the applied economic model, answering these questions depends on (i) whether there were increasing returns to scale realized over the full range of KV size; (ii) whether there were increasing returns to scale realized up to a certain minimum efficient KV size, and vanishing beyond that size; or (iii) whether there was an optimal firm size implying a U-shaped pattern indicating, in particular, decreasing returns to scale if growing beyond optimal size. While, regarding (ii), it would have been simply no disadvantage to grow beyond the MES, it would have been a disadvantage regarding (iii).

In fact, the small KVs were, first and foremost, exposed to actuarial risk stemming from pension provision. Depending on one's perspective, findings either point to a minimum efficient pension fund size of approximately 5 000 contributors – a size at which actuarial risk appears to have been minimized – or to an optimal size of slightly fewer contributors – a situation in which growing considerably *was* connected with rising actuarial risk. It is important to note, however, that this is not to say that the risk itself was zero after the MES or when the OS was reached. A residual positive risk due to pure chance – and accumulation risks! – always remained. The usual assessment of contemporaries that KVs needed – by all means – to be even larger than they were to ensure financial stability cannot be proved regarding pension insurance.

Unfortunately, regarding our historical case, findings are not as clear as one might wish. Indeed, we have to weigh two alternative formulations stemming from the contradictory results of the piecewise-constant and quadratic models.

On the one hand, KVs did, indeed, not gain further actuarial advantage by being larger than the MES. In this case, if the KVs, the contemporary observers and the regulators were sufficiently informed about how large the MES were, they all would have likely been – as rational actors – indifferent towards "growing" KVs even larger. This, of course, holds under the assumption that no other motives for additional growth kicked in at this point. In fact, all actors had no estimate of the MES in mind that was as low as 5 000 contributors. Hence, from our ex-post perspective, they encountered an "MES delusion." This delusion, however, was not that problematic: Rather than experiencing decreasing returns to scale after passing the MES, there simply were no increasing returns left. Thus, there was enough maneuvering room left for other motives to kick in, such as minimizing operating costs.

On the other hand, by growing beyond the OS, KVs encountered an actuarial disadvantage – even if actuarial risk did not increase that fast. In this case, all actors encountered an "OS delusion" that mattered. What is more, small KVs sized well below the OS "jumped, to some degree, out of the frying pan and into the fire" if the post-merger KV was larger than the OS.

I tend to accept formulation one, not the least because the rate at which actuarial risk increased in the pension-insurance section after passing the OS was so low that even a really large KV, such as the *Allgemeine KV Bochum*, would not have suffered too much from extraordinary growth in size. It follows that a merger, as a device to improve efficiency, definitely helped KVs out of an actuarial trap as long as they were notably smaller than the MES/OS. If the effects of mergers on the exposure of large KVs to actuarial risk were not negative, they were, at best, zero. In all, against the background of a persistently low median size, the findings suggest that in every year under observation, more than one half of all KVs operated on a scale too small to deal adequately with actuarial risk. Thus, the "fragmentation" design flaw, in combination with the "insufficient merger activity" design flaw, mattered – but not to the same extent that contemporaries perceived it to matter.

Third, integrating sickness insurance into the picture, the alleged "different efficient size" or, respectively, "compound insurance" design flaw of *Knappschaft* insurance – the bundled provision of pension and sickness benefits until 1906/1907 – was empirically relevant. Indeed, contemporaries legitimately pointed that out. However, the question of whether there was true a trade-off between the size of the pension-insurance section and the size of the sickness-insurance section – the existence of which, if maintained, would have made it even more problematic to care for both insurance-sections simultaneously – depends on whether there existed either an MES in both sections or an

OS. Evidence strongly makes the case for an MES in the sickness-insurance section of only about 1 000 contributors or, alternatively, an OS of slightly more contributors. Only at the MES/OS was the exposure to actuarial risk at its minimum. Yet, again, we have to weigh contradictory results.

On the one hand, the piecewise-constant model states that, between approximately 1 000 and 10 000 contributors, a KV did not encounter a disadvantage regarding its exposure to actuarial risk stemming from provision of sickness benefits. In particular, this implies that a KV was – up to those 10 000 contributors – not required to split its membership into numerous self-standing sub-funds with a maximum of 1 000 contributors. Numerous KVs sized below that upper threshold were *not* wrong to ignore the opportunity to split their membership that the 1865 regulation gave them. Rather, we might ask what it had additionally cost the few KVs that used the opportunity to implement and maintain the split. However, the actuarial disadvantage kicked in after passing a size of 10 000, rendering the strategy to split membership into such sub-funds more useful the larger the KV grew. What about the question of whether there was a trade-off between minimizing actuarial risk in the sickness-insurance section and minimizing it in the pension-insurance section? Actually, there was no such trade-off if we accept the MES in both sections: A KV that had 1 000 contributors had minimized its actuarial risk in the sickness-insurance section. Without actuarial disadvantage, the KV could grow further, to a size of 5 000, and minimize actuarial risk in the pension-insurance section, too.

In contrast, the quadratic model on the other hand suggests an OS of about the same scale as the MES. However, after surpassing that size, a KV faced decreasing returns to scale. Up to a size of about 145 000 contributors, the actuarial risk would increase less-than-proportionately and more-than-proportionately afterwards. In this scenario, there existed a trade-off between efficient sickness insurance and efficient pension insurance: On the one hand, actuarial risk in the sickness-insurance section was reduced if a KV size of about 1 000 contributors was implemented, although actuarial risk in the pension-insurance section was still not minimized. On the other hand, actuarial risk in the pension-insurance section was minimized if a KV size of 5 000 contributors was implemented. But this implies that KVs operated with diseconomies of scale, considering actuarial risk in the sickness-insurance section.

As with pension insurance, I favor the MES scenario as adequately describing the empirical case because, again, I think that we should not adhere too much to the pure technical implications of the models. Consequently, the "different efficient size" or, respectively, "compound insurance" design flaw mattered inasmuch as the MES were not the same in the pension- and sickness-insurance sections. However, a KV size of between 5 000 and 10 000 contributors was compatible with minimization of actuarial risk in both sections. Whether a negative effect on the exposure to actuarial risk regarding sickness insurance did kick in beyond 10 000 is, again, a matter of weighing. Definitely, we can reject the null hypothesis for the benefit of the alternative hypothesis.

Arguing for the adequacy of the MES scenario, what are the broader implications concerning the effectiveness of mergers? Contemporary observers strongly recommended mergers among KVs as a strategy to realize the appropriate fund size and thereby eliminate the many small funds that, they claimed, were exposed to too much variability. There was no doubt that many KVs had still not "found" that appropriate operational scale. Unfortunately, there was no contemporary attempt to measure that size. Given the MES estimates, mergers were an appropriate strategy to "grow" KVs that were smaller than 5 000 contributors to precisely that scale. By this standard, a number of absorptions were compatible with the strategy – even though contemporary actors might not have perceived them as a real success. The bundled provision of sickness and pension benefits – perceived as a major obstacle to implementing the strategy – actually was no obstacle at all. At any rate, the concentration process was unsuccessful insofar as economies of scale were not exploited optimally for numerous KVs that remained well below 5 000 contributors until their liquidation or until 1920. Once one KV or another reached the MES of 5 000 contributors, mergers no longer helped to implement the mentioned strategy since actuarial risk was minimized. This means that mergers among two or more KVs that had already reached the MES were ineffective with regard to actuarial risk due to the accumulation risks inherent in the KVs' operations. However, they might not have been ineffective at all since they might have helped to realize the MES regarding operating costs. This consideration leads us straight to the following hypothesis test.

Concerning the null hypothesis that a KV's average operating costs reduced steadily alongside a growing collective of insurants.

The question of whether KVs needed a certain size to minimize operating costs – or, put differently, of whether it made any sense to focus on the issue of operating costs at all – certainly played a role in discussing the situation of a particular KV – i.e., the *Allgemeine KV Bochum*, the single-most important KV. However, the question played no role in the broader discussion on appropriate KV size – a discussion that was aimed at proposing reforms of institutional design that would affect every KV. Taking into account evidence on the relationship between KV size and operating costs, too, calls another, maybe more important, trade-off into play – namely the trade-off between minimizing actuarial risk and minimizing operating costs: If a KV had reached a size of approximately 5 000 contributors, actuarial risk would have been optimally reduced with regard to both pension and sickness benefits. But what if operating costs were minimized at a scale much larger than that MES? If this was an OS, there was a true trade-off, indeed. If, in contrast, there was an MES, as in the case of actuarial risk, such a trade-off would not have existed.

For the reasons mentioned in Chapter 4, I think that only the models on the pre-reform period are adequate in accommodating the historical data. Again, we have to weigh two views: Is it adequate to accept economies of scale over the full range of observable KV sizes – i.e., should we keep the null hypothesis? Or should we accept the alternative hypothesis of an MES of about 33 000 contributors (or 42 000 members)? I tend towards the latter. However, the implications regarding the potential trade-off between minimizing actuarial risk and minimizing operating costs and those regarding the effectiveness of mergers as a reform strategy are actually the same. As long as we accept that there existed an MES that, if realized, would have meant that actuarial risk was minimized in both insurance sections simultaneously, there was no trade-off, but only the potential to further improve efficiency. In this case, forming large KVs of at least 33 000 contributors – while keeping a share of pensioners in overall membership that would be equal to the population mean – would have "killed two birds with one stone"! Consequently, arguing on the grounds of the KVs' multidimensional objective function – minimize actuarial risk in the pension-insurance section, minimize it in the sickness-insurance section, and minimize operating costs per member – contemporaries were, indeed, right to advocate large KVs. However, owing to the complexity of the historical setting, the socially desired institutional design – i.e., every KV would have been so large that certain actuarial and financing problems did not occur – was definitely not realized since numerous KVs remained below a size that would have minimized actuarial risk only in the sickness-insurance section. To an extent, desired mergers did not take place. This leads us straight to the test of the hypothesis that the mergers *that effectively occurred* among KVs were conducted for the rescue of KVs in actuarial and financial trouble – i.e., those mergers were largely compatible with the contemporary critics' strategy to reform the insurance system.

Concerning the null hypothesis that mergers fulfilled the purpose of rescuing KVs in trouble.

Complementing the analysis of KV size, this study also asks what might have motivated actors to enter into that long-term concentration process that we see in the data. This historical concentration process between 1861 and 1920 was driven, to an extent, by external growth, but also by liquidations. The determinants of those two "exit routes" are an important source of information because they enable us to assess the question of why certain KVs were not absorbed, although they deserved to be. I fully admit that my point of view is somewhat restricted by the fact that I build and interpret a statistical model that can only indirectly explore KVs' motivations to absorb another fund, refuse to be absorbed, or liquidate themselves. There is, of course, a straightforward reason for liquidating a KV: If, as a consequence of the structural decline of a particular mining area, membership in the responsible KV also declined, the KV's liquidation was the only remaining option.

In light of the statistical evidence, is it adequate to maintain the null hypothesis that absorptions served the purpose of rescuing KVs that showed existing actuarial deficiencies or that were overburdened with pensioners, sick days and operating costs? My reading of the contemporary, but also the more recent, literature is that this reasoning was implicitly accepted as explaining the phenomenon of concentration among KVs – implying that actors had a good deal of common sense and knew how to "live" solidarity. Strictly speaking, in interpreting the evidence, we must clearly distinguish between a KV's intention to rescue another KV and actions – in the form of absorbing a KV – that seem to be compatible with those intentions. Statistical evidence cannot *prove* the existence of those intentions at all, of course, but can only highlight observable patterns in the data that imply or that, respectively, are consistent with such intentions. In this sense, a rescue absorption is assumed if it had the effect of rescuing another KV. Indeed, part of the findings derived in Chapter 5 support the null hypothesis that mergers *were compatible* with the idea of a rescue measure: In the presence of "liquidation" as a serious exit strategy, the conditional probability of being absorbed (i) ceteris paribus increased with an increase in the SCR; (ii) was ceteris paribus generally higher for KVs that, initially, were small *and* faced a high ICR at the same time; (iii) ceteris paribus increased with an increase in operating costs per member; and (iv) was ceteris paribus higher if a net deficit occurred.

However, while I expected to find that the ICR as the main measure of financial distress, and as a proxy measure of unsustainability, is significant in explaining the concentration process from the perspective of the KVs that exited one way or another, it is not. This leads us to the conclusion that the (potentially) serious consequences of changes in the existing burden with invalids (i) were, as such, acknowledged, but ignored; (ii) were misinterpreted as not being serious and, thus, as not being a cause for action; or (iii) were unknown and, therefore, irrelevant for decisions from the start. Additionally, evidence on size-related variables suggests that size was not relevant for the decision to absorb a KV.

There are other indications that KVs in trouble were less likely to be absorbed, implying that they were not very attractive. If a KV's exposure to actuarial risk played a role for other KVs' decision to absorb it, we would expect to see another KV selecting it as a target for absorption – which was then a rescue absorption. Indeed, a high actuarial risk linked with pension provision, rather, was a crucial factor in preventing the absorption of numerous KVs that needed to be absorbed. Moreover, the higher both the ratio of young, established contributors to old contributors and the entrepreneurs' financing share in total claims costs were, the more likely a KV was to be absorbed; the effects were much higher than those supporting the "rescue" hypothesis as the null. In addition, the more diversified a KV was over mining subsectors, the more likely it was to be absorbed.

The picture that the statistical evidence draws, though diverse, tends to suggest rejection of the null. Taking evidence on liquidations into account, the "self-interest" hypothesis finds even more support: In the presence of the alternative "being merged into another KV," KVs exhibiting a bad age structure – indicating recruitment problems – and a smaller or diminishing size were liquidated with significantly larger probability. This implies that those KVs were not attractive enough to be rescued by absorption.

Hence, the analysis suggests that the concentration process among the Prussian miners' social insurance funds was, indeed, a combination of a selection process sorting out unsound funds and a process of "cherry picking." However, it was not a process that was just driven by acts of cross-KV solidarity. Thus, favoring the alternative hypothesis over the null appears adequate.

Now, after shedding light on the issue of KV size from different perspectives, we can form a consistent enough picture from the evidence to answer the central research question. Did the process we observe make the whole *Knappschaft* institution somewhat better off? What lessons can we draw about the costs and benefits of size from the KVs' quest for optimal scale?

The KVs had to act according to a reasonable multidimensional objective function to pursue and realize several, perhaps conflicting, aims simultaneously: (i) minimize actuarial risk stemming from the provision of sickness benefits; (ii) minimize actuarial risk stemming from the provision of invalidity and survivorship pensions; (iii) minimize operating costs per member; (iv) maintain a sustainable implicit rate of return, keeping intergenerational balance and encouraging the confidence in mandatory contributions; and (v) realize other aims that link up with social, societal and political processes. In this sense, the task of KV boards was complex, required the making of decisions under uncertainty, and – if the KV was to complete the task – required a basic understanding of the underlying economic relationships. The contemporary discussion on the appropriate KV size and related issues shows that actors, including the regulators, *did* think that KVs' provision of insurance coverage was a sphere in which purely economic matters mattered. The reform agenda was defined, essentially, as a quest to find the institutional design that would help KVs realize some socially desired aims.

As concluded above, mergers – in their effect on the size distribution of KVs – mattered. Some but, by far, not all mergers were helpful. Some mergers that should have been conducted were not. Evidence suggests that, in general, a KV was not required to be extremely large to operate efficiently in one way or another. Mastering certain challenges, such as the aging of the memberships, was not even a matter of size. There is, in particular, no indication that only one large state- or *Reich*-level fund covering all miners was the *conditio sine qua non* for an efficiently-working insurance system like that of Prussian miners. However, one has to bear in mind that this was what contemporaries – "on average" – identified as the socially desired outcome.

Regarding the central research question, the answer, thus, has to be: The empirical evidence suggests two possible views on the KVs' experience that lead me to reject the hypothesis that strictly monotonically increasing returns to scale existed. One may find the MES-scenario more adequate, or the stricter OS-scenario. Regardless of which is more adequate, the baseline is that mergers, in this study's framework, were efficiency-enhancing up to a KV size that was not as large as I, or the contemporary observers, expected. Regarding this point, the study of KVs clearly highlights how the presence of accumulation risks endogenously determines the optimal fund size. Liquidations were certainly a "necessary evil" when mergers were not an attractive strategy. Given that there were disincentives for voluntary – or, better, solidary – mergers that would have improved the stability of the whole insurance system and would have helped to realize the socially desired outcome, could not the regulators have provided a solution to this dilemma? After all, the regulators had once implemented those legal entitlements that were actually lost if a KV ended up in liquidation. The regulators, in fact, seem to have been very passive overall, which clearly is consistent with the ideas of economic liberalism that emerged in Prussia and Germany in the middle of the nineteenth century, but not with implementing the right incentives to improve efficiency as it was perceived in the observation period. Indeed, it could have been reasonable for the regulators to intervene if a clear market failure (a lack of security of supply) had been detected. KVs did not really reflect a market with market forces at work, and something like the market mechanism, working towards eliminating inefficient KVs and forming efficient KVs operating at the minimum efficient scale or above, did not really exist. I emphasize "not really" since, above, I identified the free spatial movement of miners as a mechanism by which KVs might have been forced into some competition. So, if there was no real market mechanism to automatically bring out the efficient KV, strong regulators would have been of use. This is especially so if we consider the evidence from Chapter 5, highlighting that individual rationality – and not just feelings of solidarity – regarding which merger to conduct played a role in what is good for the whole insurance system. It seems to me as if the regulators were either too tolerant, too uninformed, or too weak. Stated differently, there seem to be a "liberal policy argument," an "inappropriate actuarial knowledge argument," and a "weakness argument." The latter, especially, would account for the fact that employers, who arguably dominated the business decisions of the KVs, might have successfully lobbied the regulators not to intervene rigorously. This holds, first and foremost, for the period before the reform of 1906. This might have applied to employers involved in potential absorber KVs, who did not want to share their prosperity, but also to employers involved in smaller and/or stagnating (often firm-related) KVs, who might have feared losing control over their employees.

Many KVs persistently operated below a minimum efficient scale, and, thus, a potential for improving efficiency remained untapped. In the absence of market mechanisms, it seems appropriate to assign responsibility for that issue to the regulators. Although it was not the aim of this study to conduct an in-depth analysis of the mining administration as the visible regulatory authority, evidence on the historical concentration process among KVs and the actuarial discussion of the appropriate KV size suggest regulatory failure. It is up to future research to bring this study's evidence, the implied "regulatory-failure hypothesis," and a broader historical analysis of the regulators' position together.

Appendix

Tables

Table A.1: The population of Prussian KVs, 1861–1920

No.	KV name	Observation period	Size in first observed year/ size in last observed year
	Mining administration region Bonn		
1	Arnsberger KV	1861–1916 (merged into no. 39)	233/1 264
2	Briloner KV	1861–1916 (merged into no. 39)	2 279/1 189
3	Brühler KV	1861–1920	874/25 734
4	Burbacher Hütte KV	1861–1920	581/5 915
5	Cottenheimer KV	1862–1916 (liquidated)	167/493
6	Deutz KV	1861–1916 (liquidated)	2 919/1 426
7	Dillinger Hütten KV	1861–1920	710/7 952
8	Eifel KV	1861–1911 (merged into no. 3)	449/561
9	Emser KV	1867–1912 (merged into no. 21)	1 028/920
10	Eschweiler Pümpchen KV	1861–1917 (liquidated)	333/235
11	Eschweiler KV	1861–1917 (liquidated)	2 044/2 113
12	Goffontaine KV	1861–1872 (liquidated)	80/0
13	Günnersdorfer KV	1861–1885	662/0
14	HalbergerKV	1861–1920	292/3 329
15	Heller KV	1861–1916 (merged into no. 39)	1 613/6 683
16	Hohenzollernsche Lande KV	1861–1920	44/23
17	Holzappeler KV	1867–1912 (merged into no. 21)	881/462
18	Hostenbach KV	1861–1916 (merged into no. 38)	467/797
19	Ichenberger KV	1861–1874 (merged into no. 11)	372/324
20	Krupp KV	1867–1872 (merged into no. 29)	170/139
21	Lahn KV	1912–1920	8 451/10 689
22	Lendersdorfer KV	1861–1876 (liquidated)	1 050/129
23	Mariahütte KV	1861–1920	65/337
24	Mayener KV	1862–1920	293/2 183
25	Meinerzhagener KV	1861–1908 (merged into no. 3)	1 969/1 526
26	Mosel KV	1862–1911 (merged into no. 3)	683/12
27	Münster am Stein KV	1861–1920	17/8

Table A.1 continued

No.	KV name	Observation period	Size in first observed year/ size in last observed year
28	Müsener KV	1861–1916 (merged into no. 39)	1 364/835
29	Nassau Allgemeiner KV	1869–1912 (merged into no. 21)	6 942/5 610
30	Neunkirchener KV	1861–1920	1 265/6 840
31	Niedermendinger KV	1862–1916 (liquidated)	192/201
32	Oberbergischer KV	1861–1885 (merged into no. 36)	935/1 498
33	Olpe KV	1861–1916 (merged into no. 39)	343/684
34	Quinter KV	1861–1920	512/416
35	Rheinböller Hütte KV	1861–1915 (liquidated)	522/208
36	Rheinischer KV	1861–1920	1 899/6 082
37	Rheinpreußen KV	1867–1920	113/8 478
38	Saarbrücker KV	1861–1920	12 896/63 607
39	Siegener KV	1861–1920	1 482/19 420
40	Stolberger KV	1861–1920	3 235/10 637
41	Stromberger Hütte KV	1861–1915 (liquidated)	185/401
42	St. Goar KV	1861–1915 (liquidated)	229/560
43	St. Wendel KV	1861–1906 (liquidated)	228/0
44	Theodorshalle KV	1887–1920	31/28
45	Thommer KV	1876–1920	73/102
46	Werl KV	1861–1918 (liquidated)	52/24
47	Westernkottener KV	1861–1920	16/8
48	Wetzlarer KV	1861–1912 (merged into no. 21)	993/1 554
49	Wied KV	1861–1880 (liquidated)	400/504
50	Wittgensteiner KV	1861–1920	84/394
51	Wurm KV	1861–1920	2 970/17 708
	Mining administration region Breslau		
52	Muskauer KV	1861–1885 (merged into no. 35)	226/313
53	Niederschlesischer KV	1861–1920	5 083/45 735
54	Oberschlesischer KV	1861–1920	18 895/178 227
55	Plesser KV	1862–1920	512/7 178
	Mining administration region Clausthal		
56	Casseler KV	1870–1920	719/7 212
57	Clausthaler KV	1870–1920	5 704/23 842
58	Hannoverscher KV	1874–1886 (merged into no. 57)	784/2 565
59	Hessischer KV	1870–1886 (merged into no. 57)	1 267/663
60	Hohensteinischer KV	1870–1886 (merged into no. 57)	140/71
61	Ilseder Hütte KV	1873–1920	784/8 377
62	Schaumburger KV	1870–1886 (merged into no. 57)	1 273/1 738
63	Schmalkaldener KV	1870–1886 (merged into no. 57)	158/192
64	Unterharzischer KV	1873–1920	780/1 594
	Mining administration region Dortmund		
65	Allgemeiner KV zu Bochum	1890–1920	136 314/411 585
66	Altenbekener KV	1861–1895 (liquidated)	16/0

Table A.1 continued

No.	KV name	Observation period	Size in first observed year/ size in last observed year
67	Borgloh-Oeseder KV	1867–1890 (liquidated)	329/0
68	Essen-Werdenscher KV	1861–1890 (merged into no. 69)	10 424/36 314
69	Georgs-Marien–Hütte KV	1867–1920	1 200/3 410
70	Gottesgabener KV	1861–1920	28/9
71	Ibbenbürener KV	1861–1920	673/2 413
72	Königsborner KV	1861–1877 (merged into no. 73)	175/102
73	Märkischer KV	1861–1890 (merged into no. 69)	18 228/86 027
74	Minden-Ravensberger KV	1861–1920	244/560
75	Mülheimer KV	1861–1890 (merged into no. 69)	2 070/3 730
76	Neusalzwerker KV	1861–1920	78/19
77	Piesberger KV	1867–1904 (merged into no. 69)	521/917
78	Rothenfelde KV	1867–1920	38/23
79	Salzkottener KV	1861–1907 (liquidated)	22/14
80	Sassendorfer KV	1861–1920	71/20
	Mining administration region Halle		
81	Arternscher KV	1861–1876 (liquidated)	200/130
82	Brandenburgischer KV	1872–1920	3 221/24 879
83	Brandenburg-Pommern KV	1861–1872 (merged into no. 82)	1 400/1 310
84	Berliner KV	1861–1873 (liquidated)	89/92
85	Dürrenberger KV	1861–1908 (merged into no. 90)	308/219
86	Erfurter KV	1861–1873 (merged into no. 95)	87/89
87	Finowkanal Werke KV	1861–1915 (liquidated)	189/1 453
88	Halberstädter KV	1861–1920	2 662/29 310
89	Halle Saline KV	1861–1913 (liquidated)	66/44
90	Hallescher KV	1907–1920	25 543/51 905
91	Henneberger KV	1861–1869 (merged into no. 102)	34/17
92	Kamsdorfer KV	1861–1869 (merged into no. 102)	112/1
93	Lauchhammer KV	1861–1920	600/7 113
94	Mansfelder KV	1861–1920	4 665/18 929
95	Neupreußen KV	1861–1907 (merged into no. 90)	2 922/15 572
96	Niederlausitzer KV	1861–1872 (merged into no. 82)	391/2 011
97	Rüdersdorfer KV	1861–1916 (liquidated)	664/662
98	Saalkreiser KV	1861–1907 (merged into no. 90)	1 378/7 538
99	Schönebecker KV	1861–1875 (merged into no. 88)	394/530
100	Stolberg KV	1862–1873 (merged into no. 88)	149/0
101	Tangerhütte KV	1861–1920	206/943
102	Thüringer KV	1869–1920	91/392
103	Wernigeroder KV	1887–1920	472/362

Notes: The allocation of KVs to the respective mining administration regions displays the status quo as of 1861. In case that the first year of observation is 1862 or later, it is equal to the KVs founding date; otherwise we cannot be sure. When it reads that a KV was merged in year *t* into another one, this precisely means "at the turn of year *t* to year *t*+1;" this means that year *t* is the last observed year of operation. This holds for liquidations as well. KVs that survived the year 1920 are likely to have been merged into the *Reichsknappschaft* in 1923. A number of KVs in the mining administration regions of Bonn and Halle were renamed between 1867 and 1881: The *KV für die Reviere Meschede und Stadtberge* was renamed *Brilon KV* (2) in 1867; the *Gemünder KV* was renamed *Eifel KV* (8) in 1868; the *KV für die Asbacher und Gräfenbacher Hütte* was renamed *Halberger KV* (14) in 1875; the *KV für die Reviere Kirchen, Daaden und Burbach* was renamed *Heller KV* (15) in 1872; the *KV für das Revier Ründeroth und die Herrschaft Wildenburg* was renamed *Obergischer KV* (32) in 1875; the *KV für die Reviere Unkel und Hamm* was renamed *Rheinischer KV* (36) in 1881; the *KV für die Königliche Eisengiesserei Berlin* was renamed *Berliner KV* (84) in 1867; and the *KV für das Königliche Steinsalzbergwerk Erfurt* was renamed *Erfurter KV* in 1866.

Sources: Ministerium für Handel, Gewerbe und öffentliche Arbeiten (1859–1878), Ministerium für öffentliche Arbeiten (1879–1889), and Ministerium für Handel und Gewerbe (1890–1922).

Table A.2: The population of Bavarian *Knappschaften*, 1871–1920

No.	KV name	Observation period
A) Provision of both sickness and pension insurance		
	Mining administration region Bayreuth	
104	Abbach KV[a]	1871–1902 (merged into no. 108)
105	Amberg KV[b]	1871–1917 (merged into no. 106)
106	Bayerischer Knappschaftsverband	1917–1920
107	Bayerischer Landesknappschaftsverein	1918–1920
108	Bayreuth KV	1902–1920 (merged into no. 107)
109	Bodenwöhr KV[b]	1871–1917 (merged into no. 106)
110	Burglengenfeld KV[b]	1871–1902 (merged into no. 108)
111	Fichtelberg KV	1871–1909 (liquidated)
112	Goldene Adlerhütte KV	1874–1902 (merged into no. 108)
113	Goldkronach KV	1871–1902 (merged into no. 108)
114	Hof KV	1875–1902 (merged into no. 108)
115	Kissinger Saline KV	1871–1904 (liquidated)
117	Klingenberg KV	1871–1920 (merged into no. 107)
118	Königshütte KV[b]	1873–1877 (liquidated)
119	Obereichstätt KV[b]	1871–1917 (merged into no. 106)
120	Pullenreuth KV[b]	1874–1902 (merged into no. 108)
121	Sonthofen	1917–1917 (merged into no. 106)
122	Stadtsteinach KV	1871–1902 (merged into no. 108)
123	Steben KV	1871–1902 (merged into no. 108)
124	Stockheim KV	1871–1920
125	Sulzbach KV[b]	1873–1920 (merged into no. 107)
126	Vilseck KV[b]	1884–1902 (merged into no. 108)
127	Weiherhammer	1917–1917 (merged into no. 106)
128	Wunsiedel KV	1871–1921 (merged into no. 106)
	Mining administration region München	
129	Achthal KV	1871–1920 (merged into no. 107)
130	Bergen KV	1871–1917 (merged into no. 106)
131	Berchtesgaden KV	1871–1917 (merged into no. 106)
132	Bodenmais KV	1871–1917 (merged into no. 106)
133	Hohenpeißberg KV	1871–1917 (merged into no. 106)
134	Irsee	1884–1905 (merged into no. 135)
135	Miesbach KV	1871–1920 (merged into no. 107)
136	Mittenwald KV	1875–1905 (merged into no. 135)
137	Murnau KV	1875–1881 (merged into no. 135)
138	Penzberg KV	1871–1872 (merged into no. 135)
139	Reichenhall KV	1871–1917 (merged into no. 106)
140	Rosenheim KV	1871–1917 (merged into no. 106)
141	Schneckenbach KV	1871–1875 (merged into no. 104)
142	Tölz KV	1871–1872 (merged into no. 135)
143	Traunstein KV	1871–1913 (merged into no. 140)
	Mining administration region Zweibrücken	
144	Breitenbach KV	1871–1920
145	Consolidiertes Nordfeld KV	1892–1905 (merged into no. 148)

Table A.2 continued

No.	KV name	Observation period
146	Donnersberg KV	1885–1920
147	Dürckheim KV	1879–1903 (liquidated)
148	Frankenholz KV	1873–1920
149	Glanmühlbach KV	1871–1875 (merged into no. 150)
150	Godelhausen KV	1876–1920
151	Hellenberg KV	1887–1920
152	Hoof KV	1871–1920
153	Hüffler KV	1871–1875 (merged into no. 150)
154	Odenbach und Roth KV	1871–1876 (merged into no. 159)
155	Reiffelbach KV	1871–1902 (liquidated)
156	Selchenbach KV	1873–1920
157	Steinbach KV	1871–1920
158	Steinbrück KV	1871–1871 (merged into no. 157)
159	St. Ingbert KV	1871–1917 (merged into no. 106)
160	Wolfstein	1871–1920
B) Since 1917 listed as KVs providing exclusively for sickness		
161	Eichhofen KV	1917–1920
162	Gustav KV	1917–1920
163	Haidhof KV	1917–1920
164	Klardorf KV	1917–1920
165	Marienstein KV	1917–1920
166	Schmidgaden-Schwarzenfeld KV	1919–1920
167	Untergriesbach KV	1918–1920

Notes: The allocation of KVs to the respective mining administration regions displays the status quo as of 1883 after the mining administration region of Regensburg had been dissolved. KVs that had belonged to that region were allocated either to the mining administration region of Bayreuth or that of Munich. KVs marked with a superscript "a" belonged to the mining administration region of Munich until 1879 and to the mining administration region of Regensburg between 1880 and 1882. KVs marked with a superscript "b" belonged to the mining administration region of Regensburg until 1882. In case that the first year of observation is 1872 or later, it is equal to the KVs founding date; otherwise we cannot be sure. When it reads that a KV was merged in year t into another one, this precisely means "at the turn of year t to year t+1;" this means that year t is the last observed year of operation. This holds for liquidations as well. KVs that survived the year 1920 are likely to have been merged into the *Reichsknappschaft* in 1923. The *Bayerische Landesknappschaftsverein* (107) was named "Munich KV" in 1917 and 1918. In the KV statistics, the *Bayerische Knappschaftsverband* (106) is referred to as the finance network (*Finanzverbund*) of the participating KVs. The participating KVs are, therefore, listed here.
Sources: Oberbergamt München (1873–1920).

Table A.3: The population of Saxon *Knappschaften*, 1868–1920

No.	KV name	Observation period
	Mining administration region Chemnitz (hard coal)	
168	KV der Bockwa-Hohndorfer Steinkohlen-AG	1873–1891 (merged into no. 260)
169	KV des Steinkohlenwerks Concordia	1871–1884 (liquidated)
170	KV des Steinkohlenbauvereins Deutschland	1871–1891 (merged into no. 260)
171	KV des Fürstlich Schönburgschen Steinkohlenwerks	1868–1891 (merged into no. 260)
172	KV des Gersdorfer Steinkohlenbauvereins	1872–1891 (merged into no. 260)
173	KV des Steinkohlenbauvereins Gottes Segen	1869–1891 (merged into no. 260)
174	KV des Steinkohlenbauvereins Hohendorf	1872–1891 (merged into no. 260)
175	KV des Hohendorf-Bernsdorfer Steinkohlenbauvereins	1871–1878 (liquidated)
176	KV des Steinkohlenbauvereins Kaisergrube	1871–1891 (merged into no. 260)
177	KV des Kirchberger Steinkohlenaktienvereins	1868–1878 (liquidated)
178	Königsgrube KV	1871–1875 (liquidated)
179	KV des Lugauer Steinkohlenbauvereins	1870–1891 (merged into no. 260)
180	KV des Lugau-Niederwürschnitzer Steinkohlenbauvereins	1868–1891 (merged into no. 260)
181	Neue Westphalia KV	1869–1869 (liquidated)
182	KV des Niederwürschnitzer Steinkohlenbauvereins	1868–1876 (liquidated)
183	KV der Oelsnitzer Bergbaugewerkschaft	1869–1891 (merged into no. 260)
184	Oelsnitzer Frisch Glück KV	1871–1876 (liquidated)
185	KV der Gewerkschaft Rhenania	1872–1891 (merged into no. 260)
186	Reichszeche KV	1872–1872 (liquidated)
187	Saxonia KV	1870–1875 (liquidated)
188	Sewald und Compagnie KV	1869–1869 (liquidated)
189	KV des Seewaldschen Steinkohlenwerks	1868–1878 (liquidated)
190	Tagesstrecke KV	1870–1874 (liquidated)
191	Teutonia KV	1872–1878 (liquidated)
192	Vaterlandsgrube KV	1874–1874 (liquidated)
193	Vereinsglück KV	1871–1872 (liquidated)
	Mining administration region Chemnitz (brown coal)	
194	Auferstehung zu Brandis KV	1869–1869 (liquidated)
195	Saxonia KV	1870–1874 (liquidated)
	Mining administration region Dresden (hard coal)	
196	KV des Dresden-Possendorfer Steinkohlenbauvereins	1868–1869 (liquidated)
197	KV der Fiscalischen Werke im Plauenschen Grunde	1868–1884 (liquidated)
198	KV der Freiherrlich von Burgkschen Steinkohlenwerke	1868–1891 (merged into no. 260)
199	KV des Hänicher Steinkohlenbauvereins	1868–1891 (merged into no. 260)
200	KV des Königlichen Steinkohlenwerks	1885–1891 (merged into no. 260)
201	Morgenröte KV	1868–1882 (liquidated)
202	KV des Potschappler Aktienvereins	1868–1878 (liquidated)
203	Schaarschuh KV	1873–1874 (liquidated)
204	Verehelichte Neumann KV	1868–1869 (liquidated)

Table A.3 continued

No.	KV name	Observation period
	Mining administration region Dresden (brown coal)	
205	August Gerlach KV	1868–1877 (liquidated)
206	August Lorenz KV	1868–1874 (liquidated)
207	C. August Posselt KV	1869–1884 (liquidated)
208	Ch. Gottfried Burkhardt KV	1869–1878 (liquidated)
209	Chr. Burghardt KV	1871–1872 (liquidated)
210	Ebermann KV	1879–1884 (liquidated)
211	E. G. Heidrich KV	1868–1879 (liquidated)
212	Ehregott Süß KV	1868–1868 (liquidated)
213	Erdmann Bartel KV	1868–1870 (liquidated)
214	G. Posselt KV	1875–1878 (liquidated)
215	G. A. Schubert KV	1868–1884 (liquidated)
216	Gärtners Erben KV	1868–1868 (liquidated)
217	Germania KV	1876–1876 (liquidated)
218	Gottlieb Regedly KV	1868–1869 (liquidated)
219	J. Gottlieb Schröter KV	1869–1884 (liquidated)
220	Gottes Segen KV	1874–1884 (liquidated)
221	Gottfried Bischoff und Schneider KV	1868–1870 (liquidated)
222	Gottlob Bischoff KV	1870–1872 (liquidated)
223	Gottlob Thiele KV	1868–1870 (liquidated)
224	Heidrichs in Türchau KV	1878–1884 (liquidated)
225	Heinrich Wagner KV	1868–1870 (liquidated)
226	Hennigs KV	1878–1879 (liquidated)
227	Hilfe Gottes KV	1876–1884 (liquidated)
228	Hoffnung Gottes KV	1868–1884 (liquidated)
229	L. Wagner KV	1868–1878 (liquidated)
230	Zum heiteren Blick KV	1879–1881 (liquidated)
231	K. Renner	1872–1873 (liquidated)
232	Lindner und Neumann KV	1869–1869 (liquidated)
233	Radisch und Wagner KV	1868–1884 (liquidated)
234	KV des Reichenberger Kohlenbauvereins	1868–1875 (liquidated)
235	Rönsch und Waurich KV	1869–1875 (liquidated)
236	KV des Wittgensdorfer Kohlenbauvereins	1868–1868 (liquidated)
237	Verwittwete Scholze KV	1868–1870 (liquidated)
238	Zur guten Hoffnung KV	1868–1879 (liquidated)
	Mining administration region Freiberg and Altenberg (ores)	
239	Altenberg KV	1868–1891 (merged into no. 260)
240	Freiberg KV	1868–1891 (merged into no. 260)
	Mining administration region Marienberg (ores)	
241	Annaberg KV	1868–1870 (merged into no. 243)
242	Geyer und Ehrenfriedersdorf KV	1868–1870 (merged into no. 243)
243	Marienberg KV	1868–1891 (merged into no. 260)
	Mining administration region Schwarzenberg (ores)	
244	Johanngeorgenstadt KV	1868–1891 (merged into no. 260)
245	Scheibenberg KV	1868–1891 (merged into no. 260)

Table A.3 continued

No.	KV name	Observation period
246	Schneeberg KV	1868–1891 (merged into no. 260)
247	Voitgsberg KV	1868–1891 (merged into no. 260)
	Mining administration region Zwickau (hard coal)	
248	Alte Kasse des Himmelsfürstschachtes KV	1868–1869 (liquidated)
249	KV der von Arnimschen Werke	1868–1919 (merged into no. 260)
250	Bockwa-Oberhohndorfer Knappschaftsverband	1869–1891 (merged into no. 260)
251	KV des Erzgebirgischen Steinkohlen-Aktienvereins	1868–1891 (merged into no. 260)
252	KV des Fortunaschachtes	1868–1874 (liquidated)
253	Himmelsfürst KV	1869–1872 (liquidated)
254	KV des Oberhohndorfer Forststeinkohlenbauvereins	1868–1891 (merged into no. 260)
255	KV des Oberhohndorfer Torf-Steinkohlenbauvereins	1868–1868 (liquidated)
256	KV des Zwickauer Steinkohlenbauvereins	1868–1891 (merged into no. 260)
257	KV des Aktienvereins Zwickauer Bürgergewerkschaft	1868–1891 (merged into no. 260)
258	KV des Steinkohlenbauvereins Zwickau-Brückenberg	1868–1920
259	KV des Zwickau-Oberhohndorfer Steinkohlenbauvereins	1868–1891 (merged into no. 260)
	Covering mining administration regions Altenberg, Chemnitz, Dresden, Freiberg, Marienberg, Schwarzenberg and Zwickau (ores and hard coal)	
260	Allgemeine Knappschafts-Pensionskasse Sachsen	1891–1920

Notes: Displayed are *Knappschaft* pension funds. Regarding the mining administration regions Altenberg, Freiberg, Marienberg, and Schwarzenberg, KVs relate to mining areas; the *Allgemeine Knappschafts-Pensionskasse Sachsen* was an area-related fund, too. Though, the other KVs belonged to a particular mine or company. "AKP" is the abbreviation for the Allgemeine Knappschafts-Pensionskasse Sachsen (260). "n.a." abbreviates "not available." KVs that survived the year 1920 are likely to have been merged into the *Reichsknappschaft* in 1923. A number of KVs appeared with at least two alternative names: *Mörgenröte KV* (201) as *KV des Rittergutsbesitzers Brendel*; *Hoffnung Gottes KV* (228) as *Ernst Geißler KV*; *L. Wagner* (229) as *Verwittwete Kelchen KV* or *Carlsfeld KV*, respectively; *Radisch und Wagner KV* (233) as *Glückauf KV*; *August Gerlach KV* (205) as *Schlösschen KV*; *Gottes Segen KV* (220) as *Riedel KV* or *Schnitter KV*, respectively; and *Zur guten Hoffnung KV* (238) as *Ernst Eduard Krause KV*.

Sources: Königliches Finanzministerium Sachsen (1870–1872) and (1873–1921).

References

Primary sources and contemporary secondary literature

Achenbach, Heinrich von (1871), Die deutschen Bergleute der Vergangenheit, in: *Zeitschrift für Bergrecht* 12, pp. 80–118.

Allgemeiner deutscher Knappschaftsverband (1907), *Bericht über die Versammlung zur Begründung einer knappschaftlichen Rückversicherungs-Anstalt auf Gegenseitigkeit am 30. Oktober 1907, vormittags 11 Uhr, im Weinhaus „Rheingold" zu Berlin*, Berlin.

Allgemeiner Knappschafts-Verein zu Bochum (1891–1920), *Verwaltungsbericht*, Bochum.

Bankewitz, Hermann (1908), *Die Entwicklung des Knappschaftswesens in Anhalt, nebst einer Statistik des Anhaltischen Knappschaftsvereins von seiner Gründung bis zum Jahr 1907*, Wernigerode.

Beckmann (1900), *Versicherungstechnische Prüfung der finanziellen Lage der Pensionskasse des Saarbrücker Knappschaftsvereins*, St. Johann a. d. Saar.

Behm, Karl (1863), *Die Kranken- und Invalidenversicherung – Zum Gebrauche bei Errichtung von Kranken- und Invalidenkassen*, Leipzig.

Bertrams, Ferdinand (1912), Die Sicherstellung der Leistungen der preußischen Knappschaftsvereine durch das Knappschaftsgesetz vom 19. Juni 1906, in: *Glückauf* 48, pp. 1 411–1 425, 1 459–1 471, 1 499–1 513, 1 538–1 554 and 1 584–1 597.

Böhmert, Victor (1879), Zur Statistik der Knappschaftskassen im Königreiche Sachsen, in: *Zeitschrift des Königlich-Sächsischen Statistischen Bureaus* 25, pp. 153–174.

Böhmert, Victor (1885), Die Arbeitslöhne auf den fiskalischen Steinkohlenwerken Sachsens von 1869–1885, in: *Zeitschrift des Königlich-Sächsischen Statistischen Bureaus* 31, pp. 201–204.

Borchard, Hans-Heinrich (1929), *50 Jahre preussisches Ministerium für Handel und Gewerbe 1879–1929*, Berlin.

Brämer, Karl (1880), Der finanzielle Zustand der preußischen Knappschaftsvereine, in: *Zeitschrift des königlich-preußischen statistischen Bureaus* 20, pp. 289–314.

Brassert, Heinrich (1858), *Bergordnungen der Preußischen Lande*, Köln.

Bülow, Wilhelm (1905), *Das Knappschaftswesen im Ruhrkohlenbezirk bis zum allgemeinen preußischen Berggesetz vom 24. Juni 1865*, Leipzig.

Bülow, Wilhelm (1910), Die geschichtliche Entwicklung des Allgemeinen Knappschafts-Vereins, in: Allgemeiner Knappschaftsverein Bochum (ed.), *Denkschrift zur Einweihung des neuen Verwaltungsgebäudes des Allgemeinen Knappschafts-Vereins*, Dortmund.

Carnall, Rudolf von (1854–1858), Statistik der Knappschaftsvereine des preussischen Staates, in: *Zeitschrift für das Berg-, Hütten- und Salinenwesen in dem preussischen Staate* 1–5.

Caron, Albert (1881), *Die Berechnung der Beiträge bei der obligatorischen Arbeiterversicherung*, Berlin.

Caron, Albert (1882), *Die Reform des Knappschaftswesens und die allgemeine Arbeiterversicherung*, Berlin.

Dannenberg, W. (1901), *Das Allgemeine Berggesetz für das Königreich Sachsen vom 16. Juni 1868*, Leipzig.

Elsholz, Ulrich (1910), *Die Entwicklung des sächsischen Knappschaftswesens*, Dresden.

Friedrich Wilhelm IV. (1854), *Das neue Knappschaftsgesetz*, Charlottenburg.

Gebhard, Hermann, and Paul Geibel (1889), *Führer durch das Gesetz, betreffend die Invaliditäts- und Altersversicherung vom 22. Juni 1889 sowie Anleitung für die Anwendung desselben*, Altenburg.

Gottschalk, Hans (1912), Die Knappschaftsnovelle vom 3. Juni 1912, in: *Glückauf* 29, pp. 1 162–1 170 and 1 199–1 205.

Halbach, Hermann (1906), *Die Einwirkung der Arbeiterversicherungsgesetze auf die Knappschaftsvereine und ihre Einrichtungen*, Leipzig.

Heyden-Rynsch, Hermann Freiherr von (1881), Ueber die Grundprinzipien des Knappschaftskassenwesens, in: *Jahrbuch für Gesetzgebung, Verwaltung und Volkswirthschaft im Deutschen Reich* 5, pp. 259–271.

Hilt (1877), Ueber die Errichtung einer allgemeinen Pensionskasse für die Knappschaftsmitglieder im Oberbergamtsbezirk Bonn, in: *Zeitschrift für das Berg-, Hütten- und Salinenwesen in dem preussischen Staate* 25, pp. 1–28.

Hiltrop, Julius (1869), Über die Reorganisation der Knappschaftsvereine, mit Hinblick auf die Bildung von Versicherungsgenossenschaften für Arbeiter anderer Gewerbe, in: *Zeitschrift des königlich-preußischen statistischen Bureaus* 9, pp. 216–241.

Hollborn, Karl (1942), *Die Geschichte des Halberger Knappschaftsvereins zu Brebach-Saar*, Saarbrücken.

Imbusch, Heinrich (1910), *Das deutsche Knappschaftswesen. Eine Darstellung seiner Entwicklung und seines heutigen Standes, unter besonderer Berücksichtigung des Knappschaftsrechts und des Wirkens der Knappschaftsvereine in der Praxis*, Köln.

Jordan, Gerhard (1905), *Die Geschichte des Knappschaftswesens im Mansfelder Revier*, Halle.

Jüngst, Ernst (1913), Die Knappschaftskassen, in: Gesellschaft für Wirtschafts- und Sozialwissenschaften (ed.), *Untersuchungen über das Versicherungswesen in Deutschland*, München, pp. 253–274.

Kah, Karl (1874), *Das Haftpflichtgesetz betreffend die Verbindlichkeit zum Schadenersatze für die bei dem Betriebe von Eisenbahnen, Bergwerken, Fabriken, Steinbrüchen und Gräbereien herbeigeführten Tödtungen und Körperverletzungen, erläutert nach dem Motiven, stenographischen Reichstagsverhandlungen, Präjudicien und anderen Quellen*, Mannheim/Straßburg.

Karwehl, Harry (1907), *Die Entwicklung und Reform des deutschen Knappschaftswesens*, Jena.

Kleeis, Friedrich (1910), *Das deutsche Knappschaftswesen und seine Reform*, Groß-Lichterfelde.

Klostermann, Rudolf (1866), *Das Allgemeine Berggesetz für die preußischen Staaten vom 24. Juni 1865, nebst Einleitung und Kommentar*, Berlin.

Klostermann, Rudolf (1879), Ueber den Reservefonds der Knappschaftsvereine, in: *Zeitschrift für Bergrecht* 20, pp. 62–70

Knappschaftliche Rückversicherungsanstalt a. G. (1908), *Satzung für die Knappschaftlichen Rückversicherungsanstalt*, Berlin.

Knappschaftliche Rückversicherungsanstalt a. G. (1910), *Erster Geschäftsbericht für das Jahr 1908*, Berlin.

Knappschaftlicher Rückversicherungsverband a. G. (1916), *Satzung für den Knappschaftlichen Rückversicherungsverband*, Bochum.

Knappschaftlicher Rückversicherungsverband a. G. (1917), *Neunter Geschäftsbericht für das Jahr 1916*, Berlin.

Köhne, August (1915), *Die deutschen Knappschaftsvereine, ihre Einrichtung und ihre Bedeutung*, Hannover.

Königliches Finanzministerium Sachsen (1870–1872), *Jahrbuch für den Berg- und Hütten-Mann*, Freiberg.

Königliches Finanzministerium Sachsen (1873–1921), *Jahrbuch für das Berg- und Hüttenwesen im Königreiche Sachsen*, Freiberg.

Küttner, W. (1881), Die Invalidität und Invaliditäts-Versicherung der Steinkohlenbergleute, in: *Zeitschrift für das Berg-, Hütten- und Salinenwesen in dem preussischen Staate* 29, pp. 135–191.

Küttner, W. (1888), Neuere Untersuchungen über die Invalidität der Steinkohlenbergleute Preussens, in: *Zeitschrift für das Berg-, Hütten- und Salinenwesen in dem preussischen Staate* 36, pp. 30–90.

Lahmeyer, Carl (1881), Die Knappschaftsvereine des Oberbergamtsbezirks Clausthal, in: *Zeitschrift für das Berg-, Hütten- und Salinenwesen in dem preussischen Staate* 29, pp. 1–23.

Laporte, Walter (1910), *Das Problem der Arbeiterpensionskassen in versicherungswissenschaftlicher Hinsicht*, doctoral dissertation, Universität Göttingen.

Martell, P. (1916), Die Entwicklung des Knappschaftswesens im deutschen Bergbau, in: *Zeitschrift für Sozialwissenschaft*, pp. 799–805.

Milde, Franz (1907), Zur Bildung eines Rückversicherungsverbandes für die preußischen Knappschaftsvereine, in: *Der Kompaß: Organ der Knappschafts-Berufsgenossenschaft für das Deutsche Reich* 22, pp. 124–129.

Milde, Franz (1916), *Die versicherungsrechtliche Entwicklung der preussischen Knappschaftsvereine nach Erlass der Reichsversicherungsordnung, des Versicherungsgesetzes für Angestellte und des preussischen Knappschaftsgesetzes*, Kattowitz.

Ministerium für Handel und Gewerbe (1890–1922), Statistik der Knappschaftsvereine des preussischen Staates, in: *Zeitschrift für das Berg-, Hütten- und Salinenwesen in dem preussischen Staate* 38–70.

Ministerium für Handel und Gewerbe (1890–1922), Statistische Mitteilungen über die beim Bergbau Preußens gezahlten Arbeitslöhne und erzielten Arbeitsleistungen, in: *Zeitschrift für das Berg-, Hütten- und Salinenwesen im preussischen Staate* 38–70.

Ministerium für Handel und Gewerbe (1890–1922), Statistischer Teil, in: *Zeitschrift für das Berg-, Hütten- und Salinenwesen im Preussischen Staate* 38–70.

Ministerium für Handel, Gewerbe und öffentliche Arbeiten (1856), Instruction vom 3. April 1855, zur Ausführung des Gesetzes vom 10. April 1854, betreffend die Vereinigung der Berg-, Hütten- und Salinenarbeiter in Knappschaften, für den ganzen Umfang der Monarchie, in: *Zeitschrift für das Berg-, Hütten- und Salinenwesen in dem preussischen Staate* 3, pp. 25–31.

Ministerium für Handel, Gewerbe und öffentliche Arbeiten (1859–1878), Statistik der Knappschaftsvereine des preussischen Staates, in: *Zeitschrift für das Berg-, Hütten- und Salinenwesen in dem preussischen Staate* 7–26.

Ministerium für Handel, Gewerbe und öffentliche Arbeiten (1861–1878), Statistischer Teil, in: *Zeitschrift für das Berg-, Hütten- und Salinenwesen im Preussischen Staate* 9–26.

Ministerium für Handel, Gewerbe und öffentliche Arbeiten (1872), Erlass vom 22. Juni 1871, die Bestrebungen auf Verbesserung der Organisation der Knappschaftsvereine betreffend, in: *Zeitschrift für das Berg-, Hütten- und Salinenwesen im preussischen Staate* 19, pp. 27–28.

Ministerium für öffentliche Arbeiten (1879–1889), Statistik der Knappschaftsvereine des preussischen Staates, in: *Zeitschrift für das Berg-, Hütten- und Salinenwesen in dem preussischen Staate* 27–37.

Ministerium für Handel, Gewerbe und öffentliche Arbeiten (1885–1889), Statistische Mitteilungen über die beim Bergbau Preußens gezahlten Arbeitslöhne und erzielten Arbeitsleistungen, in: *Zeitschrift für das Berg-, Hütten- und Salinenwesen im preussischen Staate* 33–37.

Ministerium für Handel, Gewerbe und öffentliche Arbeiten (1879–1889), Statistischer Teil, in: *Zeitschrift für das Berg-, Hütten- und Salinenwesen im Preussischen Staate* 27–37.

Ministerium für Handel, Gewerbe und öffentliche Arbeiten (1883), Ministerial-Erlass vom 1. October 1883 an die sämmtlichen Königl. Oberbergämter, betreffend Aenderung der Statuten der Knappschaftskassen auf Grund des Reichsgesetzes vom 15. Juni 1883 über die Krankenversicherung der Arbeiter, in: *Zeitschrift für das Berg-, Hütten- und Salinenwesen im preussischen Staate* 31, pp. 77–85.

Morgenbesser, A. (1882), *Versuch zur Aufstellung von Sterblichkeits- und Invaliditäts-Tafeln für preussische Bergleute*, Berlin.

N.N. (1876), *Die statutarischen Bestimmungen der Knappschaftsvereine im Oberbergamtsbezirke Bonn*, Bonn.

Nöggerath, Jacob (1873), Beiträge zur Geschichte der Bergknappen, in: *Zeitschrift für Bergrecht* 12, pp. 204–232.

Oberbergamt München (1873–1921), *Statistik der Knappschaftsvereine im bayerischen Staate*, München.

Pietsch, Georg (1911), *Das Knappschaftswesen in Elsass-Lothringen*, Köln.

Pietsch, Georg (1913), *Die Umgestaltung der Knappschaftspensionskassen zur Anpassung an das Versicherungsgesetz für Angestellte*, Berlin.

Pitschke, H. (1892), *Ueber die Entstehung und Entwicklung des Mansfelder Knappschaftsvereins und der Wohlfahrtseinrichtungen beim Mansfelder Bergbau*, Eisleben.

Ptok, Paul W. (1922), *Das Knappschaftswesen in seiner geschichtlichen Entwicklung und gegenwärtigen Gestaltung unter Einfluss der Reichsgesetzgebung mit besonderer Berücksichtigung Oberschlesiens*, doctoral dissertation, Universität Breslau.

Rabenau, Karl (1909), *Das bayerische Knappschaftswesen*, doctoral dissertation, Universität Jena, Coburg.

Reif, Heinrich (1908), Die preußische Berggesetznovelle vom 19. Juni betr. die Knappschaftsvereine, in: *Bergrechtliche Blätter* 3, pp. 30–73.

Reisel, Otto (1924), *Die Entstehung und Entwicklung des Saarbrücker Knappschaftsvereins*, doctoral dissertation, Universität Frankfurt am Main.

Reuss, Max (1916), Der knappschaftliche Rückversicherungsverband in Charlottenburg, in: *Zeitschrift für Bergrecht* 57, pp. 394–424.

Reuss, Max (1918), Der Freizügigkeitsvertrag der deutschen Knappschaftsvereine vom 1. September 1917, in: *Zeitschrift für Bergrecht* 59, pp. 84–104.

Saarbrücker Knappschaftsverein (1879–1920), *Jahresbericht des Saarbrücker Knappschaftsvereins*, Saarbrücken.

Serlo, Albert (1859), *Die Beschwerden gegen die neue Organisation der Knappschafts-Vereine im Distrikte des Königl. Ober-Bergamtes zu Dortmund*, Essen.

Schlick, Jakob (1927), *Das Knappschaftswesen im Saarbergbau unter besonderer Berücksichtigung der Satzung vom 26.–27. Februar und des ersten Nachtrags hierzu vom 18. Dezember 1926*, doctoral dissertation, Universität Würzburg.

Schneider, Robert (1901), *Die Entwicklung des Allgemeinen Knappschafts-Vereins*, Gelsenkirchen.

Simons, Peter (1890), *Geschichte und Statistik der Wurm-Knappschaft in Bardenberg bei Aachen*, Berlin.

Simons, Peter (1895), *Das deutsche Knappschaftswesen*, Mainz.

Steinbrink, Otto (1908), *Gesetz vom 19. Juni betreffend die Abänderung des Siebenten Titels im Allgemeinen Berggesetze für die preußischen Staaten vom 24. Juni 1865 – nebst Kommentar*, Berlin.

Tecklenburg, Theodor (1877), *Organisation der Knappschaftsvereine und Krankenkassen im Großherzogthum Hessen*, Darmstadt.

Verband der Bergarbeiter Deutschlands (1909), *Das deutsche Knappschaftswesen*, Bochum.

Wächter, Georg (1893), *Sächsische Knappschaftsstatistik, umfassend Untersuchungen über Sterblichkeit, Invalidität und Krankheit*, doctoral dissertation, Universität Leipzig.

Wiessner, Vinzenz (1912), *Die Knappschaftsvereine und die Reichsversicherungsordnung vom 19. Juli 1911*, Kattowitz.

Wilhelm I. (1883), Reichsgesetz vom 15. Juni 1883, betreffend die Krankenversicherung der Arbeiter, in: *Zeitschrift für das Berg-, Hütten- und Salinenwesen im preussischen Staate* 31, pp. 31–50.

Wirtz, August (1911), *Entwicklung und Organisation des Knappschaftswesens im Ruhrkohlenbezirk*, Weimar.

Wissmann, Georg (1930), *Das Reichsknappschaftsgesetz vom 1. Juli 1926 in der Praxis*, Bochum.

Woedtke, Erich von (1885), *Unfallversicherungsgesetz. Vom 6. Juli 1884*, Berlin.

Wolff, A. (1907), *Der Wetzlarer Knappschaftsverein, Geschichte seiner Gründung und Entwicklung*, Frankfurt am Main.

Other secondary literature

Ackermann, Peter (1983), Ein Vorschlag zur Messung der Betriebsgröße von Schaden- und Unfallversicherungsunternehmen, in: *Zeitschrift für die gesamte Versicherungswissenschaft* 72, pp. 87–97.

Adelmann, Gerhard (1963), Die Beziehungen zwischen Arbeitgeber und Arbeitnehmer in der Ruhrindustrie vor 1914, in: *Jahrbücher für Nationalökonomie und Statistik* 175, pp. 413–427.

Akerlof, George A. (1970), The Market for Lemons: Quality Uncertainty and the Market Mechanism, in: *The Quarterly Journal of Economics* 84, pp. 488–500.

Albrecht, Peter (1982), Gesetze der großen Zahlen und Ausgleich im Kollektiv – Bemerkungen zu Grundlagen der Versicherungsproduktion, in: *Zeitschrift für die gesamte Versicherungswissenschaft* 71, pp. 501–538.

Albrecht, Peter (1984a), Ausgleich im Kollektiv und Prämienprinzipien, in: *Zeitschrift für die gesamte Versicherungswissenschaft* 73, pp. 167–180.

Albrecht, Peter (1984b), Welche Faktoren begünstigen den Ausgleich im Kollektiv?, in: *Zeitschrift für die gesamte Versicherungswissenschaft* 73, pp. 181–201.

Albrecht, Peter (1991), Überlegungen zur Kalkulation eines risikoadäquaten Schwankungszuschlags in der Krankheitskosten-Versicherung, in: *Blätter/Deutsche Gesellschaft für Versicherungs- und Finanzmathematik* 20, pp. 151–168.

Albrecht, Peter (1992), *Zur Risikotransformationstheorie der Versicherung: Grundlagen und ökonomische Konsequenzen*, Karlsruhe.

Alter, George, and James C. Riley, (1989), Frailty, Sickness, and Death Models of Morbidity and Mortality in Historical Populations, in: *Populations Studies* 43, pp. 25–45.

Amel, Dean, Barnes, Colleen, Panetta, Fabio, and Carmelo Salleo (2003), Consolidation and Efficiency in the Financial Sector: A Review of the International Evidence, in: *CEIS Tor Vergata – Research Paper Series* 7.

Andrade, Gregor, Mitchell, Mark, and Erik Stafford (2001), New perspectives on mergers, in: *Journal of Economic Perspectives* 15, pp. 103–120.

Arps, Ludwig (1965), *Auf sicheren Pfeilern – Deutsche Versicherungswirtschaft vor 1914*, Göttingen.

Arrow, Kenneth J. (1996), The Theory of Risk-Bearing: Small and Great Risks, in: *Journal of Risk and Uncertainty* 12, pp. 103–111.

Arza, Camila, and Paul Johnson (2006), The Development of Public Pensions from 1889 to the 1990s, in: Gordon L. Clark, Alicia H. Munnell and J. Michael Orszag (eds.), *The Oxford Handbook of Pensions and Retirement Income*, Oxford, pp. 53–75.

Banken, Ralf (2003), *Die Industrialisierung der Saarregion 1815–1914 – Take-Off-Phase und Hochindustrialisierung 1850–1914*, Stuttgart.

BarNiv, Ran, and John Hathorn (1997), The Merger or Insolvency Alternative in the Insurance Industry, in: *The Journal of Risk and Insurance* 1, pp. 89–113.

Bartels, Christoph, Bingener, Andreas, Bluma, Lars, Braun, Hans-Joachim, Montz, Markus, Müller, Irmgard, Scholl, Lars Ulrich, Schulz, Stefan, Streb, Jochen, and Melanie Sulzer (2009), Vergangenheit und Zukunft sozialer Sicherungssysteme am Beispiel der Bundesknappschaft und ihrer Nachfolger. Ein Forschungsprojekt der Leibniz-Gemeinschaft, in: *Jahrbuch für Wirtschaftsgeschichte/ Economic History Yearbook* 2009/2, pp. 195–217.

Bartels, Christoph (2010), Vergangenheit und Zukunft sozialer Sicherungssysteme am Beispiel der Bundesknappschaft und ihrer Nachfolger. Ein Forschungsprojekt der Leibniz-Gemeinschaft 2007-2009, in: *Kompass: Amtliches Mitteilungsblatt der Bergbau-Berufsgenossenschaft und der Deutschen Rentenversicherung Knappschaft-Bahn-See* 120, pp. 24–27.

Bateman, Hazel, and Olivia S. Mitchell (2004), New evidence on pension plan design and administrative expenses: the Australian experience, in: *Journal of Pension Economics and Finance* 3, pp. 63–76.

Bauer, Keldon, Miles, Linda L., and Takeshi Nishikawa (2009), The effect of mergers on credit union performance, in: *Journal of Banking and Finance* 33, pp. 2 267–2 274.

Bayer, Detlef (1981), Die Knappschaftsältesten in der Geschichte der Sozialversicherung, in: *Kompass: Zeitschrift der Bergbau-Berufsgenossenschaft und der Bundesknappschaft* 91, pp. 321–326 and 353–356.

Beito, David T. (2000), *From Mutual Aid to the Welfare State: Fraternal Societies and Social Services, 1890–1967*, Chapel Hill.

Benson, John (2002), Coalowners, Coalminers and Compulsion: Pit Clubs in England, 1860–1880, in: *Business History* 44, pp. 47–60.

Bergamt Halle (1998), *225 Jahre Oberbergämter und Bergbehörden in Halle an der Saale*, Halle.

Bernholz, Peter (2005), Die Bedeutung der Geschichte für die Wirtschaftswissenschaften und der ökonomischen Theorie für die Geschichtsforschung, in: *Perspektiven der Wirtschaftspolitik* 6, pp. 131–150.

Beyersmann, Jan, and Martin Schumacher (2008), Time-dependent covariates in the proportional subdistribution hazards model for competing risks, in: *Biostatistics* 9, pp. 765–776.

Bikker, Jacob A., and Jan De Dreu (2009), Operating costs of pension funds: the impact of scale, governance, and plan design, in: *Journal of Pension Economics and Finance* 8, pp. 63–89.

Bikker, Jacob, Steenbeek, Onno, and Federico Torracchi (2010), The impact of scale, complexity, and service quality on the administrative costs of pension funds. A cross-country comparison, in: *DNB Working Papers*, No. 258.

Bingener, Andreas (2009), Armenkasten oder Knappschaftsbüchse – Der Kampf der Goslarer Bergknappen um eine eigene soziale Absicherung, in: *Der Anschnitt: Mitteilungsblatt der Vereinigung der Freunde von Kunst und Kultur im Bergbau* 61, pp. 294–301.

Bingener, Andreas (2010), Zur Vorgeschichte und zu den Auswirkungen des Preußischen Knappschaftsgesetzes von 1854, in: Christoph Bartels (ed.), *Berufliches Risiko und soziale Sicherheit. Beiträge zur Tagung Vergangenheit und Zukunft sozialer Sicherungssysteme am Beispiel der Bundesknappschaft und ihrer Nachfolger im Deutschen Bergbau-Museum Bochum, 8. und 9. Oktober 2009*, Bochum, pp. 29–46.

Bittl, Andreas, and Bernd Müller (1998), Das versicherungstechnische Risiko im Zentrum versicherungswirtschaftlicher Betätigung, in: *Zeitschrift für die gesamte Versicherungswissenschaft* 88, pp. 369–402.

Bluma, Lars (2009a), Der Hakenwurm an der Ruhr. Umwelt, Körper und soziale Netzwerke im Bergbau des Kaiserreichs, in: *Der Anschnitt: Mitteilungsblatt der Vereinigung der Freunde von Kunst und Kultur im Bergbau* 61, pp. 314–329.

Bluma, Lars (2009b), Die Hakenwurmepidemie im Ruhrkohlenbergbau (um 1890), in: *Kompass: Amtliches Mitteilungsblatt der Bergbau-Berufsgenossenschaft und der Deutschen Rentenversicherung Knappschaft-Bahn-See* 119, pp. 12–15.

Bluma, Lars (2010), Heterotope Orte: Raumhistorische Dimensionen des knappschaftlichen Krankenhauswesens im Ruhrgebiet, in: Christoph Bartels (ed.), *Berufliches Risiko und soziale Sicherheit. Beiträge zur Tagung Vergangenheit und Zukunft sozialer Sicherungssysteme am Beispiel der Bundesknappschaft und ihrer Nachfolger im Deutschen Bergbau-Museum Bochum, 8. und 9. Oktober 2009*, Bochum, pp. 67–98.

Bluma, Lars 2011), Die Hygiene des Bergmanns. Zur Biopolitik im Ruhrkohlenbergbau des Kaiserreichs, in: *Blätter für Technikgeschichte* 73, pp. 31–54.

Bluma, Lars, Schulz, Stefan, and Jochen Streb (2010), Prinzipal-Agenten-Probleme in der knappschaftlichen Krankenversicherung: Die Bekämpfung des Simulantentums durch Anreize und Kontrolle, in: *Vierteljahrschrift für Sozial- und Wirtschaftgeschichte* 97, pp. 310–334.

Borscheid, Peter, and Anette Drees (1988), *Versicherungsstatistik Deutschlands 1750–1985*, St. Katha-rinen.

Börsch-Supan, Axel (1992), Population Aging, Social Security Design, and Early Retirement, in: *Journal of Institutional and Theoretical Economics* 148, pp. 533–557.

Boyer, Josef (1995), *Unfallversicherung und Unternehmer im Bergbau – Die Knappschafts-Berufsgenossenschaft 1885–1945*, München.

Brettschneider, Heidrun (1991), *Die Rentenversicherung der Bergleute*, Sankt Augustin.

Brinkmann, Heinrich (1967), Die Geschichte der Knappschaft im Ruhrgebiet, in: *Kompass: Zeitschrift der Bergbau-Berufsgenossenschaft* 77, pp. 175–180.

Broten, Nicholas (2010), From Sickness to Death: The Financial Viability of the English Friendly Societies and Coming of the Old Age Pensions Act, 1875–1908, in: *Department of Economic History Working Paper*, no. 135/2010, London School of Economics.

Brown, Peter C. (1995), Mining Legislation, the Consultation Process and the Reform of Mining Law: Their Significance for Company Form in Ruhr Coal Mining in the 19th Century, in: Ekkehard Westermann (ed.), *Vom Bergbau- zum Industrierevier*, Stuttgart, pp. 295–316.

Brüggemeier, Franz-Josef (1983), *Leben vor Ort – Ruhrbergleute und Ruhrbergbau 1889–1919*, München.

Buehler, Stefan, Kaiser, Christian, and Franz Jaeger (2006), Merge or fail? The determinants of mergers and bankcruptcies in Switzerland, 1995–2000, in: *Economics Letters* 90, pp. 88–95.

Burghardt, Uwe (1995), *Die Mechanisierung des Ruhrbergbaus: 1890–1930*, München.

Burhop, Carsten (2011), *Wirtschaftsgeschichte des Kaiserreichs 1871–1918*, Göttingen.

Burhop, Carsten, and Thorsten Lübbers (2009), Cartels, managerial incentives, and productive efficiency in German coal mining, 1881–1913, in: *Journal of Economic History* 69, pp. 500–527.

Burhop, Carsten, and Guntram B. Wolff (2005), A compromise estimate of German net national product, 1851–1913, and its implications for growth and the business cycle, in: *Journal of Economic History* 65, pp. 613–657.

Cabo, Paula, and Joao Rebelo (2005), Why do agricultural credit cooperatives merge? The Portuguese Experience, in: *Annals of Public and Cooperative Economics* 76, pp. 491–516.

Caswell, Jerry W. (1976), Economic efficiency in pension plan administration: a study of the construction industry, in: *Journal of Risk and Insurance* 4, pp. 257–273.

Clark, Gregory, and David Jacks (2007), Coal and the Industrial Revolution, 1700–1869, in: *European Review of Economic History* 11, pp. 39–72.

Clark, Robert L. (2004), Social Security Financing: Facts, Fantasies, Foibles, and Follies, in: *American Economic Review* 94, pp. 182–186.

Companje, Karel-Peter, Hendriks, R., Veraghtert, Karel F. E., and Brigitte E. M. Widdershoven, (2009), *Two centuries of solidarity – German, Belgian and Dutch social health insurance 1770–2008*, Amsterdam.

Conrad, Christoph (1991), The Emergence of Modern Retirement: Germany in an International Comparison, in: *Population: An English Selection* 3, pp. 171–199.

Conrad, Christoph (1994), *Vom Greis zum Rentner*, Göttingen.

Cox, David R. (1972), Regression Models and Life-Tables, in: *Journal of the Royal Statistical Society (Series B)* 2, pp. 187–220.

Cox, David R. (1975), Partial Likelihood, in: *Biometrika* 62, pp. 269–276.

Cummins, David J., Tennyson, Sharon, and Mary A. Weiss (1999), Life insurance mergers and acquisitions, in: David J. Cummins and Anthony M. Santomero (eds.), *Changes in the Life Insurance Industry: Efficiency, Technology and Risk Management*, Boston/Dordrecht/London, pp. 159–185.

Cummins, David J., and Maria Rubio-Misas (2006), Deregulation, Consolidation, and Efficiency: Evidence from the Spanish Insurance Industry, in: *Journal of Money, Credit and Banking* 38, pp. 323–356.

Cummins, David J., and Xiaoying Xie (2008), Mergers and acquisitions in the US property-liability insurance industry: Productivity and efficiency effects, in: *Journal of Banking & Financing* 32, pp. 30–55.

Cutler, David M., and Richard Johnson (2004), The birth and growth of the social insurance state: Explaining old age and medical insurance across countries, in: *Public Choice* 120, pp. 87–121.

Dapprich, Gerhard (1979), Geschichtliche Entwicklung der Knappschaft, in: Bundesknappschaft (ed.), *10 Jahre Bundesknappschaft*, Bochum, pp. 19–54.

Dapprich, Gerhard (1981), Der soziale Charakter des Arbeitsrechts im mittelalterlichen deutschen Bergbau, in: *Kompass: Zeitschrift der Bergbau-Berufsgenossenschaft und der Bundesknappschaft 91*, pp. 129–131 and 181–184.

Dapprich, Gerhard (1982), Sozialleistungen im deutschen Bergbau in der Zeit vom frühen Mittelalter bis zum 19. Jahrhundert, in: *Die Sozialgerichtsbarkeit*, pp. 514–517.

Degner, Harald (2011), Do technological booms matter? New evidence on the relationship between firm size and innovativeness, in: *Cliometrica* 5, pp. 121–144.

De Moor, Tine (2008), The Silent Revolution: A New Perspective on the Emergence of Commons, Guilds, and other Forms of Corporate Collective Action in Western Europe, in: *The International Review of Social History* 53, pp. 175–208.

Deutsche Rentenversicherung Bund (2008), *Rentenversicherung in Zeitreihen*, Berlin.

Dewey, Donald (1961), Mergers and cartels: Some reservations about policy, in: *American Economic Review (Papers and Proceedings)* 52, pp. 255–262.

Diamond, Peter (1977), A framework for social security analysis, in: *Journal of Public Economics* 8, pp. 275–298.

Disney, Richard (2004), Are contributions to public pension programmes a tax on employment, in: *Economic Policy* 19, pp. 267–311.

Dobronogov, Anton, and Mamta Murthi (2005), Administrative fees and costs of mandatory private pensions in transition economies, in: *Journal of Pension Economics and Finance* 4, pp. 31–55.

Doherty, Neil A. (1961), The Measurement of Output and Economies of Scale in Property-Liability Insurance, in: *The Journal of Risk and Insurance* 48, pp. 390–402.

Eghigian, Greg (2000), *Making Security Social: Disability, Insurance, and the Birth of the Social Entitlement State in Germany*, Ann Arbor.

Eisen, Roland (1991), Market Size and Concentration: Insurance and the European Internal Market 1992, in: *The Geneva Papers on Risk and Insurance* 16, pp. 263–281.

Emery, J. C. Herbert (1994), The rise and fall of fraternal methods of social insurance: A case study of the Independent Order of Oddfellows of British Columbia Sickness Insurance, 1874–1951, in: *Business and Economic History* 23, pp. 10–15.

Emery, J. C. Herbert (1996), Risky Business? Nonactuarial Pricing Practices and the Financial Viability of Fraternal Sickness Insurers, in: *Explorations in Economic History* 33, pp. 195–226.

Emery, J. C. Herbert (2006), From Defining Characteristic to Vitiation of Principle – The History of the Odd Fellows Stipulated Sick Benefit and Its Implications for Studying American Fraternalism, in: *Social Science History* 30, pp. 479–500.

Emery, George, and J. C. Herbert Emery (1999), *A Young Mans Benefit – The Independent Order of Odd Fellows and Sickness Insurance in the United States and Canada, 1860–1929*, Montreal et al.

Esteve-Pérez, Silviano, Sanchis-Llopis, Amparo, and Juan A. Sanchis-Llopis (2010), A competing risks analysis of firms exit, in: *Empirical Economics* 38, pp. 281–304.

Estrin, Alexander (1988), Administrative Costs for Social Security Programs in Selected Countries, in: *Social Security Bulletin* 51, pp. 29–31.

Feinstein, Charles F., and Mark Thomas (2002), *Making History Count: A Primer in Quantitative Methods to Historians*, Cambridge.

Feldman, Gerald D., and Klaus Tenfelde (1989), *Arbeiter, Unternehmer und Staat im Bergbau – Industrielle Beziehungen im internationalen Vergleich*, München.

Fishback, Price V. (1992), *Soft coal, hard choices – The economic welfare of bituminous coal miners, 1890–1930*, New York et al.

Fine, Jason P., and Robert J. Gray (1999), A proportional Hazards Model for the Subdistribution of a Competing Risk, in: *Journal of the American Statistical Association* 446, pp. 496–509.

Fischer, Wolfram (1961a), Das wirtschafts- und sozialpolitische Ordnungsbild der preußischen Bergrechtsreform 1851–1865, in: *Zeitschrift für Bergrecht* 102, pp. 181–189.

Fischer, Wolfram (1961b), Die Stellung der preussischen Bergrechtsreform von 1851–1865 in der Wirtschafts- und Sozialverfassung des 19. Jahrhunderts, in: *Zeitschrift für die gesamte Staatswissenschaft* 117, pp. 521–535.

Fischer, Wolfram (1961c), Die Bedeutung der preussischen Bergrechtsreform (1851–1865) für den industriellen Ausbau des Ruhrgebiets, in: *Vortragsreihe der Gesellschaft für Westfälische Wirtschaftsgeschichte e.V. Dortmund* 9.

Fischer, Wolfram (1989), *Statistik der Bergbauproduktion Deutschlands 1850–1914*, St. Katharinen.

Fischer, Wolfram (1995), *Statistik der Montanproduktion Deutschlands 1915–1985*, St. Katharinen.

Flora, Peter, and Arnold J. Heidenheimer (2003), *The Development of Welfare States in Europe and America*, 6th paperback edition, New Brunswick.

Fluck, Zsuzsanna, and Anthony W. Lynch (1999), Why Do Firms Merge and Then Divest? A Theory of Financial Synergy, in: *Journal of Business* 72, pp. 319–346.

Focarelli, Dario, Panetta, Fabio, and Carmelo Salleo (2002), Why do banks merge?, in: *Journal of Money, Credit, and Banking* 34, pp. 1 047–1 066.

Fotopoulos, Georgios, and Helen Louri (2000), Determinants of Hazard Confronting New Entry: Does Financial Structure Matter?, in: *Review of Industrial Organization* 17, pp. 285–300.

Frerich, Johannes, and Martin Frey (1993), *Handbuch der Geschichte der Sozialpolitik in Deutschland – Band 1: Von der vorindustriellen Zeit bis zum Ende des Dritten Reichs*, München.

Fürer, Gotthard (1988), Bergbau und Bergbehörde in Niedersachsen gestern und heute – 120 Jahre Oberbergamt in Clausthal-Zellerfeld, in: Oberbergamt Clausthal (ed.), *Festschrift zur Feierstunde: 200 Jahre vereinigtes Bergamt Clausthal – 120 Jahre Oberbergamt Clausthal – 434 Jahre Bergamt Clausthal, Clausthal*, pp. 16–31.

Gellhorn, Nikolaus von (1967), *Die Knappschaftsversicherung*, Stuttgart.

Geyer, Martin H. (1985), Staatliche Sozialpolitik und Knappschaftsreformen in Deutschland, 1880–1910, in: Gustav Schmidt (ed.), *Bergbau in Großbritannien und im Ruhrgebiet – Studien zur vergleichenden Geschichte des Bergbaus 1850–1930*, Bochum, pp. 96–146.

Geyer, Martin H. (1987), *Die Reichsknappschaft. Versicherungsreform und Sozialpolitik im Bergbau 1900–1945*, München.

Geyer, Martin H. (1990), Invalidität und Existenzsicherung im Bergbau 1770 bis 1870, in: Gerd Göckenjan (ed.), *Recht auf ein gesichertes Alter?: Studien zur Geschichte der Alterssicherung in der Frühzeit der Sozialpolitik*, Augsburg, pp. 181–202.

Geyer, Martin H. (1991), Soziale Rechte im Sozialstaat: Wiederaufbau, Krise und konservative Stabilisierung der deutschen Rentenversicherung 1924–1937, in: Klaus Tenfelde (ed.), *Arbeiter im 20. Jahrhundert*, Stuttgart, pp. 406–434.

Geyer, Martin H. (1992), The Miners Insurance and the Development of the German Social State, in: Klaus Tenfelde (ed.), *Sozialgeschichte des Bergbaus im 19. und 20. Jahrhundert*, München, pp. 1 046–1 065.

Geyer, Martin H. (1994), Die deutsche Knappschaftsversicherung in der Weltwirtschaftskrise: Überlegungen zur Geschichte des Wohlfahrtsstaates in der Zwischenkriegszeit, in: Otto Büsch (ed.), *Internationale Zusammenhänge der Weltwirtschaftskrise: Präsentationen aus Anlass einer Fachkonferenz*, Berlin, pp. 131–138.

Giga, Helmut (1979), Zur Entwicklung der Sicherheit im Ruhrbergbau zwischen 1865 und 1900, in: *Der Anschnitt: Mitteilungsblatt der Vereinigung der Freunde von Kunst und Kultur im Bergbau* 5, pp. 177–186.

Gilbert, Bentley (1965), The Decay of Nineteenth-Century Provident Institutions and the Coming of the Old Age Pensions in Great Britain, in: *Economic History Review* 17, pp. 551–563.

Göckenjan, Gerd, and Eckard Hansen (1993), Der lange Weg zum Ruhestand. Zur Sozialpolitik für das Alter zwischen 1889 und 1945, in: *Zeitschrift für Sozialreform* 39, pp. 725–755.

Gorsky, Martin, Harris, Bernard, and Andrew Hinde (2006), Age, Sickness, and Longevity in the Late Nineteenth and the Early Twentieth Centuries, in: *Social Science History* 30, pp. 571–600.

Gottlieb, Daniel (2007), Asymmetric information in late 19th century cooperative insurance societies, in: *Explorations in Economic History* 44, pp. 270–292.

Graf, Eberhard (1989), *Der Knappschaftsälteste – Ein Ehrenamt mit jahrhundertealter Tradition*, Bochum.

Graf, Eberhard (1998), Die Gründungsversammlung des Reichsknappschaftsvereins, in: *Kompass: Zeitschrift der Bergbau-Berufsgenossenschaft und der Bundesknappschaft* 108, pp. 251–253.

Greuer, Johannes-Traugott (1961), *Die Oberharzer Knappschaftskassen vom 16. Jahrhundert bis zur Mitte des 19. Jahrhunderts. Ein Beitrag zur Sozialgeschichte der Oberharzer Bergleute*, doctoral dissertation, Universität Göttingen.

Guinnane, Timothy W., and Jochen Streb (2011), Moral hazard in a mutual health-insurance system: German Knappschaften, 1867–1914, in: *Journal of Economic History* 71, pp. 70–104.

Guinnane, Timothy W., Streb, Jochen, Frondel, Manuel, and Christoph M. Schmidt (2011), Krank oder Simulant? Das Identifikationsproblem der Krankenkassen aus wirtschaftshistorischer Perspektive, in: *Perspektiven der Wirtschaftspolitik* 12, pp. 413–429.

Guinnane, Timothy W., Jopp, Tobias A., and Jochen Streb (2012a), Die Vorteile und Nachteile großer Invaliditäts- und Krankenversicherungen: Eine quantitative Wirtschaftsgeschichte der deutschen Knappschaftsvereine von 1854 bis zur Gründung der Reichsknappschaft, in: Christoph Bartels (ed.), *...höchst verpönte Selbst-Hülfe...: zur Entstehung der Sozialversicherung in Bergbau*, Seefahrt und Eisenbahnwesen, Bochum.

Guinnane, Timothy W., Jopp, Tobias A., and Jochen Streb (2012b), The costs and benefits of size in a mutual insurance system: German miners Knappschaften, 1854–1923, in: Bernard Harris (ed.), *Welfare and Old Age in Europe and North America: The Development of Social Insurance*, London.

Haerendel, Ulrike (2001), *Die Anfänge der Gesetzlichen Rentenversicherung in Deutschland – Die Invaliditäts- und Alterssicherung von 1889 im Spannungsfeld von Reichsverwaltung, Bundesrat und Parlament*, Speyer.

Han, Aaron, Jerry A. Hausman (1990), Flexible Parametric Estimation of Duration and Competing Risk Models, in: *Journal of Applied Econometrics* 5, pp. 1–28.

Hardach, Gerd (2003), Optionen der Altersvorsorge im 19. und 20. Jahrhundert in Deutschland, in: *Zeitschrift für Unternehmensgeschichte/ Journal of Business History* 48, pp. 5–28.

Hardach, Gerd (2011), Stabilisierung des Lebenseinkommens in einer alternden Gesellschaft: Deutschland von 1871 bis heute, in: Thomas Sokoll (ed.), *Soziale Sicherungssysteme und demographische Wechsellagen – Historisch-vergleichende Perspektiven (1500–2000)*, Münster, pp. 96–132.

Hardwick, Philip (1994), Scale and scope economies in the UK life assurance industry, in: *British Review of Economic Issues* 16, pp. 69–86.

Harris, Bernard, and Paul Bridgen (2007), *Charity and Mutual Aid in Europe and North America since 1800*, New York.

Henning, Hansjoachim (1987), *Quellensammlung zur Geschichte der deutschen Sozialpolitik 1867 bis 1914. IV. Abteilung: Die Sozialpolitik in den letzten Friedensjahren des Kaiserreichs (1905–1914). 2. Band: Das Jahr 1906*, Stuttgart.

Hennock, Ernest P. (1982), Die Ursprünge der staatlichen Sozialversicherung in Großbritannien und das deutsche Beispiel 1880–1914, in: Wolfgang J. Mommsen (ed.), *Die Entstehung des Wohlfahrtsstaates in Großbritannien und Deutschland 1850–1950*, Stuttgart, pp. 92–114.

Hennock, Ernest P. (1987), *British social reform and German precedents: the case of social insurance 1880–1914*, Oxford.

Hennock, Ernest P. (1990), Public Provision for Old Age – Britain and Germany 1880–1914, in: *Archiv für Sozialgeschichte* 30, pp. 81–103.

Hennock, Ernest P. (2007), *The Origin of the Welfare State in England and Germany, 1850–1914*, Cambridge.

Hentschel, Volker (1978), Das System der sozialen Sicherung in historischer Sicht 1880 bis 1975, in: *Archiv für Sozialgeschichte* 18, pp. 307–352.

Hermann, Christopher (1987), Die Knappschaft in der Weltwirtschaftskrise, in: *Kompass: Zeitschrift der Bergbau-Berufsgenossenschaft und der Bundesknappschaft* 97, pp. 297–301 and 341–346.

Hoffmann, Walter G., and Heinz J. Müller (1959), *Das Volkseinkommen in ausgewählten Bundesstaaten 1851–1913*, Tübingen.

Hoffmann, Walter G. (1965), *Das Wachstum der deutschen Wirtschaft seit der Mitte des 19. Jahrhunderts*, Berlin.

Holtfrerich, Carl-Ludwig (1973), *Quantitative Wirtschaftsgeschichte des Ruhrkohlenbergbaus im 19. Jahrhundert – Eine Führungssektoranalyse*, Dortmund.

Holtfrerich, Carl-Ludwig (1980), *Die deutsche Inflation 1914–1923. Ursachen und Folgen in internationaler Perspektive*, Berlin/New York.

Holtfrerich, Carl-Ludwig (1990), Relative Preise, Kapazität und Produktion in der deutschen Kohlen– und Eisenindustrie 1850–1913, in: Hansjörg Siegenthaler (ed.), *Ressourcenverknappung als Problem der Wirtschaftsgeschichte*, Berlin, pp. 107–122.

Holzmann, Robert (1988), Ageing and social security costs, in: *European Journal of Population* 3, pp. 411–437.

Homburg, Stefan (1997), Old-Age Pensions: A Theoretical Evaluation, in: Herbert Giersch (ed.), *Reforming the Welfare State*, Berlin, pp. 233–246.

Honoré, Bo E., and Adriana Lleras-Muney (2006), Bounds in competing risks models and the war on cancer, in: *Econometrica* 74, pp. 1 675–1 698.

Horgby, Per-Johann, and Lars Söderström (1998), Social risk in health insurance: A reflection on the economies of scale and scope in health insurance, in: *Australian Economic Papers* 37, pp. 185–194.

Horn, Günte, and Wolfgang Buschfort (1997), Wiederaufbau und Ende der Knappschaften in der sowjetischen Besatzungszone 1945 bis 1948, in: *Kompass: Zeitschrift der Bergbau-Berufsgenossenschaft und der Bundesknappschaft* 107, pp. 105–107.

Hosmer, David W., Lemeshow, Stanley, and Susanne May (2008), *Applied Survival Analysis – Regression Modeling of Time-to-Event Data*, Hoboken.

Huske, Joachim (1998), *Die Steinkohlenzechen im Ruhrrevier – Daten und Fakten von den Anfängen bis 1907*, Bochum.

Isachenkova, Natalia, and Melvin Weeks (2008), Acquisition, Insolvency and Managers in UK Small Companies, in: *Cambridge Working Papers in Economics*, no. 838.

Jensen, Michael C. (1988), Takeovers: Their Causes and Consequences, in: *Journal of Economic Perspectives* 2, pp. 21–48.

Jopp, Tobias A. (2010a), Ein risikoreiches Geschäft? Internes und externes Wachstum als risikopolitische Instrumente im preußischen Knappschaftswesen, 1854–1923, in: Christoph Bartels (ed.), *Berufliches Risiko und soziale Sicherheit. Beiträge zur Tagung Vergangenheit und Zukunft sozialer Sicherungssysteme am Beispiel der Bundesknappschaft und ihrer Nachfolger im Deutschen Bergbau-Museum Bochum, 8. und 9. Oktober 2009*, Bochum, pp. 189–224.

Jopp, Tobias A. (2010b), The Welfare State Evolves: German Knappschaften, 1854–1923, in: *Business and Economic History On-Line* 8.

Jopp, Tobias A. (2011a), The hazard of merger by absorption: Why some Knappschaften merged and others did not, 1861–1920, in: *Zeitschrift für Unternehmensgeschichte/Journal of Business History* 56, pp. 75–101.

Jopp, Tobias A. (2011b), Old times, better times? German miners Knappschaften, Pay-as-you-go Pensions, and Implicit Rates of Return, 1854–1913, in: *Business History* 53, pp. 1 018–1 043.

Jopp, Tobias A. (2012a), Insurance, size, and exposure to actuarial risk: Empirical evidence on nineteenth- and twentieth-century German Knappschaften, in: *Financial History Review* 19, pp. 75–116.

Jopp, Tobias A. (2012b), *Insurance, Fund size, and Concentration: Prussian Miners Knappschaften in the Nineteenth- and Early Twentieth-Century and Their Quest for Optimal Scale*, doctoral dissertation, Universität Hohenheim.

Jopp, Tobias A. (2013), On the Historical Roots of the Modern Welfare State: The Knappschaft Statistics of 1861 to 1920 as a Source for Quantitative Historical Economic and Social Research, in: *Historical Social Research* 38, pp. 311–338.

Kammertöns, Johannes (1952), *Die knappschaftliche Rentenversicherung*, Köln.

Karant-Nunn, Susan C. (1993), From Adventurers to Drones: the Saxon Silver Miners as an Early Proletariat, in: Thomas Max Safely (ed.), *The Workplace before the Factory*, New York.

Kaschke, Lars (2000), Nichts als Bettelgelder? Wert und Wertschätzung der Alters- und Invalidenrenten im Kaiserreich, in: *Historische Zeitschrift* 270, pp. 345–388.

Kaufhold, Karl Heinrich (1994), Preußische Staatswirtschaft – Konzept und Realität – 1640–1806, in: *Jahrbuch für Wirtschaftsgeschichte/ Economic History Yearbook* 1994/2, pp. 33–70.

Kellermann, Kersten (2008), Kosten der Kleinheit und die Föderalismusdebatte in der Schweiz, in: *Perspektiven der Wirtschaftspolitik* 9, pp. 196–225.

Khoudour-Castéras, David (2008), Welfare State and Labor Mobility: The Impact of Bismarcks Social Legislation on German Emigration before World War I, in: *Journal of Economic History* 68, pp. 211–243.

Kiefer, Nicholas M. (1988), Economic Duration Data and Hazard Functions, in: *Journal of Economic Literature* 26, pp. 646–679.

Kiesewetter, Hubert (1989), *Industrielle Revolution in Deutschland 1815–1914*, Frankfurt am Main.

Klein, Johann (1965), *Die Entwicklung der Sozialversicherung unter besonderer Berücksichtigung der Knappschaftsversicherung*, Wiebelskirchen-Saar.

Klenk, Tanja (2008), *Innovation und Kontinuität – Die Organisationsreform in der gesetzlichen Rentenversicherung*, Wiesbaden: Verlag für Sozialwissenschaften.

Kling, Gerhard (2006), The long-term impact of mergers and the emergence of a merger wave in pre-world-War I Germany, in: *Explorations in Economic History* 43, pp. 667–688.

Kluge, Arndt (2007), *Die Zünfte*, Stuttgart.

Koetter, Michael (2008), An assessment of bank merger success in Germany, in: *German Economic Review* 9, pp. 232–264.

Köhler, Peter A., and Hans F. Zacher (1982), *The Evolution of Social Insurance 1881–1981 – Studies of Germany, France, Great Britain, Austria and Switzerland*, Munich.

Köhler-Rama, Tim (2003), *Invaliditätsversicherung in der gesetzlichen Sozialversicherung in Deutschland: Strukturprinzipien, Defizite und Reformansätze aus sozialrechtlicher und versicherungsökonomischer Sicht*, doctoral dissertation, Technische Universität Darmstadt, Online pageview: https://tuprints.ulb.tu-darmstadt.de/338/, August 21st, 2008, 3 p.m.

Krampe, Hans Dieter (1961), *Der Staatseinfluss auf den Ruhrkohlenbergbau in der Zeit von 1800 bis 1865*, Köln.

Kuhnle, Stein, and Anne Sander (2010), The emergence of the Western welfare state, in: Francis G. Castles, Stephan Leibfried, Jane Lewis, Herbert Obinger and Christopher Pierson (eds.), *The Oxford Handbook of the Welfare State*, Oxford, pp. 61–80.

Kürble, Gunter, and Edmund Schwake (1984), Größenvorteile bei Schaden- und Unfallversicherungsunternehmen, in: *Zeitschrift für die gesamte Versicherungswissenschaft* 73, pp. 113–133.

Lammers, Frauke, and Jörg Schiller (2010), Einflussfaktoren für betrügerisches Verhalten von Versicherungsnehmern: erste experimentelle Befunde, in: *Zeitschrift für die gesamte Versicherungswissenschaft* 99, pp. 649–663.

Lauf, Ulrich (1985), Die ehemalige Reichsknappschaft in Berlin, in: *Kompass: Zeitschrift der Bergbau-Berufsgenossenschaft und der Bundesknappschaft 95*, pp. 177–181.

Lauf, Ulrich (1986), Viktor Weidtmann und das deutsche Knappschaftswesen, in: *Kompass: Zeitschrift der Bergbau-Berufsgenossenschaft und der Bundesknappschaft 96*, pp. 449–453.

Lauf, Ulrich (1987), Zur Geschichte des sächsischen Knappschaftswesen, in: *Kompass: Zeitschrift der Bergbau-Berufsgenossenschaft und der Bundesknappschaft* 96, pp. 318–396.

Lauf, Ulrich (1990), Knappschaft und Reformation, in: *Kompass: Zeitschrift der Bergbau-Berufsgenossenschaft und der Bundesknappschaft* 100, pp. 541–543.

Lauf, Ulrich (1991), Zur sozialrechtlichen Eingliederung des Saarlandes in die Bundesrepublik Deutschland in den Jahren 1957–1960, in: *Kompass: Zeitschrift der Bergbau-Berufsgenossenschaft und der Bundesknappschaft* 101, pp. 355–358.

Lauf, Ulrich (1994), *Die Knappschaft. Ein Streifzug durch tausend Jahre Sozialgeschichte*, Sankt Augustin.

Lauf, Ulrich (1996a), Die Knappschaft im Ersten Weltkrieg, in: *Kompass: Zeitschrift der Bergbau-Berufsgenossenschaft und der Bundesknappschaft* 106, pp. 586–597.

Lauf, Ulrich (1996b), Die Preußische Knappschaftsnovelle von 1906. Teil 1 und Teil 2, in: *Kompass: Zeitschrift der Bergbau-Berufsgenossenschaft und der Bundesknappschaft* 106, pp. 297–304 and 366–373.

Lauf, Ulrich (1998), Zur Ausgangslage des Reichsknappschaftsgesetzes von 1923, in: *Kompass: Zeitschrift der Bergbau-Berufsgenossenschaft und der Bundesknappschaft* 108, pp. 247–251.

Lauf, Ulrich (1999), Die Reform der Knappschaftsversicherung im Dritten Reich, in: *Der Anschnitt: Mitteilungsblatt der Vereinigung der Freunde von Kunst und Kultur im Bergbau* 51, pp. 201–216.

Lauf, Ulrich (2002), Neue Hypothesen zum Ursprung der Knappschaft, in: *Kompass: Zeitschrift der Bergbau-Berufsgenossenschaft und der Bundesknappschaft* 112 (9/10 and 11/12), pp. 22–25 and 22–26.

Lauf, Ulrich (2003a), Bruderschaft und Büchsengeld: eine Untersuchung zu den mittelalterlichen Wurzeln unserer Sozialversicherung, in: *Wege zur Sozialversicherung* 57, pp. 176–188.

Lauf, Ulrich (2003b), St. Johannis und das älteste Hospital für Bergleute – Neue Hypothesen zum Ursprung der Knappschaft, in: *Der Anschnitt: Mitteilungsblatt der Vereinigung der Freunde von Kunst und Kultur im Bergbau* 55, pp. 2–11.

Lauf, Ulrich (2004a), Die Goslarer Bruderschaft der Bergleute von 1260: Eine Annäherung, in: *Harz-Zeitschrift* 54/55, pp. 115–132.

Lauf, Ulrich (2004b), Frühe Sozialeinrichtungen im Bergbau Goslars und ihre Spitäler, in: *Kompass: Zeitschrift der Bergbau-Berufsgenossenschaft und der Bundesknappschaft* 114, pp. 3–13.

Lauf, Ulrich (2005), *Die Krankenhäuser der deutschen Knappschaftsvereine im 19. und 20. Jahrhundert*, Bochum.

Lauf, Ulrich (2006a), Die deutschen Knappschaftsvereine (1854–1923). Ein altes Sozialsystem im Rückblick, in: *Der Anschnitt: Mitteilungsblatt der Vereinigung der Freunde von Kunst und Kultur im Bergbau 58*, pp. 264–285.

Lauf, Ulrich (2006b), Der Streit um den Ärztevertrag. Konfliktbewältigung vor 90 Jahren im Bochumer Allgemeinen Knappschaftsverein, in: *Kompass: Amtliches Mitteilungsblatt der Bergbau-Berufsgenossenschaft und der Deutschen Rentenversicherung Knappschaft-Bahn-See* 116, pp. 12–15.

Lauf, Ulrich (2006c), Die Influenza-Pandemie von 1918 in den Verwaltungsberichten des Bochumer Allgemeinen Knappschaftsvereins, in: *Kompass: Amtliches Mitteilungsblatt der Bergbau-Berufsgenossenschaft und der Deutschen Rentenversicherung Knappschaft-Bahn-See* 116, pp. 14–15.

Lauf, Ulrich (2008a), Die Rentenversorgung der Arbeiterinvaliden im Ruhrbergbau um die Wende vom 19. zum 20. Jahrhundert, in: *Kompass: Amtliches Mitteilungsblatt der Bergbau-Berufsgenossenschaft und der Deutschen Rentenversicherung Knappschaft-Bahn-See* 118, pp. 14–19.

Lauf, Ulrich (2008b), NS-Terror in der Verwaltung – Zu den nationalsozialistischen Säuberungsmaßnahmen bei der Ruhrknappschaft nach der Machtergreifung 1933, in: *Kompass: Amtliches Mitteilungsblatt der Bergbau-Berufsgenossenschaft und der Deutschen Rentenversicherung Knappschaft-Bahn-See*, 118, pp. 10–13.

Lauf, Ulrich (2009a), *Der Allgemeine Knappschaftsverein zu Bochum (1890–1923) – Mythos und Wirklichkeit*, Bochum.

Lauf, Ulrich (2009b), Knappschaftsdarlehen für Zechenkolonien, in: *Kompass: Amtliches Mitteilungsblatt der Bergbau-Berufsgenossenschaft und der Deutschen Rentenversicherung Knappschaft-Bahn-See* 119, pp. 6–10.

Lauf, Ulrich (2009c), Ein früher Ärztevertrag im Ruhrbergbau aus dem Jahr 1859, in: *Kompass: Amtliches Mitteilungsblatt der Bergbau-Berufsgenossenschaft und der Deutschen Rentenversicherung Knappschaft-Bahn-See* 119, pp. 8–12.

Lauf, Ulrich (2010), Die Bundesknappschaft, in: *Der Anschnitt: Mitteilungsblatt der Vereinigung der Freunde von Kunst und Kultur im Bergbau* 62, pp. 251–260.

Lauf, Ulrich (2011), Die Zeit der Bundesknappschaft, in: *Kompass: Amtliches Mitteilungsblatt der Bergbau-Berufsgenossenschaft und der Deutschen Rentenversicherung Knappschaft-Bahn-See* 121(5/6 and 7/8), pp. 20–25 and 18–23.

Lindert, Peter (1994), The Rise of Social Spending, 1880–1930, in: *Explorations in Economic History* 31, pp. 1–37.

Lindert, Peter (1996), What limits social spending?, in: *Explorations in Economic History* 33, pp. 1–34.

Lindert, Peter (1998), Poor relief before the welfare state: Britain versus the Continent, 1780–1880, in: *European Review of Economic History* 2, pp. 101–140.

Lindert, Peter (2004), *Growing Public – Social Spending and Economic Growth Since the Eighteenth Century*, Cambridge.

Lingnau, Josef (1965), *Das System sozialer Hilfeleistung für die Bergarbeiter in der Knappschaftsversicherung des Ruhrbergbaus 1767–1961*, Köln/Opladen.

Lunn, Mary, and Don McNeil (1995), Applying Cox Regression to Competing Risks, in: *Biometrics* 51, 524–532.

Manjón-Antolín, Miguel C., and Josep-Maria Arauzo-Carod (2008), Firm survival: methods and evidence, in: *Empirica* 1, pp. 1–24.

Manow, Philip (2000), Kapitaldeckung oder Umlage: Zur Geschichte einer anhaltenden Debatte, in: Stefan Fisch and Ulrike Haerendel (eds.), *Geschichte und Gegenwart der Rentenversicherung in*

Deutschland: Beiträge zur Entstehung, Entwicklung und vergleichenden Einordnung der Alterssicherung im Sozialstaat, Berlin, pp. 145–168.

Mattil, Birgit (2006), *Pension Systems. Sustainability and Distributional Effects in Germany and the United Kingdom*, Heidelberg.

Mesa-Lago, Carmelo (2006), Private and Public Pension Systems Compared: an Evaluation of the Latin American Experience, in: *Review of Political Economy* 18, pp. 317–334.

Metzger, Herbert (2007), 2010: 750 Jahre Knappschaftswesen. Gelebte Solidarität von Bergleuten prägte die spätere Sozialversicherung, in: *Kompass: Amtliches Mitteilungsblatt der Bergbau-Berufsgenossenschaft und der Deutschen Rentenversicherung Knappschaft-Bahn-See* 117, pp. 12–13.

Mills, Catherine (2010), *Regulating health and safety in the British mining industries, 1800–1914*, Farnham/Burlington.

Mitchell, Olivia S. (1998), *Administrative Costs in Public and Private Retriemnt Systems, in: Martin Feldstein (ed.), Privatizing Social Security*, Chicago and London, pp. 403–456.

Mommsen, Wolfgang J., and Wolfgang Mock (1981), *The Emergence of the Welfare State in Britain and Germany*, London.

Montz, Markus (2010), Vergessener Vorreiter der Sozialversicherung? Betriebliche soziale Sicherung bei den preußischen Eisenbahnen, 1838–1914, in: Christoph Bartels (ed.), *Berufliches Risiko und soziale Sicherheit. Beiträge zur Tagung Vergangenheit und Zukunft sozialer Sicherungssysteme am Beispiel der Bundesknappschaft und ihrer Nachfolger im Deutschen Bergbau-Museum Bochum, 8. und 9. Oktober 2009*, Bochum, pp. 109–142.

Mordi, Obi (1985), Das Produktkonzept der Versicherung: Eine alternative Interpretation, in: *Zeitschrift für die gesamte Versicherungswissenschaft* 74, pp. 81–92.

Mühlenkamp, Holger (1995a), Größen- und Verbundvorteile in der Verwaltung der gesetzlichen Krankenversicherung, in: *Zeitschrift für Betriebswirtschaft* 65, pp. 287–308.

Mühlenkamp, Holger (1995b), Betriebsgröße und Verwaltungskosten der Allgemeinen Ortskrankenkassen, in: *IFO-Studien: Zeitschrift für empirische Wirtschaftsforschung* 41, pp. 259–270.

Murray, John E. (2003), Social Insurance Claims as Morbidity Estimates: Sickness or Absence?, in: *Journal of the Society for the Social History of Medicine* 16, pp. 225–245.

Murray, John E. (2005), Worker Absenteeism under Voluntary and Compulsory Sickness Insurance: Continental Europe, 1885–1908, in: *Research in Economic History* 23, pp. 177–207.

Murray, John E. (2007), *Origins of American Health Insurance: A History of Industrial Sickness Funds*, New Haven.

Nekby, Lena (2004), Pure vs. Mutual Health Insurance: Evidence from Swedish Historical Data, in: *The Journal of Risk and Insurance* 70, pp. 115–134.

Nekby, Lena, Arai, Mahmood, and Michael Lundholm (2001), Economies of Scale and Sorting in Pure vs. Mutual Insurance Societies: Evidence from Historical Data, in: *Research Papers in Economics*, Stockholm University.

Nektarios, Milton (2010), Deregulation, Insurance Supervision and Guaranty Funds, in: *The Geneva Papers of Risk and Insurance* 35, pp. 452–468.

Nelder, John A., and Robert W. M. Wedderburn (1972), Generalized Linear Models, in: *Journal of the Royal Statistical Society* A 135, pp. 370–383.

Nell, Martin, and Jörg Schiller (2002), Erklärungsansätze für vertragswidriges Verhalten von Versicherungsnehmern aus Sicht der ökonomischen Theorie, in: *Zeitschrift für die gesamte Versicherungswissenschaft* 91, pp. 533–556.

Oberbergamt Bonn (1966), *150 Jahre Oberbergamt in Bonn*, Bonn.

Ogilvie, Sheilagh (2004), Guilds, Efficiency, and Social Capital: Evidence from German Proto-Industry, in: *Economic History Review* 57, pp. 286–333.

Pastena, Victor, and William Ruland (1986), The merger/bankruptcy alternative, in: *The Accounting Review* 61, pp. 288–301.

Pearson, Robin (1996), The development of reinsurance markets in Europe during the nineteenth century, in: *Journal of European Economic History* 24, pp. 557–571.

Pearson, Robin (1997), Towards an historical model of services innovation: the case of the insurance industry, 1700–1914, in: *Economic History Review* 50, pp. 235–256.

Pearson, Robin (2002), Mutuality Tested: the Rise and Fall of Mutual Fire Insurance Offices in Eighteenth-Century London, in: *Business History* 44, pp. 1–28.

Pearson, Robin (2003), Who pays for pensions? Das Problem der Alterssicherung in Großbritannien im zwanzigsten Jahrhundert, in: *Zeitschrift für Unternehmensgeschichte/Journal of Business History* 48, pp. 48–57.

Penrose, Edith (1995), *The Theory of the Growth of the Firm*, New York.

Pintile, Melanie (2007), Analysing and interpreting competing risk data, in: *Statistics in Medicine* 26, pp. 1 360–1 367.

Pons Pons, Jerònia, and Margarita Vilar Rodriguez (2011), Friendly Societies, Commercial Insurance, and the State in Sickness Risk Coverage: The Case of Spain (1880–1944), in: *International Review of Social History* 56, pp. 71–101.

Pope, Ralph A. (1986), Economies of Scale in Large State and Municipal Retirement Systems, in: *Public Budgeting & Finance* 6, pp. 70–80.

Putter, Hein, Fiocco, Marta, and R. B. Geskus (2005), Tutorial in biostatistics: Competing risks and multi-state models, in: *Statistics in Medicine* 26, pp. 2 389–2 430.

Ralston, Deborah, Wright, April, and Kaylee Garden (2001), Can mergers ensure the survival of credit unions in the third millennium?, in: *Journal of Banking & Finance* 25, pp. 2 277–2 304.

Reimers, Lutz (2006), Die Gesundheitsreform 2006 aus wirtschaftspolitischer Sicht, in: *Wirtschaftsdienst* 86, pp. 588–594.

Reininghaus, Wilfried (1983), Das erste staatlich beaufsichtigte System von Krankenkassen: Preußen 1845–1869. Das Beispiel der Regierungsbezirke Arnsberg und Minden, in: *Zeitschrift für Sozialreform* 29, pp. 271–298.

Rieger, Martina (1999), *Der Bundeszuschuß zur knappschaftlichen Rentenversicherung als Bergbausubvention?*, Frankfurt am Main et al.

Riley, James C. (1997), *Sick, not dead: the health of British workingmen during the mortality decline*, Baltimore.

Ritschl, Albrecht, and Mark Spoerer (1997), Das Bruttosozialprodukt in Deutschland nach den amtlichen Volkseinkommens- und Sozialproduktsstatistiken 1901–1995, in: *Jahrbuch für Wirtschaftsgeschichte* 1997/2, pp. 27–54.

Ritter, Gerhard A. (2010), *Der Sozialstaat – Entstehung und Entwicklung im internationalen Vergleich*, third edition, München.

Rothenbacher, Franz (2002), *The European Population 1850–1945*, Houndmills et al.

Samuelson, Paul A. (1958), An exact consumption-loan model of interest with or without the social contrivance of money, in: *Journal of Political Economy* 66, pp. 467–482.

Saul, Klaus (1980), Wirtschafts- und sozialpolitische Grundlagen – Industrialisierung, Systemstabilisierung und Sozialversicherung. Zur Entstehung, politischen Funktion und sozialen Realität der Sozialversicherung des kaiserlichen Deutschland, in: *Zeitschrift für die gesamte Versicherungswissenschaft* 69, pp. 177–198.

Schary, Martha A. (1991), The probability of exit, in: *RAND Journal of Economics* 22, pp. 339–353.

Schelter, Helmut (1992), 200 Jahre Landesoberbergamt Nordrhein-Westfalen, in: *Glückauf* 128, pp. 505–510.

Schiller, Jörg (2004), Versicherungsbetrug als ökonomisches Problem: Eine vertragstheoretische Analyse, in: *Zeitschrift für die gesamte Versicherungswissenschaft* 93, pp. 835–851.

Schlegel-Voß, Lil-Christine, and Gerd Hardach (2003), Die dynamische Rente. Ein Modell der Alterssicherung im historischen Wandel, in: *Vierteljahrschrift für Sozial- und Wirtschaftsgeschichte* 90, pp. 290–315.

Schmähl, Winfried (1995), Alterung der Bevölkerung, Mortalität, Morbidität, Zuwanderung und ihre Bedeutung für die Gesetzliche Rentenversicherung – Auswirkungen, Handlungsbedarf und Handlungsmöglichkeiten, in: *Zeitschrift für die gesamte Versicherungswissenschaft* 84, pp. 617–646.

Schmähl, Winfried (2001), Umlagefinanzierte Rentenversicherung in Deutschland – Optionen und Konzepte sowie politische Entscheidungen als Einstieg in einen grundlegenden Transformationprozeß, in: Winfried Schmähl and Volker Ulrich (eds.), *Soziale Sicherungssysteme und demographische Herausforderungen*, Tübingen, pp. 123–204.

Schwake, Edmund (1988), Das versicherungstechnische Risiko als arteigenes Risiko der Versicherungsunternehmen?, in: *Zeitschrift für die gesamte Versicherungswissenschaft* 77, pp. 61–81.

Seeger, Manfred (1968), *Die Politik der Reichsbank von 1876–1914 im Lichte der Spielregeln der Goldwährung*, Berlin.

Shrieves, Ronald E., and Donald L. Stevens (1979), Bankruptcy avoidance as a motive for merger, in: *Journal of Financial and quantitative Analysis* 3, pp. 501–515.

Sniegs, Monika (1998), *Statistik als Steuerungsinstrument in der historischen Entwicklung der Invaliditäts- und Altersversicherung 1891–1911*, doctoral dissertation, Universität Bremen.

Sommariva, Andrea, and Guiseppe Tullio (1987), *German Macroeconomic History 1880–1979 – A Study of the Effects of Economic Policy on Inflation, Currency Depreciation and Growth*, New York.

Sprenger, Bernd (1991), *Das Geld der Deutschen – Geldgeschichte Deutschlands von den Anfängen bis zur Gegenwart*, Paderborn et al.

Springer, Friedrich P. (2010), Über Kameralismus und Bergbau, in: *Der Anschnitt: Mitteilungsblatt der Vereinigung der Freunde von Kunst und Kultur im Bergbau* 62, pp. 230–241.

Steenbeek, Onno, and Fieke van der Lecq (2007), *Costs and Benefits of Collective Pension Systems*, Berlin/Heidelberg.

Stein, Bruno (1991), Pay-as-You-Go, Partial Prefunding, and Full Funding in American Social Security, in: *History of Political Economy* 23, 79–83.

Steinmetz, George (1993), *Regulating the Social – The Welfare State and Local Politics in Imperial Germany*, Princeton.

Stolleis, Michael (1980), Hundert Jahre Sozialversicherung in Deutschland – Rechtsgeschichtliche Entwicklung, in: *Zeitschrift für die Gesamte Versicherungswissenschaft* 69, pp. 155–175.

Stonebraker, Robert J. (1993), Optimal church Size: The Bigger The Better?, in: *Journal for the Scientific Study of Religion* 32, pp. 231–241.

Sueyoshi, Glenn T. (1992), Semiparametric proportional hazards estimation of competing risks models with time-varying covariates, in: *Journal of Econometrics* 51, pp. 25–58.

Sulzer, Melanie (2010), Die Neuordnung sozialer Sicherungssysteme für Seeleute zum Ende des 19. Jahrhunderts, in: Christoph Bartels (ed.), *Berufliches Risiko und soziale Sicherheit. Beiträge zur Tagung Vergangenheit und Zukunft sozialer Sicherungssysteme am Beispiel der Bundesknappschaft und ihrer Nachfolger im Deutschen Bergbau-Museum Bochum, 8. und 9. Oktober 2009*, Bochum, pp. 175–188.

Tampke, Jürgen (1981), Bismarck's social legislation: a genuine breakthrough?, in: Wolfgang. J. Mommsen and Wolfgang Mock (eds.), *The Emergence of the Welfare State in Britain and Germany*, London, pp. 71–83.

Tampke, Jürgen (1982), Bismarcks Sozialgesetzgebung: Ein wirklicher Durchbruch?, in: Wolfgang J. Mommsen and Wolfgang Mock (eds.), *Die Entstehung des Wohlfahrtsstaates in Großbritannien und Deutschland 1850–1950*, Stuttgart, pp. 79–91.

Tenfelde, Klaus (1977), *Sozialgeschichte der Bergarbeiterschaft an der Ruhr im 19. Jahrhundert*, Bonn-Bad Godesberg.

Tenfelde, Klaus (2004), Die Knappschaftsversicherung und die Wurzeln der Sozialversicherung in Deutschland, in: Bundesknappschaft (ed.), *150 Jahre Preußisches Knappschaftsgesetz*, Bochum, pp. 16–37.

Tenfelde, Klaus, and Helmuth Trischler (1986), *Bis vor die Stufen des Throns. Bittschriften und Beschwerden von Bergarbeitern im Zeitalter der Industrialisierung*, München.

Tennstedt, Florian (1983), Die Errichtung von Krankenkassen in deutschen Städten nach dem Gesetz betr. Die Krankenversicherung der Arbeiter vom 15. Juni 1883. Ein Beitrag zur Frühgeschichte der gesetzlichen Krankenversicherung in Deutschland, in: *Zeitschrift für Sozialreform* 29, pp. 297–338.

Tennstedt, Florian (2011), Die Hilfs- und Unterstützungskassen in Preußen, in: Peter Collin (ed.), *Selbstregulierung im 19. Jahrhundert – zwischen Autonomie und staatlichen Steuerungsansprüchen*, Frankfurt am Main, pp. 275–291.

Tennstedt, Florian, Winter, Heidi, Roeder, Elmar, and Christian Schmitz (1999), *Quellensammlung zur Geschichte der deutschen Sozialpolitik 1867 bis 1914. I. Abteilung: Von der Reichsgründungszeit bis zur kaiserlichen Botschaft (1867–1881). 5. Band: Gewerbliche Unterstützungskassen*, Darmstadt.

Thane, Patricia (2006), The History of Retirement, in: Gordon L. Clark, Alicia H. Munnell and J. Michael Orszag (eds.), *The Oxford Handbook of Pensions and Retirement Income*, Oxford, pp. 33–51.

Thielmann, Hans (1958), *Die Knappschaftsversicherung*, Bad Godesberg.

Thompson, Steven (1997), Takeover activity among financial mutuals: An analysis of target characteristics, in: *Journal of Banking & Finance* 21, pp. 37–53.

Thullen, Peter (1980), Die Finanzierungsverfahren der sozialen Rentenversicherung aus alter und neuer Sicht, in: *Zeitschrift für die gesamte Versicherungswissenschaft* 69, pp. 89–109.

Thullen, Peter (1985), Überlegungen zur Anwendung des Rentenwertumlageverfahrens in der Rentenversicherung, in: *Blätter/Deutsche Gesellschaft für Versicherungs- und Finanzmathematik* 17, pp. 129–140.

Thullen, Peter (1986), Rentenfinanzierung durch Verfahren partieller Kapitalisation unter dynamischen demographischen und wirtschaftlichen Bedingungen, in: *Blätter/Deutsche Gesellschaft für Versicherungs- und Finanzmathematik* 17, pp. 441–455.

Tilly (1990), Richard, *Vom Zollverein zum Industriestaat*, München.

Todeskino, Heinz (1974), *Von den mittelalterlichen Knappschaften bis zur Bundesknappschaft*, Bochum.

Trischler, Helmuth (1988), Arbeitsunfälle und Berufskrankheiten im Bergbau 1851 bis 1945 – Bergbehördliche Sozialpolitik im Spannungsfeld von Sicherheit und Produktionsinteressen, in: *Archiv für Sozialgeschichte* 28, pp. 111–151.

Ullmann, Hans-Peter (1979), Industrielle Interessen und die Entstehung der deutschen Sozialversicherung 1880–1889, in: *Historische Zeitschrift* 229, pp. 574–610.

United Nations (2008), *World Population Prospects – 2008 Revision*, New York.

Van den Berg, Gerard J., van Lomwel, A. Gijsbert C., and Jan C. van Ours (2008), Nonparametric estimation of a dependent competing risk model for unemployment duration, in: *Empirical Economics* 34, pp. 477–491.

Van der Linden, Marcel (1996), *Social Security Mutualism – The Comparative History of Mutual Benefit Societies,* Bern et al.

Van Kranenburg, Hans L., Palm, Franz C., and Gerard A. Pfann (2002), Exit and Survival in a Concentrating Industry: The Case of Daily Newspapers in the Netherlands, in: *Review of Industrial Organization* 21, pp. 283–303.

Varnas, Juozas (1947), *Die Knappschaften als Vorbilder der deutschen Sozialversicherung*, doctoral dissertation, Universät Münster.

Verdugo, Richard R. (2006), Workers, workers productivity and the dependency ratio in Germany: analysis with implications for social policy, in: *Population Research & Policy Review* 25, pp. 547–565.

Verweyst, Ernst (1983), 60 Jahre Reichsknappschaftsgesetz, in: Kompass: *Zeitschrift der Bergbau-Berufsgenossenschaft und der Bundesknappschaft* 93, pp. 530–543.

Verweyst, Ernst (1998), Das Reichsknappschaftsgesetz und die Renten für Bergleute, in: *Kompass: Zeitschrift der Bergbau-Berufsgenossenschaft und der Bundesknappschaft* 108, pp. 237–246.

Wagner-Braun, Margarete (2002), *Zur Bedeutung berufsständischer Krankenkassen innerhalb der privaten Krankenversicherung in Deutschland bis zum Zweiten Weltkrieg – Die Selbsthilfeeinrichtungen der katholischen Geistlichen*, Stuttgart.

Weisbrod, Bernd (1989), Arbeitgeberpolitik und Arbeitsbeziehungen im Ruhrbergbau. Vom «Herr-im-Haus» zur Mitbestimmung, in: Gerald D. Feldman and Klaus Tenfelde (eds.), *Arbeiter, Unternehmer und Staat im Bergbau*, München, pp. 107–162.

Weizsäcker, Robert K. Frhr. von (2010), Politökonomische Aspekte der Rentenversicherung und Bildungsfinanzierung im Lichte des demographischen Wandels, in: Corinne Michaela Flick (ed.), *Das demographische Problem als Gefahr für Rechtskultur und Wirtschaft*, München/Frankfurt am Main, pp. 147–165.

Whaples, Robert, and David Buffum (1991), Fraternalism, Paternalism, the Family, and the Market: Insurance a Century Ago, in: *Social Science History* 15, pp. 97–122.

Wheelock, David C., and Paul W. Wilson (2000), Why Do Banks Disappear? The Determinants of U.S. Bank Failures and Acquisitions, in: *Review of Economics and Statistics* 82, pp. 127–138.

Whitehouse, Edward (2000), Administrative charges for funded pensions: an international comparison and assessment, in: *Social Protection Discussion Paper Series*, No. 16.

Whitehouse, Edward (2006), New indicators of 30 OECD countries pension systems, in: *Journal of Pension Economics and Finance* 5, pp. 275–298.

Wirth, Olaf (1962), *Das Preussische Ministerium für Handel, Gewerbe und öffentliche Arbeiten: ein Beitrag zu seiner Entstehungsgeschichte*, München.

Wisotzky, Klaus (1992), Der Wandel der betrieblichen Sozialpolitik im Ruhrbergbau, in: Klaus Tenfelde (ed.), *Sozialgeschichte des Bergbaus im 19. und 20. Jahrhundert*, München, pp. 1 066–1 082.

Wixforth, Harald (1995), Der Kampf um die Bergbaufreiheit. Zur Geschichte der preußischen Bergrechtsreform von 1826 bis 1846, in: Hartmut Berghoff and Dieter Ziegler (eds.), *Pionier und Nachzügler? Vergleichende Studien zur Geschichte Grossbritanniens und Deutschlands im Zeitalter der Industrialisierung. Festschrift für Sidney Pollard zum 70. Geburtstag*, Bochum 1995.

Wooldridge, Jeffrey M. (2002), *Econometric Analysis of Cross Section and Panel Data*, Cambridge.

Zeger, Scott I., Liang, Kung-Yee, and Paul S. Albert (1988), Models for Longitudinal Data: A Generalized Estimating Equation Approach, in: *Biometrics* 44, pp. 1 049–1 060.

Zweifel, Peter, and Roland Eisen (2003), *Versicherungsökonomie*, Heidelberg.

Index

www.ingramcontent.com/pod-product-compliance
Lightning Source LLC
Chambersburg PA
CBHW080933240926
51849CB00003BA/445
* 9 7 8 3 0 5 0 0 6 3 7 0 6 *